Springer Handbook of Auditory Research

Volume 53

More information about this series at http://www.springer.com/series/2506

Roderick A. Suthers • W. Tecumseh Fitch
Richard R. Fay • Arthur N. Popper
Editors

Vertebrate Sound Production and Acoustic Communication

Springer

Editors
Roderick A. Suthers
Medical Science, Jordan Hall
Indiana University
Bloomington, IN, USA

W. Tecumseh Fitch
Department of Cognitive Biology
University of Vienna
Vienna, Austria

Richard R. Fay
Marine Biological Laboratory
Woods Hole, MA, USA

Arthur N. Popper
Department of Biology
University of Maryland
College Park, USA

ISSN 0947-2657 ISSN 2197-1897 (electronic)
Springer Handbook of Auditory Research
ISBN 978-3-319-27719-6 ISBN 978-3-319-27721-9 (eBook)
DOI 10.1007/978-3-319-27721-9

Library of Congress Control Number: 2015959709

Springer Cham Heidelberg New York Dordrecht London

Printed on acid-free paper

Springer International Publishing AG Switzerland is part of Springer Science+Business Media (www.springer.com)

This volume is dedicated to the memory of our mentor and colleague, Professor Donald R. Griffin (1915–2003), whose amazing contributions to bioacoustics, including the codiscovery of bat echolocation, have shaped our scientific careers (photo by Greg Auger).

Series Preface

The following preface is the one that we published in Volume 1 of the *Springer Handbook of Auditory Research* back in 1992. As anyone reading the original preface, or the many users of the series, will note, we have far exceeded our original expectation of eight volumes. Indeed, with books published to date, and those in the pipeline, we are now set for more than 50 volumes in SHAR, and we are still open to new and exciting ideas for additional books.

We are very proud that there seems to be consensus, at least among our friends and colleagues, that SHAR has become an important and influential part of the auditory literature. While we have worked hard to develop and maintain the quality and value of SHAR, the real value of the books is very much because of the numerous authors who have given their time to write outstanding chapters and to our many coeditors who have provided the intellectual leadership to the individual volumes. We have worked with a remarkable and wonderful group of people, many of whom have become great personal friends of both of us. We also continue to work with a spectacular group of editors at Springer. Indeed, several of our past editors have moved on in the publishing world to become senior executives. To our delight, this includes the current president of Springer USA, Dr. William Curtis.

But the truth is that the series would and could not be possible without the support of our families, and we want to take this opportunity to dedicate all of the SHAR books, past and future, to them. Our wives, Catherine Fay and Helen Popper, and our children, Michelle Popper Levit, Melissa Popper Levinsohn, Christian Fay, and Amanda Fay, have been immensely patient as we developed and worked on this series. We thank them and state, without doubt, that this series could not have happened without them. We also dedicate the future of SHAR to our next generation of (potential) auditory researchers—our grandchildren—Ethan and Sophie Levinsohn, Emma Levit, and Nathaniel, Evan, and Stella Fay.

Preface 1992

The *Springer Handbook of Auditory Research* presents a series of comprehensive and synthetic reviews of the fundamental topics in modern auditory research. The volumes are aimed at all individuals with interests in hearing research including advanced graduate students, postdoctoral researchers, and clinical investigators. The volumes are intended to introduce new investigators to important aspects of hearing science and to help established investigators to better understand the fundamental theories and data in fields of hearing that they may not normally follow closely.

Each volume presents a particular topic comprehensively, and each serves as a synthetic overview and guide to the literature. As such, the chapters present neither exhaustive data reviews nor original research that has not yet appeared in peer-reviewed journals. The volumes focus on topics that have developed a solid data and conceptual foundation rather than on those for which a literature is only beginning to develop. New research areas will be covered on a timely basis in the series as they begin to mature.

Each volume in the series consists of a few substantial chapters on a particular topic. In some cases, the topics will be ones of traditional interest for which there is a substantial body of data and theory, such as auditory neuroanatomy (Vol. 1) and neurophysiology (Vol. 2). Other volumes in the series deal with topics that have begun to mature more recently, such as development, plasticity, and computational models of neural processing. In many cases, the series editors are joined by a coeditor having special expertise in the topic of the volume.

Richard R. Fay, Woods Hole, MA, USA
Arthur N. Popper, College Park, MD, USA

Volume Preface

Vocal production is a central topic in biological and evolutionary approaches to animal communication. An understanding of vocal production provides the crucial link between physics, physiology, and anatomy on the one hand and perception, neural processing, and evolution of communication signals on the other. The fundamental principles of vocal production in humans are well understood and are being increasingly extended to animal taxa. The origins of the principles of vocal production in speech science create a double barrier to entry for biologists interested in understanding acoustic communication in nonhuman species because the classic texts are now quite old, and because the traditional information sources are focused entirely on human speech, it is difficult for a newcomer to know what adjustments are necessary to adapt this body of theory and practice to different animal species. This volume aims to fill these gaps, providing easy-to-understand overviews of the various relevant theories and techniques and showing how acoustic principles can be applied to thc study of all main vertebrate groups.

The volume starts with a chapter by Tecumseh Fitch and Roderick Suthers, who provide a brief history of the origins of speech research and discuss some of the major issues that have arisen as investigators used human-based studies to better understand animal sound communication. Next, in Chap. 2, Eric Parmentier and Michael Fine consider issues of sound production in fishes, the largest of all vertebrate groups. They point out that the generally wide diversity among fishes carries over to mechanisms of sound production in different species. In Chap. 3, Kaitlen Colafrancesco and Marcos Gridi-Papp deal with the bioacoustics of amphibians and reptiles and show that these species vary greatly in the extent to which they use acoustic communication. Christopher James Clark, in Chap. 4, argues that motion-induced sound is a byproduct of essentially all behaviors, and he provides a focus on how this is the case for flight behavior. Birds continue to be considered in Chap. 5 where Daniel Düring and Coen Elemans apply a new approach toward integrative studies of birdsong.

In Chap. 6, Christian Herbst provides an overview of vocal production in mammals, while in Chap. 7, Peter Narins, Angela Stoeger, and Caitlin O'Connell-Rodwell provide an overview of seismic and infrasonic communication.

Mammalian sound production is again considered in Chap. 8 by Anna Taylor, Benjamin Charlton, and David Reby as they examine evidence for and against the hypothesis that social vocal signals may convey information about the fitness of the sender to the receiver. This is followed in Chap. 9 by Peter Tyack that reviews and discusses vocal plasticity and vocal leaning in a wide range of species.

Finally, in Chap. 10, Tecumseh Fitch provides an overview of the many unusual, and sometimes bizarre, modifications of the vocal apparatus that are known among terrestrial vertebrates and how much is still to be learned about vertebrate vocal production.

Sound production has not been covered extensively in SHAR, although the general topic of comparative bioacoustics has been part of a number of volumes that are in the subset of books in the series that deal with hearing of individual groups of animals. Most notably and most closely related to this volume is *Acoustic Communication* (Vol. 16, 2003; edited by Andrea Megela Simmons et al.). Additional books on comparative hearing include *Comparative Hearing: Mammals* (Vol. 4, 1993; edited by Richard R. Fay & Arthur N. Popper); *Hearing by Bats* (Vol. 5, 1995; edited by Arthur N. Popper & Richard R. Fay); *Comparative Hearing: Insects* (Vol. 10, 1998; edited by Ronald R. Hoy et al.); *Comparative Hearing: Fish and Amphibians* (Vol. 11, 1998; edited by Richard R. Fay & Arthur N. Popper); *Hearing by Whales and Dolphins* (Vol. 12, 2000; edited by Whitlow W. L. Au et al.); *Comparative Hearing: Birds and Reptiles* (Vol. 13, 2000; edited by Robert J. Dooling et al.); *Hearing and Sound Communication in Amphibians* (Vol. 28, 2007; edited by Peter M. Narins et al.); *Fish Bioacoustics* (Vol. 32, 2008; edited by Jacqueline F. Webb et al.); *Insights from Comparative Hearing Research* (Vol. 49, 2014; edited by Christine Kőppl et al.); and *Biosonar* (Vol. 51, 2014; edited by Annemarie Surlykke et al.).

Roderick A. Suthers, Bloomington, IN, USA
W. Tecumseh Fitch, Vienna, Austria
Richard R. Fay, Woods Hole, MA, USA
Arthur N. Popper, College Park, MD, USA

Contents

Contributors

Benjamin D. Charlton School of Biology and Environmental Science, Science Centre West, University College Dublin (UCD), Dublin 4, Ireland

Christopher James Clark Department of Biology, University of California, Riverside, Riverside, CA, USA

Kaitlen Colafrancesco Department of Biological Sciences, University of the Pacific, Stockton, CA, USA

Daniel N. Düring Department of Biology, University of Southern Denmark, Odense M, Denmark

Coen P.H. Elemans Department of Biology, University of Southern Denmark, Odense M, Denmark

Michael L. Fine Department of Biology, Virginia Commonwealth University, Richmond, VA, USA

W. Tecumseh Fitch Department of Cognitive Biology, University of Vienna, Vienna, Austria

Marcos Gridi-Papp Department of Biological Sciences, University of the Pacific, Stockton, CA, USA

Christian T. Herbst Voice Research Laboratory, Department of Biophysics, Palacký University, Olomouc, Czech Republic

Peter M. Narins Department of Integrative Biology and Physiology, University of California Los Angeles, Los Angeles, CA, USA

Department of Ecology & Evolutionary Biology, University of California Los Angeles, Los Angeles, CA, USA

Caitlin O'Connell-Rodwell Department of Otolaryngology, Head & Neck Surgery, Stanford University School of Medicine, Stanford, CA, USA

Eric Parmentier Laboratoire de Morphologie Fonctionnelle et Evolutive, Institut de Chimie, Université de Liège, Sart Tilman, Liège, Belgium

David Reby School of Psychology, University of Sussex, Falmer, UK

Angela S. Stoeger Department of Cognitive Biology, UZA 1, Biologiezentrum, Vienna, Austria

Roderick A. Suthers Medical Science, Indiana University, Bloomington, IN, USA

Anna M. Taylor School of Psychology, University of Sussex, Falmer, UK

Peter L. Tyack Sea Mammal Research Unit, Scottish Oceans Institute, St Andrews, Scotland

School of Biology, University of St Andrews, St Andrews, Scotland

Chapter 1
Vertebrate Vocal Production: An Introductory Overview

W. Tecumseh Fitch and Roderick A. Suthers

Abstract Vocal production is a central topic in biological and evolutionary approaches to animal communication, linking physics, physiology, and anatomy, on the one hand, with perception, neural processing, and evolution of communication signals, on the other. Understanding of vertebrate vocal production has increased greatly in the last two decades, mainly by building on an understanding of the physics and physiology of human vocal production initially developed by speech scientists. There is an increasing feeling among specialists in bioacoustics that this discipline has entered a new scientific era where the broad theoretical and physical underpinnings of vocal production, in a wide variety of vertebrate species, are solid and well-understood. Unfortunately, the origins of this understanding in speech science pose a problem for many nonspecialists, as the founding texts are highly technical and mathematical treatments written by and for engineers. This chapter, like the volume of which it is part, aims to provide a nontechnical introduction and overview of vertebrate vocal production written by and for biologists interested in vocal communication. The chapter provides a historical overview of the origins of two critical bodies of theory, the source-filter theory of vocal production and the myo-elastic aerodynamic theory of the voice source, and details how these theories were gradually applied to nonhuman animal vocalizations. The chapter ends with a summary of the rest of the chapters in the volume.

Keywords Animal bioacoustics • Animal communication • Formants • Functional morphology • Linear predictive coding • Source-filter theory • Spectrogram • Speech science • Voice science

W.T. Fitch (✉)
Department of Cognitive Biology, University of Vienna,
14 Althanstrasse, Vienna 1020, Austria
e-mail: tecumseh.fitch@univie.ac.at

R.A. Suthers
Medical Science, Indiana University, 1001 E. Third St., Bloomington, IN 47405, USA
e-mail: suthers@indiana.edu

R.A. Suthers et al. (eds.), *Vertebrate Sound Production and Acoustic Communication*, Springer Handbook of Auditory Research 53,
DOI 10.1007/978-3-319-27721-9_1

1.1 Introduction

Vocal production is a central topic in biological and evolutionary approaches to animal communication. An understanding of vocal production provides the crucial link between physics, physiology, and anatomy, on the one hand, and perception, neural processing, and evolution of communication signals, on the other.

The fundamental principles of vocal production in humans are well understood, stemming originally from engineering research on the human voice, and these principles are being increasingly extended to animal taxa, including birds, reptiles, and mammals. Unfortunately, the origins of the principles of vocal production in speech science create a double barrier to entry for biologists interested in understanding acoustic communication in nonhuman species. First, the classic texts are now quite old, and most are highly mathematical (e.g., van den Berg 1958; Fant 1960) and thus difficult for many biologists to locate, read, and understand. Second, because the traditional information sources are focused entirely on human speech, it is difficult for a newcomer to know what adjustments are necessary to adapt this body of theory and practice to different animal species. At present, many biologists eager to study acoustic communication in animals find these barriers so daunting that they turn to other fields or (worse) do a poor job studying animal vocalizations.

This volume aims to fill this gap, providing easy to understand overviews of the various relevant theories and techniques and showing how acoustic principles can be applied to the study of vocal production in all of the main vertebrate groups. The volume assembles chapters from the world's leading researchers on vertebrate vocal communication; the chapters are designed to be intelligible to a wide audience of biologists, including those with no background in engineering or human voice science. They highlight what is known and illustrate how to implement useful techniques and methodologies, but also identify and summarize current gaps in our knowledge. The hope is that this volume will serve both as a tutorial introduction for newcomers and a useful summary for experts, as well as a springboard for further research for all scientists interested in understanding animal acoustic signals and their role in communication. The editors and authors thus expect it to serve as a catalyst to help solidify and enrich the field as a whole.

The first part of this chapter provides a brief historical overview of the origins of speech science, by which the fundamental physical and physiological principles of vertebrate vocal production in one particular species—*Homo sapiens*—were uncovered. The chapter then turns to the slow process by which these principles were gradually applied to different taxa of nonhuman animals before providing a brief overview of the current status of contemporary vertebrate bioacoustics, where principles of human vocal production are becoming increasingly widely accepted as principles of vocal production in most (and perhaps all) other terrestrial vertebrates. Finally, the chapter ends with a brief overview of each of the other chapters in the volume.

1.2 The Rise of Speech Science

The study of vocal acoustics has, since its beginnings up to the last few decades, been focused on understanding human speech, and has always been more closely linked to engineering than to biology. The earliest systematic attempts to understand the acoustic nature of speech involved the creation of "speaking" machines. For example, in 1769 the Imperial Academy of Russia posed the question of the acoustic nature of vowels in the competition for its annual prize, asking "Can an instrument be constructed… which accurately expresses the sounds of these vowels?" and shortly afterward Wolfgang von Kemepelen created his famous speaking machine that could understandably imitate whole phrases in French or Italian (Dudley and Tarnoczy 1950; Linggard 1985). But despite this engineering success, von Kempelen did not understand the acoustic nature of vowels and conflated fundamental frequency and formants. Although the basic idea that vowels involve vocal tract resonances was already clear in the nineteenth century, Helmholtz believed that certain vowels such as /o/ and /u/ possessed a single resonance, whereas other vowels such as /i/ had two. Alexander Graham Bell correctly deduced that vowel quality is determined by two formants (Linggard 1985), and by 1930 Paget had, purely by ear, produced a relatively accurate chart of the lowest three formants for the English vowels (Paget 1930). Paget also experimented with excised larynges, had relatively accurate ideas about consonants, and recognized that singers often "tune" their formants to the pitch they are currently singing (cf. Sundberg 1975).

But real progress in this area did not attain takeoff velocity until the invention of the oscilloscope in the 1930s and especially the sound spectrograph in the 1940s, which provided objective, replicable, and concrete measurements of speech sounds that could be reproduced and published. From the confusing morass of speculation and debate that preceded it, the two main pillars of modern speech science emerged clearly and quite rapidly. These are source-filter theory regarding formant frequencies, and the myoelastic-aerodynamic (MEAD) theory regarding vocal fold vibration and fundamental frequency (f_0).

Source-filter theory was the first to cohere into common knowledge (Chiba and Kajiyama 1941; Fant 1960). The basic insights were already clearly stated, concerning independence of source and filter and the role of the larynx as source and vocal tract shape as filter, in 1941 by Chiba and Kajiyama in Japan (Chiba and Kajiyama 1941). These fundamental ideas, published during World War II, seem to have had little impact in the West until around 1950, when their importance was appreciated by some of the founding fathers of speech science at MIT, where Roman Jakobson's copy of Chiba and Kajiyama's book was circulating. Gunnar Fant, who was an electrical engineer visiting MIT at the time, read the book and rapidly recognized the fundamental importance of these ideas (Fant 2001), as did Kenneth Stevens and James Flanagan, who were students at the MIT Acoustics Lab. This group made use of the new measurement tools to verify and extend Chiba and Kajiyama's ideas, which culminated in Fant's magnum opus (Fant 1960) that essentially established source-filter theory as the standard modern model of speech production. The development and codification of the powerful spectral modeling technique of linear predictive coding (LPC) in the mid-1970s introduced source-filter theory to an entire generation of electrical engineers (Markel and Gray 1976), so that by the 1980s it was common knowledge among speech scientists.

Clarification of the MEAD theory took somewhat longer (van den Berg 1958, 1968; Titze 2006). In the 1940s it was already clear to most speech researchers that vibration of the vocal folds occurs passively. Thus, slow-changing physical parameters such as subglottal pressure and vocal fold tension are actively determined by the brain, but individual glottal pulses are generated by the physics of the system. However, in 1950 Husson advanced his alternative neurochronaxic theory, proposing that during human phonation each glottal pulse corresponds to a neural impulse traveling down the recurrent laryngeal nerve to incite a muscular contraction in the thyroarytenoid muscle (Husson 1955), implying that no airflow is required to vocalize (cf. Rubin 1960). Given that scientists had successfully generated realistic vocal fold vibrations in excised larynges for centuries by simply blowing air into them, it is surprising that this theory was ever taken very seriously regarding the human voice. However, it is now known that for a small subset of vocalizations in other vertebrates (e.g., cat purring) this neurochronaxic theory is indeed an accurate description: When cats purr, each individual sound pulse corresponds to a centrally driven muscular contraction (Remmers and Gautier 1972; Frazer Sissom et al. 1991). But for speech, by the late 1950s van den Berg could confidently state the essential principles of the MEAD theory (van den Berg 1958), and by the early 1960s the neurochronaxic theory was considered to have been convincingly refuted with respect to humans (Rubin 1960).

The coherence of speech science around these two key bodies of theory led to an explosion of research in speech technology, speech perception, and speech medicine (e.g., the development of vocal prosthetics) that continues today. In particular, the development of formant synthesizers provided an important new tool to investigate perceptual processes and to develop theories of speech perception. Although effective electronic synthesis of speech had already been accomplished by Dudley in 1936, by the mid-1950s speech synthesizers became important engines of speech perception research. In particular, the pattern playback machine devised at Haskins laboratories enabled researchers to control formants flexibly and generate synthetic speech stimuli with arbitrary formant patterns (Cooper et al. 1952). This new capacity for formant synthesis in turn led to important breakthroughs in understanding speech *perception*, such as the discovery of categorical perception (Liberman et al. 1957; Harnad 1987).

This brief history of the birth of speech science illustrates that once a solid understanding of the physics and physiology of speech production was in place, technological developments opened up a whole range of new research questions concerning not only the production or transmission of speech, but also its perception. As will be seen in Sect. 1.4, a similar pattern has unfolded much more recently in the animal bioacoustics literature.

1.3 Applying Principles of Speech Science to Nonhuman Animals

For obvious practical reasons, the first animal species in which the principles of vocal production were intensively studied, and understood, was our own: *Homo sapiens*. In turning to the application of these fundamental principles to vocal production in other vertebrates, there is a clear contrast with the steady and rapid progress that typified speech science. Rather, over many decades, a series of relatively isolated scientists made important discoveries for individual clades (especially bats, birds, and anurans) independently. It is only relatively recently that more general overarching principles, applicable to most vertebrates, have begun to be discerned and potential exceptions isolated.

One of the first groups in which vocal production was intensively studied were bats (Au and Suthers 2014). The discovery of bat echolocation in the early 1940s (Griffin and Galambos 1941) energized a large research community that attempted to understand the acoustic and neural basis of this remarkable ability (cf. Busnel and Fish 1980). Understanding how microchiropteran bats make their loud ultrasonic vocalizations was one important aspect of this system (Suthers and Fattu 1973), and by the late 1950s it was already clear that echolocation sounds were made in the larynx by vibrations of very thin vocal membranes, tensed by the cricothyroid muscles.

This conclusion was reached via multiple converging sources of data: cannulating the trachea, slitting the vocal membranes, cutting the superior laryngeal nerve to denervate the cricothyroid, and recording cricothyroid electromyograms simultaneously with ultrasound recording (Griffin 1958; Novick and Griffin 1961; Suthers 1988). Whether these laryngeal sounds are emitted from the mouth depends on the type of bat, with some bats being mouth emitters (e.g., members of the Vespertilionidae such as *Myotis*) and others being nostril emitters (e.g., members of the Rhinolophidae such as *Rhinolophus*) (Griffin 1958). That the source-filter theory applied to bat echolocation, and that there was an absence of source-tract coupling, was confirmed by placing bats in heliox and observing that, despite formant shifts, their f_0 does not shift (Pye 1967; Roberts 1973; Hartley and Suthers 1988). Thus by the late 1960s Pye was able to affirm explicitly that vocal production in bats follows the same essential principles as vowel production in humans (Pye 1967, p. 53ff.), with formants filtering a harmonic sound produced by tissue vibrations within the larynx. During the 1980s, detailed experimental work on bat vocal production was carried out by Roderick Suthers and his colleagues that verified and extended the applicability of source-filter theory to bats (Hartley and Suthers 1988; Suthers 1988; Suthers et al. 1988).

Another important focus of research was echolocation in toothed whales, the odontocetes. As for bats, much of this research was funded by the U.S. Navy in the hope of developing artificial echolocation systems that could perform at the remarkable level of echolocating animals (a goal that remains elusive today). Odontocetes produce two major classes of sounds: communicative sounds such as whistles and echolocating clicks that are typically uttered in bursts called click trains (Caldwell and Caldwell 1972; Evans 1973). A long debate about where these sounds were produced—in the larynx as for most mammals or in the highly complex and novel nasal sac system—was not definitively resolved until recently (Au and Suthers 2014). Although early work based on cineradiography (Dormer 1979) strongly suggested that the source of both sound types was the nasal sac system, even earlier work had suggested an alternative: that clicks might be produced in the nasal sacs and whistles in the larynx via some sort of whistle mechanism. It took heliox experiments to demonstrate that "whistle" is a misnomer for these vocalizations, as they are not produced using an aerodynamic whistle mechanism (Amundin 1991; Madsen et al. 2012). More recent work confirms that, at least in those species studied to date, both whistles and clicks are produced in the nasal sac system, with clicks produced in the larger right half of the system and whistles in the left (cf. Schenkkan 1973; Cranford et al. 1996). The odontocete nasal production system is thus analogous to the two-voice syrinx of songbirds: Both systems involve novel sound-producing sources, capable of producing two simultaneous sounds under separate control, and both involve novel musculature and neural control mechanisms. This reliance on a newly-evolved voice source may be relevant to the fact that the capacity for vocal learning appears to be pervasive in both of these clades, but is unusual elsewhere among vertebrates.

Application of source-filter theory to other mammals was more erratic. Although Philip Lieberman had discussed formants in the vocalizations of nonhuman primates (Lieberman 1968), his subsequent work suggesting that nonhuman primates have a restricted capacity to vary these formants (Lieberman et al. 1969) seems to have had the effect of dampening interest in primate formants until the mid-1990s (exceptions include Andrew 1976; Richman 1976). Edward Carterette investigated feline vocalization from a source-filter viewpoint in the late 1970s (Carterette et al. 1979), applying the then-new tool of linear predictive coding (LPC) to cat meows, and did some further work with colleagues on meow production (Carterette et al. 1984; Shipley et al. 1991), but this interesting line of research does not appear to have been picked up by other researchers or applied to other species until much later (cf. Owren and Bernacki 1998).

Finally, for rodents, there was some early interest in the mechanisms underlying ultrasound production, but this interest seems to have fizzled, which is all the more surprising because (Roberts 1975) suggested, based on heliox experiments, that rodent ultrasounds are made using a whistle mechanism in which source f_0 and tract are tightly coupled. If Roberts is correct, rodents would be the *only* known animals that use a wind instrument–like mechanism to produce vocalizations. In conclusion, work on mammalian vocal production remained relatively sporadic, except for bats and cetaceans, for several decades until the 1990s.

For vocal production in anurans (frogs and toads) the story is rather similar. Despite a long interest in the perception and evolution of vocalizations of this prominent vertebrate group (Blair 1963; Capranica 1968), research on vocal production has been quite limited until recently (e.g., Gridi-Papp et al. 2006;

Kime et al. 2013). Paulsen was able to produce high-speed films of vocal production in excised frog larynges in the 1960s (Paulsen 1967), and a series of highly insightful excised experiments on toad vocal production in the early 1970s by William Martin (1971, 1972; Martin and Gans 1972) seems to have fostered a slowly increasing interest in anuran production (Schmidt 1972; Gans 1973; Weber 1977). The unusual laryngeal mechanism used in underwater sound production by *Xenopus* became a well-studied system (Yager 1982, 1996; Kelley and Tobias 1999). But for "normal" terrestrial frogs, it was not until quite recently that vocal production again became a major focus of interest (see Colafrancesco and Gridi-Papp, Chap. 3 for a review).

Finally, in birds there has been a more consistent research focus on problems of vocal production, but many questions are more difficult to answer because of both the small body size of many laboratory species and the relative inaccessibility of the avian vocal source, the syrinx (located at the base of the trachea, deep within the torso). Much of the earliest work was on larger species such as ducks, chickens, and geese (Paulsen 1967; Lockner and Youngren 1976; Gaunt and Gaunt 1977) or based on careful spectrographic description rather than direct experiment (e.g., Greenewalt 1968). In the early 1970s, a review by Gaunt and Wells still expressed great caution about what the essential mode of avian vocal production might be, or even whether this mode is the same for all species (Gaunt and Wells 1973).

Although Greenewalt (1968) did not think that vocal tract filtering played any important role in avian vocal production, some researchers in the 1980s already recognized the importance of the filter, and source-filter theory, in at least some bird species (e.g., Suthers 1988). A prominent advance in this regard involved the use of heliox to demonstrate that birds singing in light gases do not shift f_0, but do show "unmasking" of harmonics that are normally suppressed by vocal tract filtering (Hersch 1966; Nowicki 1987). This work, along with later verification (Ballintijn and ten Cate 1998; Riede et al. 2004), showed that the essential features of the source-filter hypothesis also appear to apply to avian vocal production. The relatively "pure" signal typical of some bird calls, and much bird song, is achieved by vocal tract filtering and suppression of harmonics that were present in the glottal source (cf. Düring and Elemans, Chap. 5).

Important experimental research on avian vocal production continued through the 1980s and 1990s (Suthers and Hector 1988; Hartley and Suthers 1989; Goller and Suthers 1996). Nonetheless, even the fundamental question of what, exactly, vibrates within the syrinx to produce this source remained uncertain and debated, and it was not until Franz Goller and Ole Larsen's pioneering work with endoscopy that these tissues were finally visualized directly (Goller and Larsen 1997a, b; Larsen and Goller 2002). To the surprise of many, the equivalent of the vocal fold in songbirds is *not* the medial tympaniform membrane (MTM), as had been traditionally supposed based on anatomy alone, but rather the more massive syringeal labia (the medial and lateral labia). Thus, even for birdsong, in which there has been sustained interest, we have only recently reached the point where the fundamental principles of vocal production seem reasonably clear.

1.4 Contemporary Bioacoustics: Entering the Modern Era

Since the late 1990s progress in understanding vertebrate vocal production has been rapid. In birds, the delineation of the avian source mechanism led to the discovery of superfast muscles in the avian syrinx (Elemans et al. 2004), the mechanisms of avian vocal tract filtering have been elucidated (Riede et al. 2004, 2006), and high-resolution imaging has provided a new level of insight into syringeal anatomy (Düring et al. 2013). Songbirds, especially the zebra finch (*Taeniopygia guttata*), are now *the* model species for understanding the neuroscience of vocal learning (Doupe and Kuhl 1999; Marler and Slabbekoorn 2004; Catchpole and Slater 2008), and vocal control plays an important role in this body of research (Elemans 2014).

For anurans, excised larynx experiments have uncovered the mechanism of ultrasound production in certain frogs (Suthers et al. 2006), insights into sound filtering and radiation have been applied across multiple species (Mason and Narins 2002; Gridi-Papp 2008), and sophisticated computer models have been developed (Kime et al. 2013). For more detail see Colafrancesco and Gridi-Papp (Chap. 3).

Progress has been particularly impressive in understanding mammalian vocalization (see Taylor et al., Chap. 8). After many years of uncertainty, the production mechanism underlying "vocal" production in toothed cetaceans has been clarified: A complex novel organ in the nasal cavities (Cranford et al. 1996; Madsen et al. 2003). The hypothesis that dolphin "whistles" are produced by a whistle-like mechanism has been falsified by using heliox to show that, as for most other mammals, f_0 does not shift when formants are modified (Madsen et al. 2012). For primates, the realization that formant frequencies can play important roles signaling size (Fitch 1997, 2000) or individual identity (Rendall et al. 1998; Rendall 2003) has sparked a renewed interest in the role of formants in primate vocal production (Riede et al. 2005; Ey et al. 2007; Ghazanfar et al. 2007), including both heliox experiments (Koda et al. 2012) and playback experiments in which formant frequencies are experimentally manipulated (Fitch and Fritz 2006). Similar insights have been applied to both production and perception in red deer (*Cervus elaphus*; Fitch and Reby 2001; Reby and McComb 2003; Charlton et al. 2007), fallow deer (*Dama dama*; McElligott et al. 2006; Vannoni and McElligott 2007), and koalas (*Phascolarctos cinereus*; Charlton et al. 2011, 2012), along with a variety of other mammalian species (McComb et al. 2003; Frey et al. 2007; Sanvito et al. 2007).

The applicability of MEAD to nonhuman species has taken longer to become a research focus. Despite the clear and early recognition of the fundamental principles of vocal fold vibration in humans (van den Berg 1968), it is only recently that researchers have begun to ask how widely applicable these principles are across the animal kingdom. A number of early excised larynx studies focused mainly on finding a suitable animal "model" for simulating human vocal fold physiology (Cox et al. 1999; Kim et al. 2004; Alipour and Jaiswal 2009). More recently, animal laryngeal dynamics have become a focus of interest in their own right, for example in tigers (*Panthera tigris*; Titze et al. 2010) or African elephants *Loxodonta africana* (Herbst et al. 2012). Models initially developed for understanding human laryngeal vibrations (Steinecke and Herzel 1995) have been applied to a wide range of nonhuman species (Mergell et al. 1999; Zaccarelli et al. 2006).

An area of particular recent interest has been so-called "nonlinear phenomena" (NLP)—that class of sounds that show irregular "noisy" spectra, along with changes in source vibration frequency such as period doubling or tripling. It first became clear that nonlinear dynamics and chaos theory applied to mammalian voices in the context of human baby crying (Mende et al. 1990), and similar techniques were rapidly applied to adult human speech (Herzel et al. 1991; Herzel and Wendler 1991; Herzel 1993). However, because NLP are relatively rare in the speech of healthy adults, they seem to be of rather limited interest for non-clinicians. In contrast, for animals such phenomena are often much more common, or even pervasive (Wilden et al. 1998; Riede et al. 2000; Fitch et al. 2002). For many bioacousticians the discovery of these phenomena opened the door to analysis of a large class of vocalizations that had previously gone unstudied, across a range of vertebrates from frogs and lizards (Feng et al. 2009; Labra et al. 2013) to birds (Fee et al. 1998; Owren and Rendall 2003; Suthers et al. 2006) and primates (Berry et al. 1996; Brown et al. 2003; Owren and Rendall 2003). NLP thus appear to be an example of a field in which animal bioacoustics is now leading the way and advancing basic bioacoustics beyond what had been previously studied for human speech (e.g., Blumstein et al. 2010).

As highlighted in Sect. 1.3, several clades of vertebrates have evolved novel sound-producing sources. These include all of the sounds produced by fish, except perhaps lungfish, which have a primitive larynx and are reputed to produce squeaks, burps, and "vocal sounds" (M'Donnell 1860). All other fish sounds are produced non-laryngeally by a wide variety of novel mechanisms of diverse mechanical functions and anatomical origins. Among tetrapods, the avian syrinx and odontocete nasal sac system are examples of novel vocal sources, as mentioned in Sect. 1.3. Although the details of aerodynamic and mechanical function remain imperfectly understood in these systems, neither system exhibits an appreciable fundamental frequency shift during vocalization in a heliox atmosphere by cetaceans (Amundin 1991; Madsen et al. 2012) or birds (Hersch 1966; Nowicki 1987; Ballintijn and ten Cate 1998). This indicates that at least basic source-filter independence is conserved in these systems, as for other tetrapods (Beil 1962; Brauer et al. 1966; Rand and Dudley 1993). The relative inaccessibility of the avian and cetacean source makes their direct endoscopic viewing a challenge, as does their small size in birds, so excised preparations or in vivo experiments on anesthetized animals may be the simplest way to gain deeper insights into the operation of the syrinx and/or nasal sacs. But at least for birds, current understanding and models are clearly consistent with the idea that the MEAD theory applies to the vibrating membranes of the avian syrinx (Gardner et al. 2001; Mindlin and Laje 2005; Zaccarelli et al. 2006).

1.5 Overview of This Volume

With the historical development of the essential principles of vertebrate vocal production now clarified, this introduction ends with a brief overview of the rest of the chapters in this book.

Chapter 2 focuses on sound production in fishes. Eric Parmentier and Michael L. Fine point out that a restricted number of fish families produce sound for social communication but find that the mechanisms for sound production are so variable that it is not possible to classify them into useful subcategories. Although all fishes can hear sound, only a minority can produce it. The authors suggest that acoustic communication in fishes came about via exaptation, in which a structure that produced sound as a byproduct of other activity was then over taken over or co-opted for a new use that enhanced acoustic communication.

In Chap. 3, Kaitlen Colafrancesco and Marcos Gridi-Papp report that reptiles and amphibians vary greatly in the extent to which they use acoustic communication. Among the most prominent vocalists are frogs that call in choruses. Sound is produced by air flowing between a pair of vocal cords in the larynx. Four external laryngeal muscles control the position and tension of the vocal cords. Male frogs raised in isolation produce normal calls as adults, and there is no evidence of vocal learning in anurans. Salamanders produce soft hissing or clicks produced by snapping their mouth shut, but these sounds appear to have little or no communicative value. More interesting vocal behavior is displayed by some reptiles including tortoises, during courtship and mating. Crocodilians, along with some lizards (e.g., geckos), also have a rich vocal repertoire produced mainly during sexual and agonistic behavior.

In Chap. 4, Christopher James Clark argues that some motion-induced sound is a byproduct of nearly all motor behavior. The term "sonation" was proposed by Bostwick and Prum (2003) to mean nonvocal modulated sounds that have evolved specializations for communication. Clark discusses some of the problems that arise when attempting to apply this term rigorously. Rigor is also needed when using vision and hearing to determine the relationship between sound and motion in a flight display. Clark describes a flight pendulum display that is performed by Allen's hummingbird in which the bird flips its tail up and down in synchrony with rapid song notes. Previous investigators assumed the sound was produced by the synchronized tail flips. However, a tail feather source of the song was ruled out when it was shown that removing portions of tail feathers did not affect the song. Clark solved the mystery when he showed that normal speed video is too slow to follow the wing movements; high-speed video showed that wing beats were synchronized with the song.

In Chap. 5, Daniel N. Düring and Coen P. H. Elemans apply a promising new approach toward the integrative study of birdsong. Instead of focusing predominantly on either neural aspects (the central song system) or on the peripheral vocal system, as typical of much past work, they advocate a systems view of embodied motor control. The authors argue that recent emphasis on biological research stresses the need for a systems view of motor control that includes both mechanical and neural feedback and takes account of the fact that muscle forces depend on body motion through nonlinear force–length and force–velocity properties of the muscles.

Christian T. Herbst is a biophysicist studying both the human voice and vocal mechanisms in other mammals, and in Chap. 6 he provides an overview of the physics of vocal production in mammals. It has been shown over the past several decades that laryngeal vocalizations do not always consist of simple harmonics and their overtones. Instead the mammalian larynx is capable of producing complex irregular vocalizations collectively referred to as nonlinear phenomena. These complex vocalization can include bifurcations, subharmonics, or deterministic chaos. Their function remains unclear, but it has been suggested that such phenomena may provide the mammal with a complex vocal repertoire without requiring it to evolve correspondingly complex neural motor circuits to control production. Nonlinear dynamics are also involved in "normal" MEAD sound production and vibration of coupled nonlinear laryngeal oscillators, and thus clearly play important roles in vocal production.

In Chap. 7, Peter M. Narins, Angela S. Stoeger, and Caitlin O'Connell-Rodwell provide an overview of seismic and infrasonic communication in terrestrial mammals. The airborne range for infrasonic frequency is less than 20 Hz. During seismic communication, sound energy is transmitted through solid substances such as the ground. A particularly interesting example of seismic communication is observed in blind mole rats, which both send and detect substrate-borne vibrations, received using a unique middle ear morphology and a jaw-based listening device. Despite their seemingly tactile nature, their communication signals are processed mainly by the auditory system instead of the somatosensory system. There is evidence that blind mole rats can estimate physical properties of underground obstacles based on self-generated reflections, which has been called "seismic echolocation."

In Chap. 8, Anna M. Taylor, Benjamin D. Charlton, and David Reby again consider mammalian vocalization, reviewing evidence for and against the hypothesis that social vocal signals may convey information about the fitness of the sender to the receiver. Nonhuman animals use vocal interactions to resolve many social interactions. Receivers of these interchanges may benefit from the opportunity to evaluate "honest" information that results from biomechanical constraints on acoustic variables. An essential assumption of the source-filter theory is that the vibration frequency of the source is independent from the filter (feedback from the vocal tract filter does not control the fundamental frequency of the glottal source). Vocalizations may thus contain information concerning body size, sex, and/or age, and so forth, and the outcome of vocal interactions may often depend on "honest" biomechanical constraints associated with vocal production.

In Chap. 9, Peter L. Tyack reviews and discusses vocal plasticity and vocal learning. Vocal learning varies in complexity but can generally be regarded as modification of one's vocalizations by reference to auditory feedback. Striking examples of vocal learning can be found in the ability of some nonhuman animals to imitate human speech. Examples include mynah birds, which can learn to match both the fundamental frequency and consonants of their owners' speech accurately, or a male Asian elephant named "Koshik" that was able to raise the normally low frequencies of his large vocal tract and modify formants to generate speech by inserting the tip of his trunk into his mouth. His speech imitation precisely matched the first two formants of the trainer's voice. Despite these examples, vocal learning is not well developed in nonhuman primates: Winter et al. (1973) found that infant squirrel monkeys raised with a muted mother developed calls that were essentially identical to those of normal mothers. In songbirds, the "template model" of song learning separates the formation of auditory learning from the process of auditory feedback. Some mammals (e.g., harbor seals) appear to have developed a template model similar to that of songbirds, in which the vocal motor learning phase does not develop until sexual maturity.

Tyack also discusses the extent to which many animals vary their vocal output to compensate for fluctuation in auditory noise during acoustic communication. Vocal output can be modulated based on auditory evaluation of environmental noise, to balance the need for effective communication against associated costs and risks. How does an animal select a relationship between auditory input and vocal motor output when communicating in varying noise? The best known compensation mechanism is the Lombard effect, which was first described in humans who attempted to compensate for background noise by speaking more loudly, lengthening their utterances and increasing the sound frequencies. This Lombard effect has subsequently been found in birds and other mammals.

Finally, in Chap. 10, W. Tecumseh Fitch provides an overview of the many unusual, and sometimes bizarre, modifications of the vocal apparatus that are known among terrestrial vertebrates. The chapter illustrates how much is left to learn about vertebrate vocal production, in part because the advances in bioacoustics and voice science, reviewed in Sects. 1.2–1.4, occurred long after the "golden age" of comparative anatomy during which most vocal modifications were discovered and described. Unusual modifications of the vocal source include giant larynges, like those of howler monkeys (*Alouatta* spp.), or of the hammerhead bat (*Hypsignathus monstrosus*), where the larynx has enlarged so much that it fills the thoracic cavity. Modifications of the vocal tract are also common, for example, the elongated trachea seen in many bird species. Fitch also provides an overview of laryngeal air sacs. Although such air sacs are quite common in mammals, their acoustic function remains mysterious. This final chapter illustrates that despite major advances in vocal bioacoustics, documented in this volume, there are many intriguing phenomena concerning vertebrate vocal production that remain poorly understood and in some cases essentially unstudied.

1.6 Conclusions

This chapter has provided a concise overview of the rapid rise of speech science in the 1960s as soon as the core theoretical insights—source-filter theory and myoelastic aerodynamic theory—became available. Although the application of these insights to nonhuman animals has taken much longer, this is perhaps unsurprising given both the great diversity of animal vocal production systems and the much smaller amount of research funding available for any single species. In the last two decades, however, the situation has improved substantially and the field of animal bioacoustics seems to be entering a "golden age" perhaps equivalent to that seen for human speech science in the 1960s and 1970s.

In certain bioacoustic domains, for example the study of nonlinear phenomena, we can expect animal studies to lead the way because of the more frequent occurrence of such phenomena in animal communication. We can also expect that, among the many species that have not yet been studied, there will be new surprises and perhaps even exceptions to the "standard" theories (e.g., concerning whistle-like ultrasonic vocalization in rodents; see Sect. 1.3). Finally, playback experiments in which increased understanding of vocal production allows behavioral biologists to modify animal vocal sounds in very precise ways, or even synthesize vocalizations "from scratch," can certainly be expected to provide a crucial new source of understanding of animal communication and its evolution. This volume, it is hoped, will provide an accessible introduction to the study of vertebrate vocal production that will help fuel such progress.

References

Alipour, F., & Jaiswal, S. (2009). Glottal airflow resistance in excised pig, sheep, and cow larynges. *Journal of Voice, 23*(1), 40–50.

Amundin, M. (1991). Helium effects on the click frequency spectrum of the harbor porpoise, *Phocoena phocoena. Journal of the Acoustical Society of America, 90*, 53–59.

Andrew, R. J. (1976). Use of formants in the grunts of baboons and other nonhuman primates. *Annals of the New York Academy of Sciences, 280*, 673–693.

Au, W. W. L., & Suthers, R. A. (2014). Production of biosonar signals: Structure and form. In A. Surlykke, P. E. Nachtigall, R. R. Fay, & A. N. Popper (Eds.), *Biosonar* (pp. 61–105). New York: Springer.

Ballintijn, M., & ten Cate, C. (1998). Sound production in collared doves: A test of the "whistle" hypothesis. *Journal of Experimental Biology, 201*, 1637–1649.

Beil, R. G. (1962). Frequency analysis of vowels produced in a helium-rich atmosphere. *Journal of the Acoustical Society of America, 34*, 347–349.

Berry, D. A., Herzel, H., Titze, I. R., & Story, B. (1996). Bifurcations in excised larynx experiments. *Journal of Voice, 10*, 129–138.

Blair, W. F. (1963). Acoustic behaviour of Amphibia. In R. Busnel (Ed.), *Acoustic behavior of animals* (pp. 694–708, 803–804). Amsterdam: Elsevier.

Blumstein, D. T., Davitian, R., & Kaye, P. D. (2010). Do film soundtracks contain nonlinear analogues to influence emotion? *Biology Letters, 6*(6), 751–754.

Bostwick, K. S., & Prum, R. O. (2003). High-speed video analysis of wing-snapping in two manakin clades (Pipridae: Aves). *Journal of Experimental Biology, 206*, 3693–3706.

Brauer, R. W., Jennings, R. A., & Poulter, T. C. (1966). *The effect of substituting helium and oxygen for air on the vocalization of the California sea lion,* Zalophus californianus. Paper presented at the Third Annual Conference on Biological Sonar and Diving Mammals, Fremont, CA.

Brown, C. H., Alipour, F., Berry, D. A., & Montequin, D. (2003). Laryngeal biomechanics and vocal communication in the squirrel monkey (*Saimiri boliviensis*). *Journal of the Acoustical Society of America, 113*, 2114–2126.

Busnel, R.-G., & Fish, J. F. (Eds.). (1980). *Animal sonar systems*. New York: Plenum Press.

Caldwell, M. C., & Caldwell, D. K. (1972). Vocal mimicry in the whistle mode by an Atlantic bottlenosed dolphin. *Cetology, 9*, 1–8.

Capranica, R. R. (1968). Vocal repertoire of the bullfrog (*Rana catesbeiana*). *Behavior, 31*, 302–325.

Carterette, E. C., Shipley, C., & Buchwald, J. (1979). Linear prediction theory of vocalization in cat and kitten. In B. Lindblom & S. Ohman (Eds.), *Frontiers in speech communication research* (pp. 245–257). New York: Academic Press.

Carterette, E. C., Shipley, C., & Buchwald, J. S. (1984). On synthesizing animal speech: The case of the cat. In G. Bristow (Ed.), *Electronic speech synthesis: techniques, technology, and applications* (pp. 292–302). New York: McGraw-Hill.

Catchpole, C. K., & Slater, P. J. B. (2008). *Bird song: Themes and variations* (2nd ed.). New York: Cambridge University Press.

Charlton, B. D., Ellis, W. A. H., Brumm, J., Nilsson, K., & Fitch, W. T. (2012). Female koalas prefer bellows in which lower formants indicate larger males. *Animal Behaviour, 84*(6), 1565–1571.

Charlton, B. D., Ellis, W. A. H., McKinnon, A. J., Cowin, G. J., Brumm, J., Nilsson, K., et al. (2011). Cues to body size in the formant spacing of male koala (*Phascolarctos cinereus*) bellows: Honesty in an exaggerated trait. *Journal of Experimental Biology, 214*, 3414–3422.

Charlton, B. D., Reby, D., & McComb, K. (2007). Female perception of size-related formants in red deer, *Cervus elaphus*. *Animal Behaviour, 74*, 707–714.

Chiba, T., & Kajiyama, M. (1941). *The vowel: Its nature and structure*. Tokyo: Tokyo-Kaiseikan.

Cooper, F. S., Delattre, P. C., Liberman, A. M., Borst, J. M., & Gerstman, L. J. (1952). Some experiments on the perception of synthetic speech sounds. *Journal of the Acoustical Society of America, 24*, 597–606.

Cox, K. A., Alipour, F., & Titze, I. R. (1999). Geometric structure of the human and canine cricothyroid and thyroarytenoid muscles for biomechanical applications. *Annals of Otology, Rhinology and Laryngology, 108*, 1151–1158.

Cranford, T. W., Amundin, M., & Norris, K. S. (1996). Functional morphology and homology in the odontocete nasal complex: Implications for sound generation. *Journal of Morphology, 228*, 223–285.

Dormer, K. J. (1979). Mechanisms of sound production and air recycling in delphinids: Cineradiographic evidence. *Journal of the Acoustical Society of America, 65*, 229–239.

Doupe, A. J., & Kuhl, P. K. (1999). Birdsong and human speech: Common themes and mechanisms. *Annual Review of Neuroscience, 22*, 567–631.

Dudley, H., & Tarnoczy, T. H. (1950). The speaking machine of Wolfgang von Kempelen. *Journal of the Acoustical Society of America, 22*(2), 151–166.

Düring, D. N., Ziegler, A., Thompson, C. K., Ziegler, A., Faber, C., Müller, J. et al. (2013). The songbird syrinx morphome: A three-dimensional, high-resolution, interactive morphological map of the zebra finch vocal organ. *BMC Biology, 11*(1). doi:10.1186/1741-7007-11-1. Retrieved from http://www.biomedcentral.com/1741-7007/11/1.

Elemans, C. P. H. (2014). The singer and the song: The neuromechanics of avian sound production. *Current Opinion in Neurobiology, 28*, 172–178.

Elemans, C. P. H., Spierts, I. L. Y., Müller, U. K., van Leeuwen, J. L., & Goller, F. (2004). Bird song: Superfast muscles control dove's trill. *Nature, 431*, 146.

Evans, W. E. (1973). Echolocation by marine delphinids and one species of fresh-water dolphin. *Journal of the Acoustical Society of America, 54*, 191–199.

Ey, E., Pfefferle, D., & Fischer, J. (2007). Do age- and sex-related variations reliably reflect body size in non-human primate vocalizations? A review. *Primates, 48*(4), 253–267.

Fant, G. (1960). *Acoustic theory of speech production*. The Hague, Netherlands: Mouton & Co.

Fant, G. (2001). T Chiba and M Kajiyama: Pioneers in speech acoustics. *Quarterly Progress and Status Report, 42*(1), 59–60.

Fee, M. S., Shraiman, B., Pesaran, B., & Mitra, P. P. (1998). The role of nonlinear dynamics of the syrinx in the vocalizations of a songbird. *Nature, 395*, 67–71.

Feng, A. S., Riede, T., Arch, V. S., Yu, Z., Xu, Z.-M., Yu, X.-J., et al. (2009). Diversity of the vocal signals of concave-eared torrent frogs (*Odorrana tormota*): Evidence for individual signatures. *Ethology, 115*, 1015–1028.

Fitch, W. T. (1997). Vocal tract length and formant frequency dispersion correlate with body size in rhesus macaques. *Journal of the Acoustical Society of America, 102*(2), 1213–1222.

Fitch, W. T. (2000). Skull dimensions in relation to body size in nonhuman mammals: The causal bases for acoustic allometry. *Zoology, 103*(1–2), 40–58.

Fitch, W. T., & Fritz, J. B. (2006). Rhesus macaques spontaneously perceive formants in conspecific vocalizations. *Journal of the Acoustical Society of America, 120*(4), 2132–2141.

Fitch, W. T., Neubauer, J., & Herzel, H. (2002). Calls out of chaos: The adaptive significance of nonlinear phenomena in mammalian vocal production. *Animal Behaviour, 63*, 407–418.

Fitch, W. T., & Reby, D. (2001). The descended larynx is not uniquely human. *Proceedings of the Royal Society of London B: Biological Sciences, 268*(1477), 1669–1675.

Frazer Sissom, D. E., Rice, D. A., & Peters, G. (1991). How cats purr. *Journal of Zoology (London), 223*, 67–78.

Frey, R., Gebler, A., Fritsch, G., Nygrén, K., & Weissengruber, G. E. (2007). Nordic rattle: The hoarse vocalization and the inflatable laryngeal air sac of reindeer (*Rangifer tarandus*). *Journal of Anatomy, 210*, 131–159.

Gans, C. (1973). Sound production in the Salientia: Mechanism and evolution of the emitter. *American Zoologist, 13*, 1179–1194.

Gardner, T., Cecchi, G., Magnasco, M., Laje, R., & Mindlin, G. B. (2001). Simple motor gestures for birdsongs. *Physical Review Letters, 87*, 28101–28104.

Gaunt, A. S., & Gaunt, S. L. L. (1977). Mechanics of the syrinx in *Gallus gallus*: II Electromyographic studies of *ad libitum* vocalizations. *Journal of Morphology, 152*, 1–20.

Gaunt, A. S., & Wells, M. K. (1973). Models of syringeal mechanisms. *American Zoologist, 13*, 1227–1247.

Ghazanfar, A. A., Turesson, H. K., Maier, J. X., van Dinther, R., Patterson, R. D., & Logothetis, N. K. (2007). Vocal-tract resonances as indexical cues in rhesus monkeys. *Current Biology, 17*(5), 425–430.

Goller, F., & Larsen, O. N. (1997a). A new mechanism of sound generation in songbirds. *Proceedings of the National Academy of Sciences of the U S A, 94*(26), 14787.

Goller, F., & Larsen, O. N. (1997b). *In situ* biomechanics of the syrinx and sound generation in pigeons. *Journal of Experimental Biology, 200*(16), 2165–2176.

Goller, F., & Suthers, R. A. (1996). Role of syringeal muscles in gating airflow and sound production in singing brown thrashers. *Journal of Neurophysiology, 75*(2), 867–876.

Greenewalt, C. H. (1968). *Bird song: Acoustics and physiology*. Washington, DC: Smithsonian Institution Press.

Gridi-Papp, M. (2008). The structure of vocal sounds produced with the mouth closed or with the mouth open in treefrogs. *Journal of the Acoustical Society of America, 123*, 2895–2902.

Gridi-Papp, M., Rand, A. S., & Ryan, M. J. (2006). Complex call production in the túngara frog. *Nature, 441*, 38.

Griffin, D. R. (1958). *Listening in the dark*. New Haven, CT: Yale University Press.

Griffin, D. R., & Galambos, R. (1941). The sensory basis of obstacle avoidance by flying bats. *Journal of Experimental Zoology, 86*, 481–506.

Harnad, S. R. (1987). *Categorical perception: The groundwork of cognition*. Cambridge, England: Cambridge University Press.

Hartley, D. J., & Suthers, R. A. (1988). The acoustics of the vocal tract in the horseshoe bat, *Rhinolophus hildebrandti. Journal of the Acoustical Society of America, 84*, 1201–1213.

Hartley, R. S., & Suthers, R. A. (1989). Airflow and pressure during canary song: Direct evidence for mini-breaths. *Journal of Comparative Physiology, A: Sensory, Neural, and Behavioral Physiology, 165*, 15–26.

Herbst, C. T., Stoeger, A. S., Frey, R., Lohscheller, J., Titze, I. R., Gumpenberger, M., et al. (2012). How low can you go? Physical production mechanism of elephant infrasonic vocalizations. *Science, 337*(6094), 595–599.

Hersch, G. L. (1966). *Bird voices and resonant tuning in helium air mixtures*. PhD dissertation, University of California, Berkeley.

Herzel, H. (1993). Bifurcations and chaos in voice signals. *Applied Mechanics Reviews, 46*, 399–413.

Herzel, H., Steinecke, I., Mende, W., & Wermke, K. (1991). Chaos and bifurcations during voiced speech. In E. Mosekilde & L. Mosekilde (Eds.), *Complexity, chaos and biological evolution* (pp. 41–50). New York: Plenum Press.

Herzel, H., & Wendler, J. (1991). Evidence of chaos in phonatory samples. *Proceedings Eurospeech Genova, 1991*, 263–266.

Husson, R. (1955). Données expérimentales nouvelles concernant la physiologie de la phonation. *Exposés Annuelles d'Otorhinolaryngolie, 1955*, 287.

Kelley, D. B., & Tobias, M. L. (1999). Vocal communication in *Xenopus laevis*. In M. D. Hauser & M. Konishi (Eds.), *The design of animal communication* (pp. 9–35). Cambridge, MA: MIT Press/Bradford.

Kim, M. J., Hunter, E. J., & Titze, I. R. (2004). Comparison of human, canine and ovine laryngeal dimensions. *Annals of Otology, Rhinology and Laryngology, 113*, 60–68.

Kime, N. M., Ryan, M. J., & Wilson, P. S. (2013). A bond graph approach to modeling the anuran vocal production system. *Journal of the Acoustical Society of America, 133*(6), 4133–4144.

Koda, H., Nishimura, T., Tokuda, I. T., Oyakawa, C., Nihonmatsu, T., & Masataka, N. (2012). Soprano singing in gibbons. *American Journal of Physical Anthropology, 149*(3), 347–355.

Labra, A., Silva, G., Norambuena, F., Velásquez, N., & Penna, M. (2013). Acoustic features of the weeping lizard's distress call. *Copeia, 2013*(2), 206–212.

Larsen, O. N., & Goller, F. (2002). Direct observations of syringeal muscle function in songbirds and a parrot. *Journal of Experimental Biology, 203*, 25–35.

Liberman, A. M., Harris, K. S., Hoffman, H. S., & Griffith, B. C. (1957). The discrimination of speech sounds within and across phoneme boundaries. *Journal of Experimental Psychology, 53*, 358–368.

Lieberman, P. (1968). Primate vocalization and human linguistic ability. *Journal of the Acoustical Society of America, 44*(6), 1574–1584.

Lieberman, P. H., Klatt, D. H., & Wilson, W. H. (1969). Vocal tract limitations on the vowel repertoires of rhesus monkeys and other nonhuman primates. *Science, 164*, 1185–1187.

Linggard, R. (1985). *Electronic synthesis of speech*. Cambridge, England: Cambridge University Press.

Lockner, F. R., & Youngren, O. M. (1976). Functional syringeal anatomy in the mallard. I. In situ electromyograms during ESB elicited calling. *Auk, 93*, 324–342.

M'Donnell, R. (1860). Observations on the habits and anatomy of the *Lepidosiren annexes*. *Natural History Review, 7*, 93–112.

Madsen, P. T., Carder, D. A., Au, W. W. L., Nachtigall, P. E., Møhl, B., & Ridgway, S. H. (2003). Sound production in neonate sperm whales. *Journal of the Acoustical Society of America, 113*(6), 2988–2991.

Madsen, P. T., Jensen, F. H., Carder, D., & Ridgway, S. H. (2012). Dolphin whistles: A functional misnomer revealed by heliox breathing. *Biology Letters, 8*, 211–213.

Markel, J. D., & Gray, A. H. (1976). *Linear prediction of speech*. New York: Springer.

Marler, P., & Slabbekoorn, H. (2004). *Nature's music: The science of birdsong*. New York: Academic Press.

Martin, W. F. (1971). Mechanics of sound production in toads of the genus *Bufo*: Passive elements. *Journal of Experimental Zoology, 176*, 273–294.

Martin, W. F. (1972). Evolution of vocalizations in the genus *Bufo*. In W. F. Blair (Ed.), *Evolution in the genus Bufo* (pp. 279–309). Austin, TX: University of Texas Press.

Martin, W. F., & Gans, C. (1972). Muscular control of the vocal tract during release signaling in the toad *Bufo valliceps*. *Journal of Morphology, 137*, 1–28.

Mason, M., & Narins, P. M. (2002). Vibrometric studies of the middle ear of the bullfrog (*Rana catesbeiana*) I: The extrastapes. *Journal of Experimental Biology, 205*, 3153–3165.

McComb, K., Reby, D., Baker, L., Moss, C., & Sayialel, S. (2003). Long-distance communication of social identity in African elephants. *Animal Behaviour, 65*, 317–329.

McElligott, A. G., Birrer, M., & Vannoni, E. (2006). Retraction of the mobile descended larynx during groaning enables fallow bucks (*Dama dama*) to lower their formant frequencies. *Journal of Zoology, 270*, 340–345.

Mende, W., Herzel, H., & Wermke, K. (1990). Bifurcations and chaos in newborn infant cries. *Physics Letters A, 145*(8–9), 418–424.
Mergell, P., Fitch, W. T., & Herzel, H. (1999). Modeling the role of non-human vocal membranes in phonation. *Journal of the Acoustical Society of America, 105*, 2020–2028.
Mindlin, G. B., & Laje, R. (2005). *The physics of birdsong*. New York: Springer.
Novick, A., & Griffin, D. R. (1961). Laryngeal mechanisms in bats for the production of orientation sounds. *Journal of Experimental Zoology, 148*(2), 125–145.
Nowicki, S. (1987). Vocal tract resonances in oscine bird sound production: Evidence from birdsongs in a helium atmosphere. *Nature, 325*, 53–55.
Owren, M. J., & Bernacki, R. H. (1998). Applying linear predictive coding (LPC) to frequency-spectrum analysis of animal acoustic signals. In S. L. Hopp, M. J. Owren, & C. S. Evans (Eds.), *Animal acoustic communication: Sound analysis and research methods* (pp. 130–162). New York: Springer.
Owren, M. J., & Rendall, D. (2003). Salience of caller identity in rhesus monkey (*Macaca mulatta*) coos and screams: Perceptual experiments with human (*Homo sapiens*) listeners. *Journal of Comparative Psychology, 117*, 380–390.
Paget, R. A. S. (1930). *Human speech*. London: Kegan Paul, Trench, Trubner and Co.
Paulsen, K. (1967). *Das Prinzip der Stimmbildung in der Wirbeltierreihe und beim Menschen*. Frankfurt, Germany: Akademische Verlagsgesellschaft.
Pye, J. D. (1967). Synthesizing the waveforms of bat's pulses. In R.-G. Busnel (Ed.), *Animal sonar systems* (pp. 69–83). Jouy-en-Josas, France: Laboratoire de Physiologie Acoustique.
Rand, A. S., & Dudley, R. (1993). Frogs in helium: The anuran vocal sac is not a cavity resonator. *Physiological Zoology, 66*(5), 793–806.
Reby, D., & McComb, K. (2003). Anatomical constraints generate honesty: acoustic cues to age and weight in the roars of red deer stags. *Animal Behaviour, 65*, 519–530.
Remmers, J. E., & Gautier, H. (1972). Neural and mechanical mechanisms of feline purring. *Respiration Physiology, 16*(3), 351–361.
Rendall, D. (2003). Acoustic correlates of caller identity and affect intensity in the vowel-like grunt vocalizations of baboons. *Journal of the Acoustical Society of America, 113*, 3390–3402.
Rendall, D., Owren, M. J., & Rodman, P. S. (1998). The role of vocal tract filtering in identity cueing in rhesus monkey (*Macaca mulatta*) vocalizations. *Journal of the Acoustical Society of America, 103*(1), 602–614.
Richman, B. (1976). Some vocal distinctive features used by gelada monkeys. *Journal of the Acoustical Society of America, 60*, 718–724.
Riede, T., Beckers, G. J. L., Blevins, W., & Suthers, R. A. (2004). Inflation of the esophagus and vocal tract filtering in ring doves. *Journal of Experimental Biology, 207*, 4025–4036.
Riede, T., Bronson, E., Hatzikirou, H., & Zuberbühler, K. (2005). Vocal production mechanisms in a non-human primate: Morphological data and a model. *Journal of Human Evolution, 48*, 85–96.
Riede, T., Herzel, H., Mehald, D., Seider, W., Trumler, E., Böhme, G., Tembrock, G (2000). Nonlinear phenomena in the natural howling of a dog-wolf mix. *Journal of the Acoustical Society of America, 108*, 1435–1442.
Riede, T., Suthers, R. A., Fletcher, N. H., & Blevins, W. E. (2006). Songbirds tune their vocal tract to the fundamental frequency of their song. *Proceedings of the National Academy of Sciences of the U S A, 103*(14), 5543–5548.
Roberts, L. H. (1973). Cavity resonances in the production of orientation cries. *Periodicum Biologorum, 75*, 27–32.
Roberts, L. H. (1975). The rodent ultrasound production mechanism. *Ultrasonics, 13*(2), 83–88.
Rubin, H. J. (1960). The neurochronaxic theory of voice production—A refutation. *Archives of Otolaryngology, 71*(6), 913–920.
Sanvito, S., Galimberti, F., & Miller, E. H. (2007). Vocal signaling of male southern elephant seals is honest but imprecise. *Animal Behaviour, 73*(2), 287–299.
Schenkkan, E. J. (1973). On the comparative anatomy and function of the nasal tract in odontocetes (Mammalia: Cetacea). *Bijdragen tot de Dierkunde, 43*(1), 127–159.

Schmidt, R. S. (1972). Action of intrinsic laryngeal muscles during release calling in leopard frog. *Journal of Experimental Zoology, 181*, 233–244.
Shipley, C., Carterette, E. C., & Buchwald, J. S. (1991). The effects of articulation on the acoustical structure of feline vocalizations. *Journal of the Acoustical Society of America, 89*(2), 902–909.
Steinecke, I., & Herzel, H. (1995). Bifurcations in an asymmetric vocal-fold model. *Journal of the Acoustical Society of America, 97*(3), 1874–1884.
Sundberg, J. (1975). Formant technique in a professional female singer. *Acustica, 32*, 89–96.
Suthers, R. A. (1988). The production of echolocation signals by bats and birds. In P. E. Nachtigall & P. W. B. Moore (Eds.), *Animal sonar: Processes and performance* (pp. 23–45). New York: Plenum Press.
Suthers, R. A., & Fattu, J. M. (1973). Mechanisms of sound production in echolocating bats. *American Zoologist, 13*, 1215–1226.
Suthers, R. A., Hartley, D. J., & Wenstrup, J. J. (1988). The acoustic role of tracheal chambers and nasal cavities in the production of sonar pulses by the horseshoe bat, *Rhinolophus hildebrandtii. Journal of Comparative Physiology A, 162*, 799–813.
Suthers, R. A., & Hector, D. H. (1988). Individual variation in vocal tract resonance may assist oilbirds in recognizing echoes of their own sonar clicks. In P. E. Nachtigall & P. W. B. Moore (Eds.), *Animal sonar: Processes and performances* (pp. 87–91). New York: Plenum Press.
Suthers, R. A., Narins, P. M., Lin, W.-Y., Schnitzler, H.-U., Denzinger, A., Xu, C.-H., Feng, A. S. (2006). Voices of the dead: Complex nonlinear vocal signals from the larynx of an ultrasonic frog. *Journal of Experimental Biology, 209*, 4984–4993.
Titze, I. R. (2006). *The myoelastic aerodynamic theory of phonation*. Denver, CO: National Center for Voice and Speech.
Titze, I. R., Fitch, W. T., Hunter, E. J., Alipour, F., Montequin, D., Armstrong, D. L., et al. (2010). Vocal power and pressure-flow relations in excised tiger larynges. *Journal of Experimental Biology, 213*(22), 3866–3873.
van den Berg, J. (1958). Myoelastic-aerodynamic theory of voice production. *Journal of Speech and Hearing Research, 1*, 227–244.
van den Berg, J. (1968). Sound production in isolated human larynges. *Annals of the New York Academy of Sciences, 155*, 18–27.
Vannoni, E., & McElligott, A. G. (2007). Individual acoustic variation in fallow deer (*Dama dama*) common and harsh groans: A source-filter theory perspective. *Ethology, 113*, 223–234.
Weber, E. (1977). Effects of extirpation of laryngeal muscles and vocal cords on vocalizations of *Bufo viridis* (Amphibia, Anura). *Israel Journal of Zoology, 26*, 230–247.
Wilden, I., Herzel, H., Peters, G., & Tembrock, G. (1998). Subharmonics, biphonation, and deterministic chaos in mammal vocalization. *Bioacoustics, 9*, 171–196.
Winter, P., Handley, P., Ploog, D., & Schott, D. (1973). Ontogeny of squirrel monkey calls under normal conditions and under acoustic isolation. *Behaviour, 47*(3/4), 230–239.
Yager, D. D. (1982). A novel mechanism for underwater sound production in *Xenopus borealis*. *American Zoologist, 22*, 887.
Yager, D. D. (1996). Sound production and acoustic communication in *Xenopus borealis*. In R. Tinsely & H. R. Kobel (Eds.), *The biology of Xenopus*. Oxford, England: Oxford University Press.
Zaccarelli, R., Elemans, C. P. H., Fitch, W. T., & Herzel, H. (2006). Modelling bird songs: Voice onset, overtones and registers. *Acta Acustica, 92*(5), 741–748.

Chapter 2
Fish Sound Production: Insights

Eric Parmentier and Michael L. Fine

Abstract In addition to briefly reviewing sound-producing mechanisms, this chapter focuses on an under-appreciated evolutionary process, exaptation, which could aid in understanding the independent origins and high diversity of sound-producing mechanisms in fishes. Existing anatomical structures first used in non-voluntary sound production provide advantages that result in further selection and refinement of sophisticated sonic organs. Moreover, comparisons of the relationships between fish size and spectral features in multiple not phylogenetically related species highlight two acoustic patterns. In species using superfast muscles, the slope of the relationship between fish size and sound frequency is weak (1°–5°) so that emitter size is unlikely inferred from call frequency. In other species that stridulate or use bones or tendons to stimulate the swimbladder, the high slopes (25°–80°) indicate major differences in the call frequencies within a species. These signals likely convey important information (size and potential fitness of the emitter) to conspecific receivers.

Keywords Acoustic • Call • Communication • Evolution • Gas bladder • Mechanism • Message • Size effect • Sonic • Sonic muscle • Stridulation • Swim bladder • Teleost

2.1 Introduction

Although numerous sonic fishes produce many different sounds (Fig. 2.1), sound production for social communication occurs in a restricted number of families. In some taxa (Doradidae, Bagridae, Pimelodidae, Batrachoididae, Gadidae, Sciaenidae, Holocentridae, Pomacentridae, and Carapidae, for example) all, or almost all, species

---

E. Parmentier (✉)
Laboratoire de Morphologie Fonctionnelle et Evolutive, Institut de Chimie, Université de Liège Sart Tilman, Liège 4000, Belgium
e-mail: E.Parmentier@ulg.ac.be

M.L. Fine
Department of Biology, Virginia Commonwealth University, Richmond, VA 23284, USA
e-mail: mlfine@vcu.edu

R.A. Suthers et al. (eds.), *Vertebrate Sound Production and Acoustic Communication*, Springer Handbook of Auditory Research 53,
DOI 10.1007/978-3-319-27721-9_2

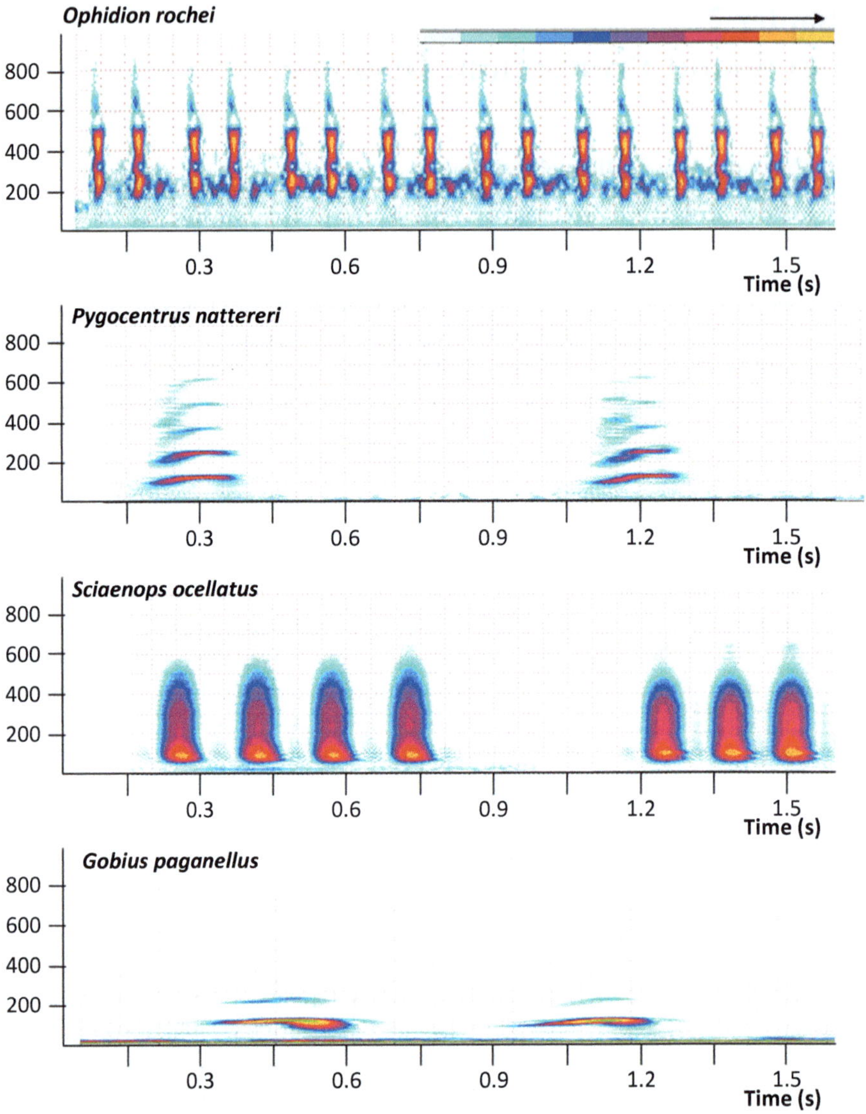

Fig. 2.1 Spectrogram of different fish sounds. Color scale: relative intensity

have the ability to call although mute species exist (Fine and Parmentier 2015). At the opposite extreme, large groups such as minnows (cyprinids) are mostly mute, but a few species produce socially relevant sounds (Johnston and Johnson 2000; Holt and Johnston 2014; Fine and Parmentier 2015). From a number of reviews devoted to the sound-producing mechanisms in fishes (Ladich and Fine 2006; Fine and Parmentier 2015) it is apparent that (1) sound-producing mechanisms have evolved independently and sporadically in various lineages (Fine and Parmentier 2015) and (2) the variety of sound-producing mechanisms is so great that it has not been possible to classify these mechanisms satisfactorily in useful subcategories (Ladich and Fine 2006).

Many studies indicate advantages and even a necessity to produce sounds. However, the question remains: why are sounds produced in some taxa but not in others? Acoustic communication likely evolved in distantly related species because this function is important to reproductive success. This assumption is reinforced by the high diversity of mechanisms that fishes have developed independently, which leads to the suggestion that morphological characters promoting acoustic communication have evolved multiple times. These recurrent selections support the importance of the acoustic function to species fitness. Although all fishes possess the hearing sense and detect the acoustic scene (Popper and Fay 2011), most fish species lack the ability to produce sounds, indicating that acoustic communication may be advantageous but is not a vital function as is swimming, feeding, breathing or eating. Interestingly, many of the structures used in these vital functions can be modified for sound production.

Exaptation refers to a functional character previously shaped by natural selection for a particular function that is co-opted for a new use that enhances fitness (Gould and Vrba 1982). The term exaptation has been used once in the fish sound-production literature (Parmentier et al. 2007) in regard to the jaw-snapping mechanism in damselfishes. Recent descriptions of different mechanisms allow the suggestion that sound production mechanisms result from numerous and varied exaptations of existing structures. The parsimony principle states that a history involving a minimum number of changes in a set of sequences likely approximates the actual evolutionary history of the sequences (Fitch 1971; Hein 1990). We postulate that sound production appeared in fish taxa that were able to take advantage of their non-voluntary sounds. This hypothesis supports both observations of numerous unrelated mechanisms of sound production in fishes and that many species do not produce sounds.

Producing sound involves a vibration that is coupled to the medium (Bradbury and Vehrencamp 1998). There are five basic mechanisms for producing sounds, all of which are present in fishes: (1) muscular vibrations of a membrane or sac (Fine et al. 2009; Millot et al. 2011), (2) stridulation (Fine et al. 1999; Parmentier et al. 2010b; Bertucci et al. 2014), (3) forced flow through a small orifice (Wahlberg and Westerberg 2003; Lagardère and Ernande 2004), (4) muscular vibration of appendages (Kratochvil 1978, 1985; Ladich et al. 1992) and (5) percussion on a substrate (Colleye et al. 2013).

Although multiple mechanisms have been described, most can be grouped into two categories: muscles that directly or indirectly connect to the swim bladder and stridulatory mechanisms involving the rubbing of bones. Additional mechanisms such as fin plucking in gouramis or jaw snapping in damselfish can be found in the literature (see Ladich and Fine 2006; Fine and Parmentier 2015). The two main groups can be split into multiple smaller categories that are quite different.

2.2 Swim Bladder Mechanisms

Many of the swim bladder–based mechanisms result from evolutionary convergence and are constructed around the same basic principle: fish have to provoke the vibration of a gas-filled structure whose base functions include buoyancy and respiration (Alexander 1966). Classically the swim bladder has been modeled as a pulsating underwater bubble (Harris 1964; van Bergeijk 1964), an omnidirectional and resonant monopole. Because of the compressibility of gas in the bladder compared with the surrounding water, an acoustic pressure wave is believed to excite the bladder into vibration that radiates particle motion to the ears (Sand and Hawkins 1973). Similarly, single muscle contractions would excite the swim bladder wall for sound production. Based on this logic, many investigators have assumed that the resonant properties of swim bladders can magnify sounds produced elsewhere in the body (Fish 1953; Demski et al. 1973; Smith and Croll 2011). Removing gas from toadfish (Tavolga 1964b), cichlid (Longrie et al. 2009), and damselfish swim bladders (Colleye et al. 2012) decreases sound amplitude but not fundamental frequency. In contrast, filling the swim bladder with fluid induces a significant decrease in pulse duration and a significant increase in dominant frequency (Colleye et al. 2012). These experiments indicate the importance of gas in the bladder to amplify and radiate movement of sonic muscles but do not support the logic of a resonant structure.

In Clark's anemonefish (*Amphiprion clarkii*), striking the ventral surface of the swim bladder with an impact hammer forces it inward, increasing pressure within the bladder. However, the compressed bladder does not rebound sufficiently to cause sound vibration, indicating that the swim bladder is an inefficient resonator (Colleye et al. 2012). Similar findings in oyster toadfish (*Opsanus tau*) and red-bellied piranha (*Pygocentrus nattereri*) (Fine et al. 2009; Millot et al. 2011) indicate that swim bladders are highly damped and thus prevented from prolonged resonant vibrations.

However, striking the rib cage of anemonefish (intimately surrounding the swim bladder) with an impact hammer generates sound waveforms similar to those of natural sounds (Colleye et al. 2012). Therefore, the vibrating ribs drive the swim bladder wall, which appears to function like a loudspeaker membrane driven by rib displacements (a forced response) rather than as an independent resonator. Furthermore, filling the swim bladder with physiological saline increased radiation mass and thus changes the properties of the swim bladder wall. Therefore, modifying the physical properties of the swim bladder should affect the vibrational properties of the rib cage. These considerations further support the notion that the swim bladder is not a resonant structure. Fine and Parmentier (2015) provide additional arguments showing conflicts between the resonant bubble model and different aspects of fish biology. Sonic muscles attached to swim bladders are among the fastest muscles in vertebrates (Skoglund 1961; Rome and Lindstedt 1998; Fine et al. 2001), yet a resonant structure such as a bell does not require extreme speed to excite it into resonance. The resonant frequency of an underwater bubble increases with depth (hydrostatic pressure) and decreases with bubble radius. Therefore swim bladder and sonic muscle size increases with fish growth could create mismatches

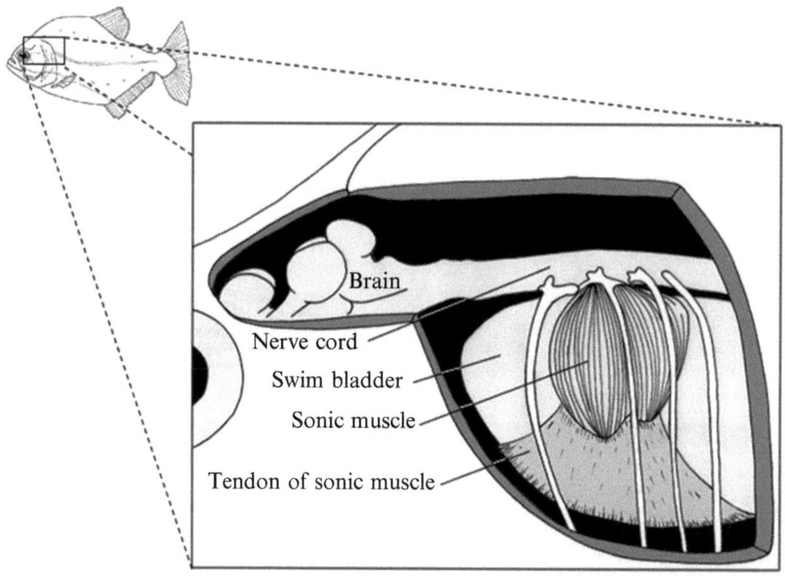

Fig. 2.2 Schematic *left* lateral view of the sound-producing mechanism (*black*) piranha (*Serrasalmus rhombeus*). Skull and vertebrae are not shown (Redrawn from Ladich and Bass 2005)

between communicating individuals. A resonant bubble will continue to oscillate after sound termination and would interfere with temporal coding of fish sounds, most of which are of short duration and pulsed. Finally, sonic swim bladders have a number of shapes (i.e., heart shaped in toadfish) or diverticula (Hawkins 1993; Barimo and Fine 1998; Ramcharitar et al. 2006; Schulz-Mirbach et al. 2013), which is not logical for an omnidirectional source (Fine 2012).

Sound production requires development of intrinsic or extrinsic muscles that deform the swim bladder to radiate sound. Mechanisms of swim bladder excitement are particularly diverse at various levels including muscle origins, insertions, ultrastructure, and contraction speed as well as in the sizes, shapes, and structures. Intrinsic muscles attach completely to large areas of the swim bladder. They are capable of producing short-duration pulsed-type sounds with single or a small number of contractions, but they are generally associated with production of long duration tonal notes. During a sustained contraction, the sonic muscle contraction rate can decrease slightly, causing a slight decrease (frequency modulation) of the fundamental frequency.

Extrinsic sonic muscles generally have their origins on various bones on the skull but also on the pectoral girdle, the ribs, or epineurals (Tavolga 1964a; Schneider 1967; Ladich and Fine 2006) and insert on the swim bladder or on a structure that attaches to the swim bladder. Sonic muscles in piranhas (Fig. 2.2) originate on the vertebral column and insert on a broad tendon that surrounds the ventral surface of the anterior chamber of the swim bladder (Markl 1971; Ladich and Bass 2005). In the red drum (*Sciaenops ocellatus*), sonic muscles are bilaterally symmetrical muscles

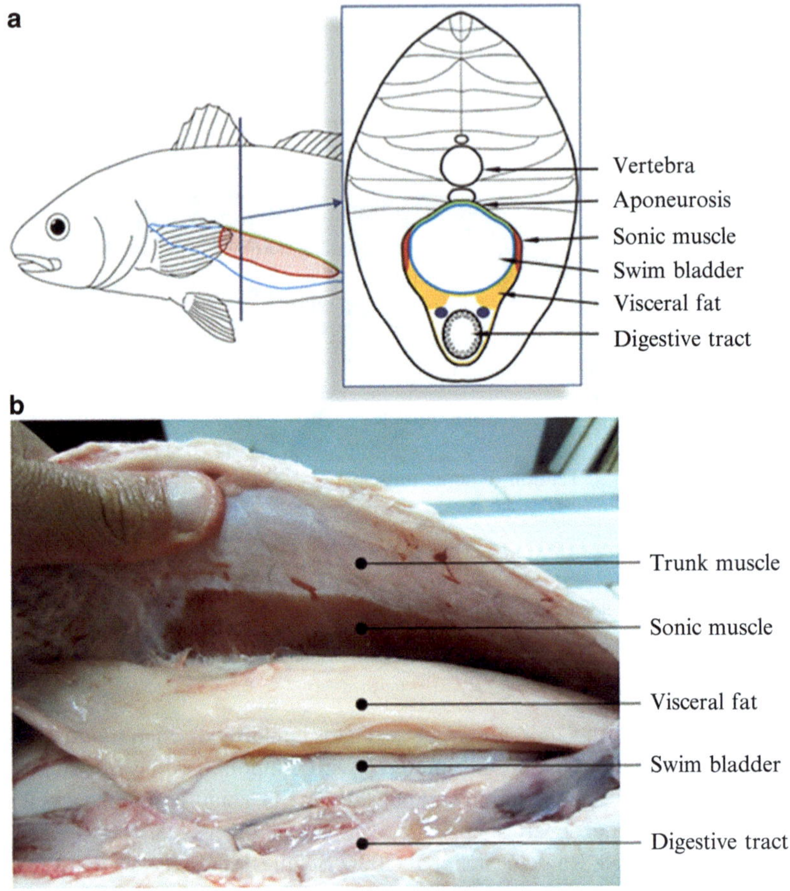

Fig. 2.3 (**a**) Schematic *left* lateral view of the sound-producing mechanism in the *red* drum (*Sciaenops ocellatus*). The *blue line* corresponds to the shape of the swim bladder and the *red lines* to the position of the sonic muscle. In the schematic cross section, *left* and *right* sonic muscles are dorsally united by the aponeurosis (in *green*). (**b**) Ventral view of the body cavity showing the relative position of the sonic muscle

that run perpendicular to the long axis of the fish. These muscles originate on the abdominal hypaxial musculature and insert on a central tendon that attaches to the dorsal swim bladder (Fig. 2.3). Alternately, extrinsic muscles can also insert between two bones, one of which is connected to the swim bladder via ligaments or connective tissue. This situation is found in some Scorpaenidae (Hallacher 1974) and Holocentridae (Parmentier et al. 2011b). In catfishes (Ladich and Fine 2006; Kaatz and Stewart 2012; Boyle et al. 2014), the sonic muscle inserts on variously derived elastic-spring mechanisms, the ramus Mülleri, a modified rib, that attaches to the bladder (Sörensen 1895; Chardon 1968). The muscle pulls the bladder forward directly or via the spring mechanism. Sound production is due to the pull and rebound from the spring mechanism and stretched bladder. In some ophidiiform fishes the swim bladder insertion can be highly modified (Parmentier et al. 2002). In *Onuxodon* (Carapidae) and some *Ophidion* (Ophidiidae), for example (Fig. 2.4), the sonic muscles insert on a lima bean–shaped hard structure protruding from the anterior wall of the swim bladder (Parmentier et al. 2006a, 2008, 2010a).

Although there is generally one symmetrical pair of sonic muscles, some species have developed additional muscles (Fig. 2.4). In some Ophidiiformes such as the pearlfish *Carapus* and *Encheliophis*, for example, a pair of ventral muscles originates on the neurocranium (in the orbit ceiling) and inserts on the dorsal part of the swim bladder. Additionally a pair of dorsal muscles originates on the neurocranium and inserts on the first epineurals (Fig. 2.4), which connect to the swim bladder by ligaments (Parmentier et al. 2003a, b). In these fishes, contraction of the sonic muscles inserted directly on the swim bladder is the driving force for sonic emission, while the muscles inserted on the epineurals probably should modify the sounds.

In other Ophidiiformes, both dorsal and ventral muscles are required to produce sounds. Sustained contractions of dorsal muscles during the entire call place the swim bladder under tension, and a series of rapid contraction/relaxation cycles from the second pair of muscles creates multiple sound pulses (Fig. 2.5) (Parmentier et al. 2010a). This system is analogous to a bow. At rest, the string and the rod are separated; the contraction of the first pair of muscles would tense the rod with the string to stretch the bow, and the contraction and relaxation of the second set of muscles pulls and releases the bow string. This mechanism is experimentally supported by electromyographic recording from both muscles (Kéver et al. 2014b). The same kind of mechanism can apply in the glaucosomatid *Glaucosoma buergeri*. This fish also possesses two pairs of sonic muscles. The anterior sonic muscles originate on the skull and insert on the outside of the anterior–dorsal region of the swim bladder. The posterior muscle inserts on the inner side of the swim bladder and has the morphology of a typical smooth muscle (which is quite exceptional). Its contraction works as an antagonist to anterior skeletal muscles that extend the anterior swim bladder. The smooth muscle presumably functions to damp vibrations from the tendon, which would drive the swim bladder to produce sound (Mok et al. 2011).

Although diversity is high due to independent evolution, the systems correspond finally just to variations on a theme, and two main kinds of mechanisms will be highlighted.

The forced-response model (Fine 2012) posits that the frequency spectrum is dictated by contraction dynamics of superfast extrinsic or intrinsic sonic muscles (Fine et al. 2001; Connaughton 2004; Millot et al. 2011). This system, the drumming muscle system, can also involve bony or ligamentous attachments to the swim bladder if each contraction cycle generates a cycle of sound waveform. Drumming requires superfast muscles (Skoglund 1961; Rome et al. 1996; Fine et al. 2001) and can be found in various species of the Batrachoididae (Tower 1908; Rice and Bass 2009), Triglidae (Connaughton 2004), Serrasalmidae (Markl 1971; Kastberger 1981a; Millot et al. 2011) and Zeidae (Onuki and Somiya 2004).

The swim bladder rebound model posits that swim bladder sounds are driven by vibration of surrounding structures such as epineurals or ribs (Parmentier et al. 2006b, 2010a; Oliver and Lobel 2013). In this case, the dominant frequency is determined by

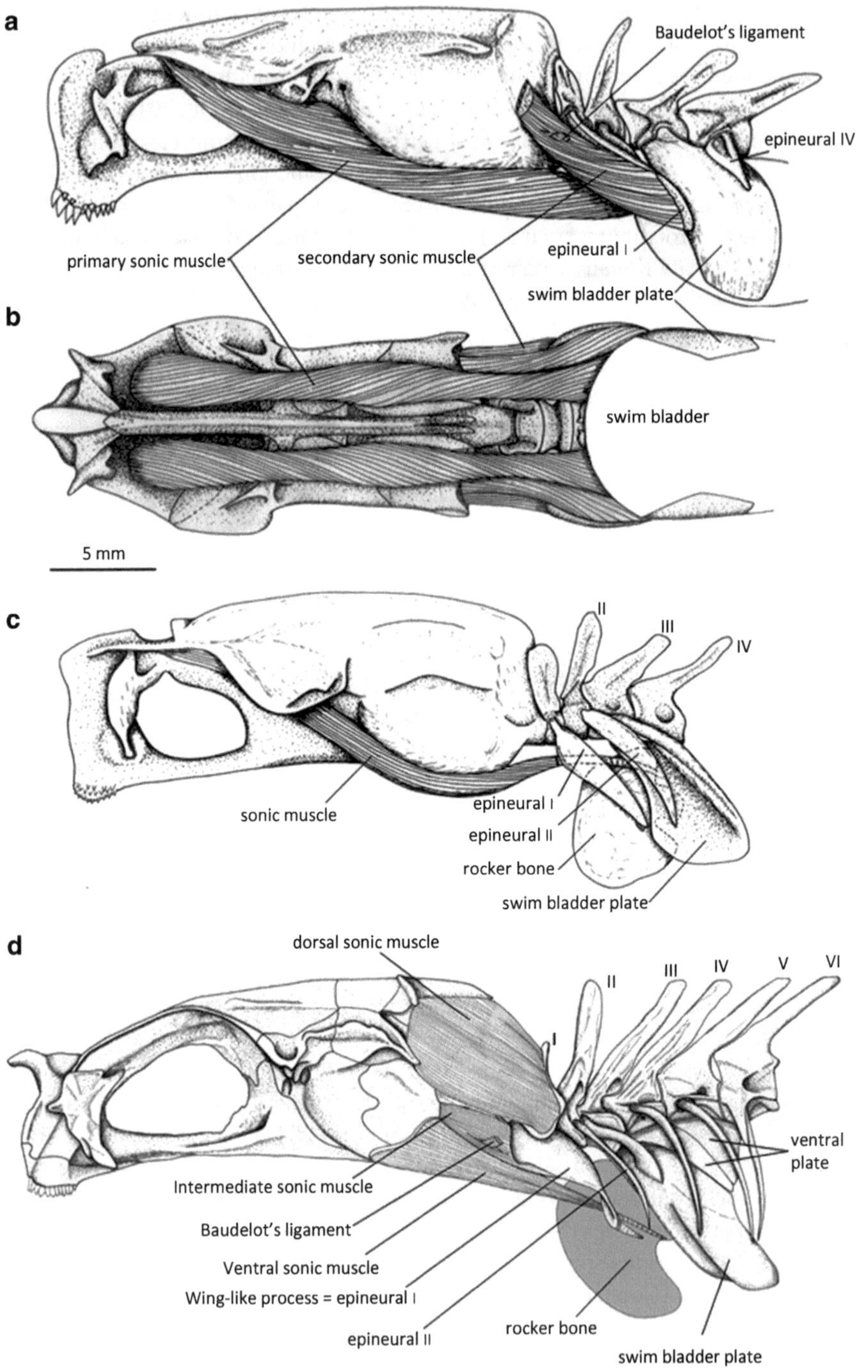

Fig. 2.4 Sound-producing mechanism in different ophidiiform fishes. *Left* lateral (**a**) and ventral (**b**) view in *Carapus boraborensis*; (**c**) *left* lateral view in *Onuxodon fowleri* and (**d**) *left* lateral view in *Ophidion barbatum*. Swim bladders are not shown in **c** and **d** (Modified from Parmentier et al. 2002, 2006a)

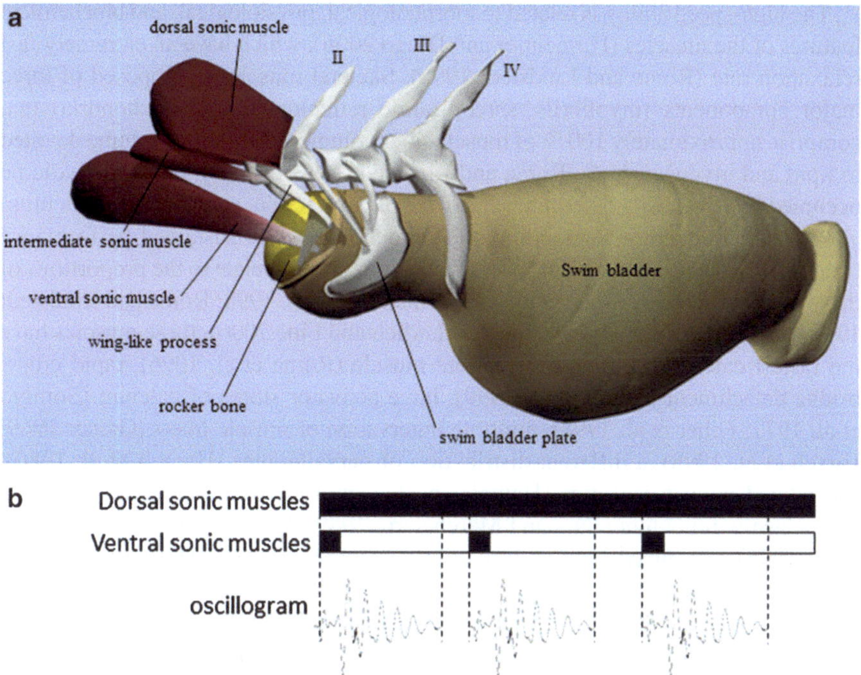

Fig. 2.5 Schematic view of the sound-producing mechanism in *Ophidion rochei* (**a**) and (**b**) schematic graph showing the muscle activity during sound production and the related oscillograms of calls. Dark areas correspond to the muscle activity

vibratory properties of the surrounding structure. Each muscle contraction causes a sound pulse but does not determine the sound's frequency spectrum.

These two systems highlight dramatic differences in sonic muscle anatomy and physiology. Skeletal sonic muscles have varying contraction speeds, and maximal rates of contraction when stimulated electrically extend from 10 to 300 Hz (Gainer et al. 1965; Fine et al. 2001; Millot et al. 2011). In drumming fishes, the muscle contraction rate sets the fundamental frequency. For example, contraction of sonic muscles at 150 Hz will drive a sound with a fundamental frequency of 150 Hz and typically harmonics at multiples of 150 Hz. In this case a muscle twitch, the time for a contraction/relaxation cycle, is short: toadfish sonic muscles require about 10 ms for a twitch (Skoglund 1961; Fine et al. 2001). Additional studies on sonic muscles in the weakfish *Cynoscion regalis* (Sciaenidae), the leopard searobin *Prionotus scitulus* (Triglidae), the hardhead catfish *Arius felis* (Ariidae), the gafftopsail catfish *Bagre marinus* (Ariidae), and the tiger perch *Terapon jarbua* (Terapontidae) place them among the "champions" of contraction speed (Schneider 1967; Sprague 2000). The yellow pyramid butterflyfish (*Hemitaurichthys polylepis*) also produces rapid pulse train sounds with extrinsic high-speed swim bladder muscles (Boyle and Tricas 2010; Boyle et al. 2013).

The high-speed ability is related to morphological, physiological, and biochemical features of the muscles (Parmentier and Diogo 2006), which have an extremely fast relaxation rate (Rome and Lindstedt 1998). Skeletal muscle is composed of three major components (myofibrils, sarcoplasmic reticulum, and mitochondria) that comprise approximately 100 % of muscle fiber volume (less a small volume devoted to lipid and glycogen fuel) (Rome and Lindstedt 1998). Briefly stated, the volume occupied by myofibrils determines the force of contraction, sarcoplasmic reticulum allows high frequency contraction, and mitochondria fuel sustained performance (fatigue resistance). Functional specializations in muscle relate to the proportions of these three structures (Rome et al. 1996; Lindstedt et al. 1998; Rome and Lindstedt 1998). In comparison to white muscles (Ladich and Fine 2006), these muscles have the fastest calcium spike in a vertebrate muscle (Rome et al. 1996), rapid cross-bridge detachment (Rome et al. 1999), huge activator stores of calcium (Somlyo et al. 1977; Feher et al. 1998), multiple innervation of muscle fibers (Gainer 1969; Hirsch et al. 1998), a different distribution of parvalbumins (Hamoir et al. 1980), specialized myosin isoforms (Hamoir and Focant 1981), an increased volume of sarcoplasmic reticulum (Bass and Marchaterre 1989; Appelt et al. 1991; Schaeffer et al. 1996), a reduced fiber and myofibril diameter (Evans 1973; Ono and Poss 1982; Kéver et al. 2014b), a higher content of mitochondria (Eichelberg 1977; Bass and Marchaterre 1989; Parmentier et al. 2013), multiple capillaries surrounding muscle fibers that maximize the supply of oxygen and exchange of other metabolites (Lewis et al. 2003).

The rebound system has been described mainly in ophidiiform and glaucosomatid fishes. In the rebound system, the mechanism of some carapid species utilizes slow muscles that tetanize at about 10 Hz (Parmentier et al. 2006b). Thus there is one slow muscle contraction for each sound pulse, and a resonant response (multiple cycles in the sound waveform) seems to be driven by a bone, the swim bladder plate, rather than the swim bladder. With slow muscles each muscle contraction generates a pulse but not the frequency within a pulse. Within the subfamily Ophidiinae sounds have been recorded from two species: *Ophidion marginatum* (Mann et al. 1997; Sprague and Luczkovich 2001) and *Ophidion rochei* (Parmentier et al. 2010a; Kéver et al. 2012, 2014a). Calls from the striped cusk-eel *Ophidion marginatum* have peak frequencies above 1 kHz (Mann et al. 1997; Sprague and Luczkovich 2001), which should be impossible even with superfast swim bladder muscles because twitches would have to occur in less than 1 ms, faster than any known direct muscle. As in *Carapus* species, calls would result from a release mechanism that utilizes three steps. The contraction of the dorsal muscle first pulls the epineurals, which are in close relationships with the anterior part of the swim bladder, backward. Second, contraction of sound-producing muscle stretches the anterior part of the swim bladder rostrally, creating a tension opposed to the action of the dorsal muscle. Third, sound-producing muscle relaxation combined with the caudally-acting force cause rapid rebound of the swim bladder (Parmentier et al. 2010a).

Other indirect swim bladder systems were recently summarized in Fine and Parmentier (2015). Sound production in the Nile tilapia (*Oreochromis niloticus*) occurs by contraction of a horizontal band of muscle that initiates movements of the rib cage and the swim bladder (Longrie et al. 2009). Although sounds have been described in more than 30 cichlid species, the sound-producing mechanism is currently described only in one species. Additional studies are required to increase understanding of the mechanism(s) in this family.

2.3 Sounds and Information

Temporal patterns may be an important sound characteristic for acoustic communication in fishes, especially in noisy and/or shallow water where low frequencies do not propagate well and the spectral content of signals is easily altered (Mann 2006; Ghahramani et al. 2014). In many cases, temporal and spectral features are related to fish size, and therefore the calls can convey phenotypic differences between males: smaller individuals typically produce sounds of higher frequency and shorter duration than larger individuals (Myrberg et al. 1993; Connaughton et al. 2000).

However, the relationship between these variables and size is not invariant. In some species, the fundamental frequency may not change with fish size because muscle contraction rate determines the fundamental frequency (Skoglund 1961; Fine et al. 2001). Grunt fundamental frequency did not change with size in toadfish varying from 29 to 760 g (Fine and Waybright 2015), and choruses of toadfish in nature, composed of different sized fish, can have fundamental frequencies varying over as little as 10 Hz (Fine 1978). At the opposite extreme, fundamental frequency in *Amphiprion clarkii* decreases by 500 Hz in individuals between 30- and 90-mm total length.

In Fig. 2.6, data from published studies were collected to compare the slopes of the relationships between fish size and fundamental frequency. When the equation was given, slopes were drawn on the basis of the specimen sizes in the study. For other studies, the slopes were calculated based on data in the graphs, in which case data from the smallest and largest individuals were used. These slopes should be considered approximations. Figure 2.6a includes fishes in which the sound-producing mechanism is based on superfast muscles. In these examples sound frequencies are dependent on the sonic muscle contraction rate. The species in this group have negative slopes between 1 and 10°. Moreover, highest slopes are found in studies comparing adults and larvae in which the sound-producing system is not completely developed (Vasconcelos and Ladich 2008; Parmentier et al. 2011b). The sound production mechanism of the rock-pool blenny (*Parablennius parvicornis*) is unknown (De Jong et al. 2007). Interestingly, the slope of the relationship in this species is greater than 10°, and these blennies produce harmonic calls, suggesting mechanism involving high-speed muscles.

In the second group (Fig. 2.6b), the relationships between dominant frequency and fish size have steeper slopes, between 25 and 80°. Species in this group belong to distantly related taxa: cichlids (Amorim et al. 2003, 2004; Bertucci et al. 2012), pomacentrids (Myrberg et al. 1993; Colleye et al. 2011), and gobiids (Malavasi et al. 2003). Although it was not possible to place them on the graph because the authors used the swim bladder size or the body weight rather than body length, similar

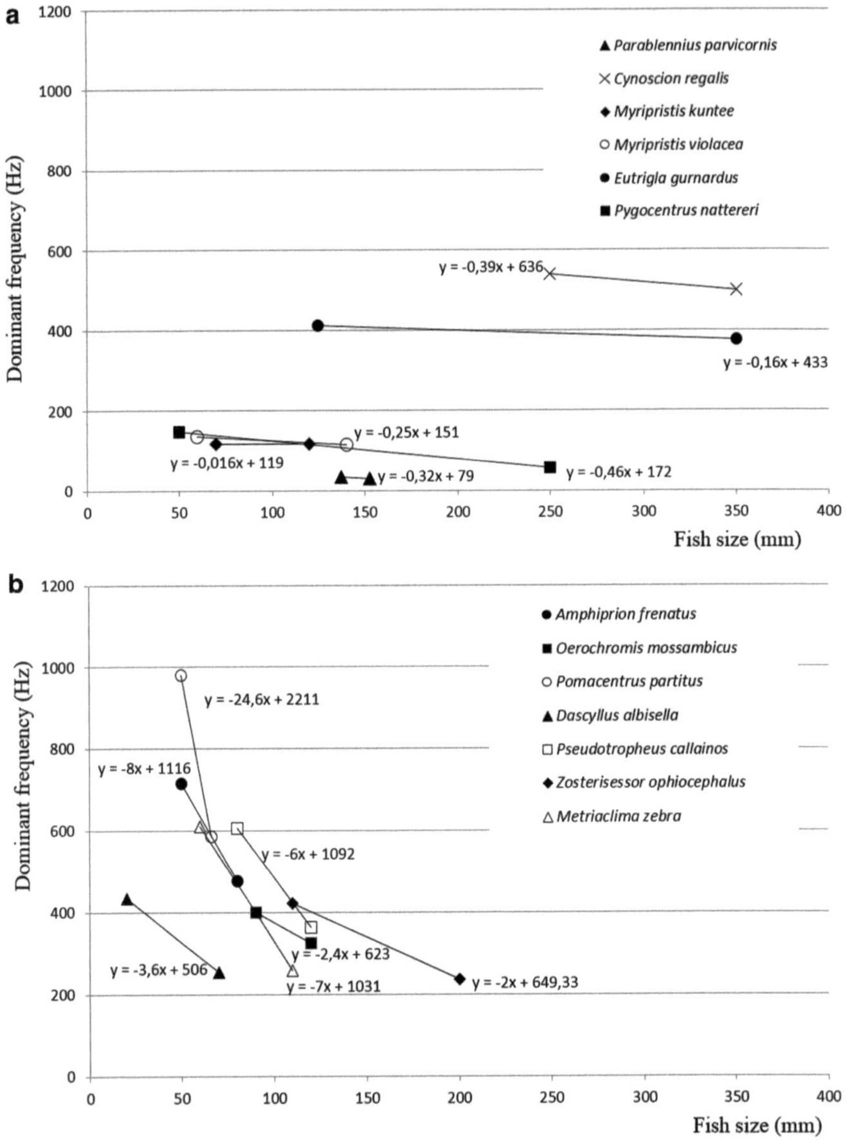

Fig. 2.6 Relationships between dominant frequency and fish size in different species. (**a**) Fishes in which the sound-producing mechanism is based on superfast muscles. The mechanism is, however, not yet known in *P. parvicornis*. All the fish species in this group have a negative slope between 1 and 5°. (**b**) Fishes in which the sound-producing mechanism is not based on fast-contracting muscles. The relationships between dominant frequency and fish size have slopes between 60 and 87°

relationships are found in the Carapidae (Parmentier et al. 2006b), Mormyridae (Crawford 1997), and Ophidiidae (Kéver et al. 2014a). Moreover, similar steep slopes are also found in fishes producing stridulatory sounds as in Osphronemidae (Henglmuller and Ladich 1999).

Comparison of Fig. 2.6a, b suggests another way that the sound-producing mechanism can be used to group fishes, namely size information in their calls. In fishes of the first group (Holocentridae, Sciaenidae, Batrachoididae), fish size would not be inferred from the main frequency by conspecifics whereas it would be possible in fishes from the second group. In the lattice soldierfish (*Myripristis violacea*), for example, 60- and 130-mm individuals produce the same main frequency (Parmentier et al. 2011b). As a comparison, the calling frequency of 60-mm skunk clownfish (*Amphiprion akallopisos*) is 700 Hz but less than 400 Hz in 130-mm specimens (Colleye et al. 2009). The high slopes in the second group, indicate that frequency can provide information on emitter size. Teleost fishes such as the black goby (*Gobius niger*, Gobiidae) and the annular seabream (*Diplodus annularis*, Sparidae) are able to discriminate tonal sounds differing in frequency by approximately 10 %, equivalent to 40 Hz for a 400-Hz sound (Fay 1988). Similar frequency discrimination occurs in bicolor damselfish (*Stegastes partitus*); females preferentially respond to lower frequency chirps of larger males (Myrberg et al. 1986). In the sergeant major (*Abudefduf saxatilis*), fish size affects auditory sensitivity (Egner and Mann 2005); all fish are most sensitive to low frequencies (100–400 Hz), but larger fish are more likely to respond to higher frequencies (800–1600 Hz). On the other hand, females of Hawaiian dascyllus (*Dascyllus albisella*) choose their mate on the basis of their courtship rate and not on phenotypic characters such as weight size, or dominant frequency (Oliver and Lobel 2013).

Despite these comments, slopes from group 1 species (Fig. 2.2) were all negative. This shared phenomenon is likely a scaling effect because longer muscles, presumably with longer fibers, from larger fish would take longer to complete a twitch (Connaughton et al. 2000).

An important set of studies in the Lusitanian toadfish (*Halobatrachus didactylus*) provides more information on this kind of vocal behavior (Amorim and Vasconcelos 2008). The acoustic features that consistently best discriminate individual toadfish are the dominant frequency of the middle tonal segment of the boat whistle and dominant frequency modulation. If scientists can recognize individuals based on their calls, it is likely that the fish can too, and evidence of individual recognition has been demonstrated in the Gulf toadfish *Opsanus beta* (Thorson and Fine 2002). However, these frequencies are related to the pulse period (i.e., the rate of muscle contraction) and not fish size. Moreover, the pulse period has low variability in this taxa, which is consistent with their vocal central pattern generator (Bass and Baker 1990; Barimo and Fine 1998; Amorim et al. 2010). In *Halobatrachus didactylus*, reproductive success appears to be determined by calling rate and calling effort (i.e., the highest percentage of time spent calling). These parameters indicate male condition (Vasconcelos et al. 2012), as reflected in sonic muscle hypertrophy and larger gonads (Amorim et al. 2010). In this case, the sonic muscle mass would allow long periods of calling but not affect main frequency. In other words, sounds would be related to male quality, that is to males that are likely to confer greater fitness on their offspring (Amorim and Vasconcelos 2008; Amorim et al. 2010) but not to its size.

2.4 Stridulation Mechanisms

These mechanisms work by friction caused by rubbing skeletal elements (teeth or bones), which produce a series of irregular pulses containing a wide range of frequencies (Tavolga 1971; Hawkins 1993). Two examples have dominated the literature: rubbing of the pharyngeal teeth and friction of the pectoral fin against the shoulder girdle (Fig. 2.7).

Likely all fishes can produce sounds using their pharyngeal teeth during feeding movements or substrate manipulations, and therefore it is not always easy to relate sound production to communication. These sounds, however, can be used by other species (interception or eaves dropping), which may be inimical to communication because the beneficiary is not the sender (Myrberg 1981). Unfortunately in many fishes without obviously distinct sound-producing elements, the sonic mechanism

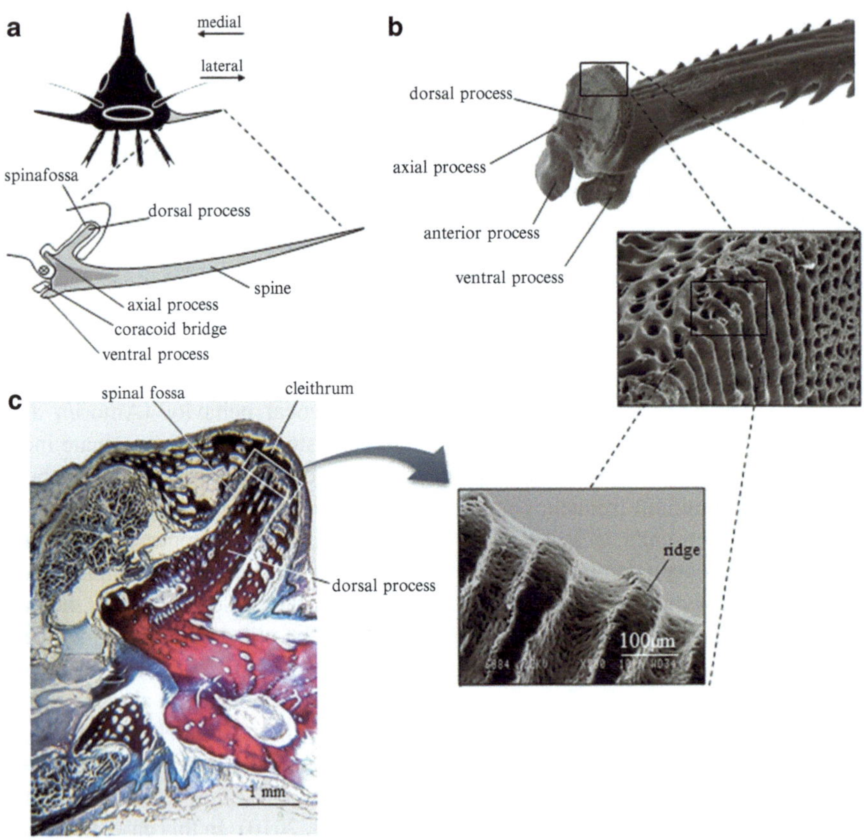

Fig. 2.7 (**a**) Frontal schematic view of the *left* pectoral spine and the position of its dorsal process in the spinal fossa in mochokid catfish. (**b**) Scanning electron micrographs of a mochokid catfish with enlarged views of the ridges situated on the dorsal process. (**c**) Cross section at the level of the dorsal process of the pectoral spine (Modified from Parmentier et al. 2010b)

has been attributed, without explicit evidence, to sounds from the pharyngeal jaws and a resonating effect of the swim bladder (Fine and Parmentier 2015). In this regard the swimbladder does not appear to radiate stridulatory sounds in the tiger-tail seahorse (Lim et al. 2015) or in channel catfish (Fine et al. 1997). To the best of our knowledge, the use of rubbing teeth in communication process can be found in haemulid grunts (Burkenroad 1930; Bertucci et al. 2014), but additional experiments are required to understand the related behaviors.

A second stridulatory mechanism utilizes pectoral spines in catfishes (Sörensen 1895; Schachner and Schaller 1981; Fine and Ladich 2003). During abduction and in some species adduction of the fins, sweep movements of the pectoral spine produce a number of discrete pulses with varying waveforms (Fine et al. 1996, 1999; Vance 2000). Sounds are produced by microscopic bony ridges on the dorsal process (Schachner and Schaller 1981; Fine et al. 1997; Parmentier et al. 2010b) coming into contact with the cleithrum (Parmentier et al. 2010b).

Seahorses produce sounds by rubbing the exoccipital bone on the back of the skull against a coronet (Colson et al. 1998; Oliveira et al. 2014; Lim et al. 2015).

Some species of croaking gouramis (*Trichopsis vittata*) provide another kind of pectoral mechanism. Rather than bone against bone, the pectoral fin has two hypertrophied tendons that rub against other fin rays, producing a double-pulsed sound (Kratochvil 1985; Ladich et al. 1992).

2.5 A Thought About Evolution of Sound Production

Although callers are present in distantly related taxa and in some basal groups of bony fishes (Fine and Parmentier 2015), there is no phylogenetic continuity, as with the syrinx of birds and the larynx of mammals. The development of acoustic communication in fishes is possible because required morphological characters were present as precursors. Quite all teleosts have bones, teeth, an air sac, inner ears with otoliths (and three semicircular canals) and more or less developed fins that constitute the raw material for the development of sound-producing structures. We note that the swim bladder may be lost in certain benthic or deep-water forms. In fact, all fishes can produce sounds if we consider ones produced by hydrodynamic movements (Moulton 1960) and chewing sounds, but the challenge is to demonstrate which sounds are incidental byproducts and which are voluntarily used to communicate (mainly agonistic and courtship behavior). Some physostome fishes, with connections between the swim bladder and the gut, can produce sound by shuttling gas from the swim bladder (Wahlberg and Westerberg 2003; Lagardère and Ernande 2004), but it is unclear if any such sounds have evolved for communication or are merely incidental, as in gut rumbling in humans.

A signal can be selected for communication if it fulfills the following conditions: (1) it can be generated at relatively low cost, (2) it can propagate to an individual able to perceive and interpret it, and (3) it elicits a response of the receiver that is advantageous for the sender (Myrberg 1981; Bradbury and Vehrencamp 1998).

Individuals able to elicit, modify, or generate informative signals should be favored through evolutionary history because they can inform conspecifics about their location, intentions, and potential fitness. Data are, however, currently insufficient to show which groups have taken advantage of this ability to evolve more rapidly or to produce a higher specific diversity. Catfishes, for instance, have many species capable of sound production, but they also have other adaptations (electro- and chemoreception), and therefore it is not possible to claim that sound production is responsible for their diversity. In terms of adaptive radiation, evolving sound production does not generally open up new avenues leading to diversification, but this question has never been studied systematically.

Production of underwater acoustic signals is subject to constraints that differ between swim bladder and stridulation mechanisms. Further, these systems likely evolved convergently to improve calling abilities. This scheme can also explain why distantly related families such as sciaenids (Connaughton et al. 2000; Ramcharitar et al. 2006; Parmentier et al. 2014), characids (Markl 1971; Kastberger 1981a; Ladich and Bass 2005; Millot et al. 2011), or batrachoidids (Fine et al. 2001; Rice and Bass 2009) share similar characteristics. In drums (Sciaenidae), the swim bladder is surrounded laterally by bilaterally symmetrical sonic muscles originating from a ventral tendon or the hypaxial musculature and inserting on a large, flattened central tendon that attaches to a large extent of the dorsal swim bladder (Fig. 2.3) (Hill et al. 1987; Parmentier et al. 2014). In piranhas, a broad tendon is not dorsal but ventral to the swim bladder (Fig. 2.2), and lateral sonic muscles insert on transverse expansions at the base of the second pair of ribs (Ladich and Bass 2005; Millot et al. 2011). In black drum (unlike in typical sciaenids), toadfishes, and searobins sonic muscles are intrinsic and attach exclusively to the swim bladder (Fine et al. 2001; Rice and Bass 2009). Despite these differences, the frequency spectrum is dictated by contraction dynamics of superfast sonic muscles acting on the damped swim bladder. Similarities in ultrastructure of nonhomologous sonic muscle fibers are particularly striking in weakfish, a sciaenid (Ono and Poss 1982), and the oyster toadfish (Fawcett and Revel 1961; Appelt et al. 1991; Fine et al. 1993) although the sciaenid is innervated segmentally by true spinal nerves and the toadfish by occipital spinal nerves. Again, evolution has produced muscles with convergent abilities to contract rapidly (Rome et al. 1996; Young and Rome 2001; Parmentier and Diogo 2006).

Similar variability occurs in stridulatory mechanisms of sound production that involve movements of pectoral and dorsal fins, pharyngeal teeth, buccal teeth, neurocranium, and so forth (see Fine and Parmentier 2015 for a review). Therefore caution is required in assuming homologous characters involved with sound-producing mechanisms in phylogenetic studies; similar functions can be produced by convergent structures whose similarity is superficial, as is often the case with morphology (Kocher et al. 1993; Rüber and Adams 2001; Frédérich et al. 2013). In other words, fishes using swim bladder mechanisms are not phylogenetically closer than ones using stridulatory mechanisms. Some taxa even employ both mechanisms. For instance some catfishes (Siluriformes) produce sounds using swim bladder muscles (Sörensen 1895; Tavolga 1977; Parmentier and Diogo 2006) or pectoral (Kaatz et al. 2010; Parmentier et al. 2010b) or dorsal spines (Mahajan 1963; de Pinna 1996). However, no catfish possesses all of these mechanisms, and some employ one, two or none of them (Fine and Ladich 2003; Parmentier and Diogo 2006). Even within swim bladder mechanisms, some catfishes have muscles connected directly to the swim bladder whereas in others they insert on a modified rib (the Springfederapparat or elastic spring mechanism), which then attaches to the bladder.

2.5.1 Concept of Exaptation

A phenotype is composed of modular units that integrate functionally related characters into units of evolutionary transformation. These units may emerge spontaneously (large-effect mutations of homeobox genes, for example) and are then acted on by natural selection (Wagner 1996). Functional modularity refers to the interactions of traits in performing one or more functions (Klingenberg 2008). The teleost head, for example, is used for prey capture and breathing. Motor patterns allowing these movements are usually based on the same mechanical principle that allows gill ventilation. In teleosts and sharks, feeding movements may be exaggerations of ones used in breathing (Hughes 1960; Liem 1985; Ferry-Graham 1999). By introducing the term exaptation, Stephen Jay Gould and Elizabeth Vrba published a provocative challenge to orthodox evolutionary theory (Larson et al. 2013). Exaptation refers to a functional character previously shaped by natural selection for a particular function that is co-opted for a new use that enhances fitness (Gould and Vrba 1982). However, the character can retain its plesiomorphic (ancestral) form while taking on a new function, thus expanding functional diversity (McLennan 2008). Exaptation has not been used widely in the biological sciences (Larson et al. 2013), mainly because few concrete examples have been properly demonstrated (Ostrom 1979; Cullen et al. 2013; Patek et al. 2013). Bird feathers are usually used to support the concept because they probably evolved for temperature regulation and display functions and later co-opted for flight. The jaws in trap-ants are typically used in rapid closing strikes for prey capture but also allow ants to propel themselves into the air (Patek et al. 2013).

The concept of exaptation is highly interesting because the mechanical units that change or incorporate new functions have the potential for rapid evolutionary change and may not require transitional forms (McLennan 2008). In fish species, sound was likely an initial byproduct of mechanical functions involved in feeding or locomotion. To be part of an operational system, sounds produced early in the evolution of the trait likely evoked modifications in the behavior of the recipient individuals (conspecific or not). In this scheme, behavioral responses would have improved the fitness of the emitter although advantages likely accrued to the recipient as well. The ability to produce sound allows the fish to enter a new adaptive zone, an environmental (not necessarily new) space that is exploitable after the acquisition of morphological and/or physiological characters. For instance, wing development permitted birds to enter a new adaptive zone (the aerial way of life), and then minor morphological modifications allowed them to colonize various milieus (Mayr 1989). This process allows the acceleration of diversification by ecological opportunity, such as dispersal into newly opened territory, extinction of competitors, or adoption of a new way of life (Simpson 1953), which for the purpose of this chapter involves sound production. Calling species do not necessarily develop new ecological opportunities but increase attraction of sexual partners, discourage predators, or improve territorial defense, thus providing evolutionary advantages. Historically, morphological modifications that permit entrance into a new adaptive zone were thought to result from one or several changes to an ancestral plan (Zeldicht and Fink 1996) or from the emergence of novelties (Futuyma 1986; Heard and Hauser 1995). However, the establishment of a relationship between exaptation and an adaptive zone might allow an adaptation such as sound production to develop rapidly because extensive modification of morphology may be unnecessary, particularly if fish sound-producing mechanisms arose from pre-existing structures adapted for other functions. Once sounds have been incorporated into a species' behavior, natural selection can reinforce calling behavior through morphological and physiological modifications. Beautiful and surprising sound-producing mechanisms arose in many taxa, notably in ophidiiform (Courtenay 1971; Parmentier et al. 2006a) and batrachoidiform fishes (Skoglund 1961; Fine et al. 2001; Rice and Bass 2009). In these taxa, a well-developed mechanism allows rapid identification of its structural components, for example, muscles on a swim bladder. On the other hand, many taxa are deprived of obvious mechanical structures that would cause sound production. Cichlids (Rice and Lobel 2003; Longrie et al. 2009), pomacentrids (Parmentier et al. 2007), gobiids (Stadler 2002; Parmentier et al. 2013), cottids (Colleye et al. 2013), chaetodontids (Boyle and Tricas 2010, 2011; Parmentier et al. 2011a), and cyprinids, for example, all include species capable of sound production, but these species do not exhibit major modifications of their Bauplan. The anatomy of these fishes is similar to that of mute relatives. Surprisingly, in some groups the responsible anatomical structures are unknown, and it is difficult to determine which parts of the body to investigate. Therefore sounds can be produced with only minor modifications of fish morphology. Sections 2.5.1.1–2.5.1.5 discuss several interesting examples highlighting taxa that have taken advantage of their incipient abilities to produce voluntary communication signals.

2.5.1.1 Damselfish (Pomacentridae)

Damselfish are well-known vocal species that produce sounds in courtship and agonistic contexts (Myrberg et al. 1978; Mann and Lobel 1998; Colleye and Parmentier 2012). The sonic mechanism appears unique among teleosts and results from teeth collisions induced by a fast jaw slam (Parmentier et al. 2007). The vibration is radiated through the ribs and induces oscillations of the swim-bladder wall (Colleye et al. 2012). Rapid mouth closure is forced by an apomorphic ceratomandibular ligament (Stiassny 1981) that connects the medial face of the lower jaw and the lateral face of the ceratohyal (Fig. 2.8). The ligament, stretched when the oral jaws

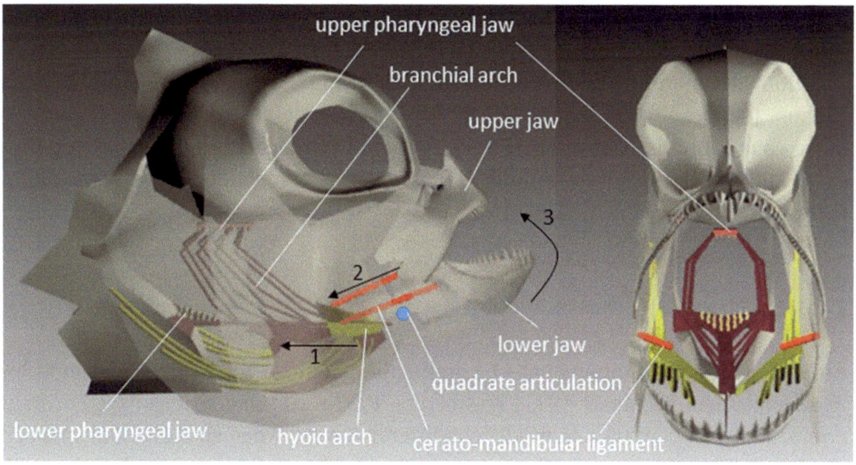

Fig. 2.8 Schematic *right* lateral view (**a**) and *rostral* view (**b**) of the sound-producing mechanism illustrating the relative movement of skeletal components in clownfish. Lowering the hyoid bar (1) stretches the sonic ligament (2), and the jaw closes the mouth (3) by rotating around the mandible articulation on the quadrate (Modified from Parmentier et al. 2007)

are opened, enables rapid closure causing teeth collisions and sound production (Parmentier et al. 2007; Colleye et al. 2012; Frédérich et al. 2014). Manual manipulation of fresh specimens indicates mouth closure is caused by the stretched ligament and does not require adductor muscle contraction, as in other teleosts (Olivier et al. 2014). Further, cutting the ceratomandibular ligaments prevents both feeding and sound production.

In the clownfish (*Amphiprion clarkii*) and the filamentous algae grazer (*Stegastes rectifraenum*), the slam occurs during feeding, likely the precursor behavior, and sound production (Olivier et al. 2014, 2015). In *Stegastes rectifraenum*, the buccal jaw slam probably plays a major role in farming activity enabling accurate strikes on small filamentous algae (Olivier et al. 2014). Kinematic analysis has demonstrated that similar jaw slams and sounds are produced during biting of filamentous algae and agonistic behavior. Similar movements are also found in sound production and biting in *Amphiprion*, and the characters and motor patterns used in feeding have been co-opted for sound production. Based on feeding movements and parsimony, the ancestral call was likely a single pulse. Different sequences of pulses are produced in different behaviors although they all utilize the c-md ligament. Sounds generally occur simultaneously with aggressive actions related to territorial defense. In all studied species, a single jaw slam occurs during fighting and produces a single pulse. The origin of sound production would come from a biting action for two reasons. (1) Fighting sounds usually occur before the display of aggressive behavior with biting (Parmentier et al. 2010c). (2) Alternatively, biting occurs during foraging activities in *Stegastes rectifraenum* (Olivier et al. 2014). Intact individuals were able to produce sounds and maintain their territorial boundaries whereas muted individuals did not deter intruders from entering their shelter sites despite appropriate visual displays (Myrberg 1997).

Initial communicative sounds were probably single pulses, which were selected because they resulted in successful territory and nest defense. Currently, one or two pulse sounds are used to deter conspecifics and heterospecifics, and courtship dips or visiting calls utilize trains of pulses that result from repetition of the same motor pattern.

2.5.1.2 Piranhas (Serrasalmidae)

Piranhas produce drumming calls by contracting high-speed swim bladder muscles in several species (Markl 1971; Kastberger 1981a, b). One species, *Pygocentrus nattereri*, has been shown to produce a different sound when an individual snapped its jaws to bite a conspecific (Millot et al. 2011). The sound has a single pulse with a dominant frequency of approximately 1740 Hz. In videos, 90 % of the sounds occurred when chasing a conspecific. Further studies are required to determine if these teeth sounds have a communication function.

2.5.1.3 Grunt (Haemulidae)

Stridulation is based on friction of skeletal elements such as teeth, fin rays, and vertebrae (Burkenroad 1930; Tavolga 1971). Haemulids produce stridulatory sounds when the upper and lower pharyngeal teeth grate against each other (Burkenroad 1930; Moulton 1958). These sounds are also made when the fish are handheld, and to the best of our knowledge, sound production under natural conditions has not been observed. Therefore the function of these sounds is unknown although the association with being grabbed suggests they are emitted in stressful situations. The detailed cyclic pharyngeal jaw movement pattern during food processing has been described in nine haemulid species (Wainwright 1989a, b). In the French grunt (*Haemulon flavolineatum*) Bertucci and colleagues found similar sounds produced during food processing and when fish are handheld (Bertucci et al. 2014). High-speed X-ray videos confirmed sounds result from the rubbing of teeth located on the upper and lower pharyngeal jaws, and the cyclic movements during sound production correspond to those made during food processing (Bertucci et al. 2014).

In the Haemulidae, as in many perciform fishes (Vandewalle et al. 1992, 1995), the motion of the upper jaw is generally greater than that of the lower jaw. During the rhythmic pharyngeal transport of food to the esophagus, the upper jaw sweeps dorsally from the posterior pharyngeal cavity forward, descends, and then returns to its initial position. The lower pharyngeal jaws move similarly. The upper and lower jaws meet during the posterior movement of the upper and anterior movement of the lower jaws, creating a sheering action. Without food, these movements provoke sound production. We hypothesize incidental sounds produced during food processing assumed a communication function that was favored over time.

2.5.1.4 Sea Horses (Syngnathidae)

Sea horses and pipefishes produce stridulation clicks during feeding strikes but also in courtship, male–male competition, and when held out of water (Fish 1953; Colson et al. 1998; Ripley and Foran 2007). Observations of head movements with high-speed video and synchronous sound recording indicate sound clicks and feeding strikes are due to a bony articulation: ridges of the supraoccipital bone slide under the groove in the coronet (bony plate at the back of the head) during rapid head elevation (Colson et al. 1998; Lim et al. 2015). In pipefishes, feeding strikes evoke click production (Ripley and Foran 2007), but the functional significance of the feeding click is unknown. The incidental byproduct hypothesis has been rejected because clicks may increase predation risk (Oliveira et al. 2014), although the balance between risk and benefit has not been studied. The benefit should be more important than the risk for the caller, and the feeding sounds may help maintain proximity between male–female pairs since these fish swim slowly. Once more, the parsimony principle implies sound production was first a byproduct of feeding strikes and has been selected for use in courtship and pair maintenance. It may have contributed to the complex courtship behavior found in many members of the family.

2.5.1.5 Catfish (Siluriformes)

Catfishes (Fig. 2.7) use large, complex, and armored pectoral spines that can be bound and locked as antipredator adaptations (Fine and Ladich 2003). A locked spine more than doubles the width of a juvenile channel catfish and complicates ingestion by gape-limited fish predators (Bosher et al. 2006; Sismour et al. 2013). In fact, dead fish, snakes, and birds have been found with spines stuck in their tissues (Sismour et al. 2013). Furthermore, spine and pectoral girdle mass have decreased in domesticated channel catfish that have experienced reduced predation for a number of generations. Selection for fast growing individuals may have also played a part in pectoral reduction (Fine et al. 2014). In addition, catfishes have evolved toxins multiple times that can be delivered by the pectoral spines (Burkhead 1972; Wright 2009).

The base of catfish spines has apomorphic dorsal, anterior, and ventral processes that are not present in related taxa such as characids and cyprinids (Hubbs and Hibbard 1951; Fine and Ladich 2003). The processes mate with complementary surfaces on the cleithrum and coracoid of a fused pectoral girdle (Brousseau 1978; Diogo et al. 2001; Miano et al. 2013) that provides support to anchor the spines (Schaefer 1984). Fossils of well-developed pectoral spines and girdles date back to the Cretaceous (Gayet and Van Neer 1990; Lundberg 1997; Gayet and Meunier 2003); therefore no direct information exists on the formation of these processes from a typical pectoral first spine. The dorsal and anterior processes likely evolved from the dorsal half of the first lepidotrich and the ventral process from the ventral half (John Lundberg and John Friel, pers. comm., 2014).

In a distress situation, pectoral spines are bound after partial and locked after complete abduction. The locked spine resists any linear motion, and unlocking requires posterodorsal rotation of the spine followed by adduction (Fine et al. 1997). The deployment of an enlarged spine provided some degree of protection. However, the spine function does not seem limited to this function in all Siluriformes. Numerous catfish species use also the dorsal process of the pectoral spine to stridulate, producing a series of pulses when grabbed by a predator (Bosher et al. 2006) or when handheld (Heyd and Pfeiffer 2000; Kaatz et al. 2010). The fused pectoral girdle, in addition to providing a rigid platform to anchor the spine (Schaefer 1984), has secondarily become specialized as a sound radiator. Species capable of sound production have developed ridges, visible with scanning electron microscopy, on the under surface of the dorsal process (Fine et al. 1997; Kaatz et al. 2010; Parmentier et al. 2010b). In these species, sounds are caused by a slip-stick mechanism: Small jerk-like motions of ridges against the cleithrum stimulate the pectoral girdle to vibrate (Parmentier et al. 2010b; Ghahramani et al. 2014; Mohajer et al. 2015). Initially the sound is of low amplitude, which then increases after termination of the jerk suggesting constructive interference, when the spine is immobile (Mohajer et al. 2015).

Most catfish species produce sound during spine abduction although a number of families have members that stridulate during abduction and adduction (Heyd and Pfeiffer 2000; Kaatz et al. 2010; Parmentier et al. 2010b). The channel catfish stridulates only during abduction although one adduction sound was videotaped out of 256 recorded (Fine et al. 1996). Therefore existing morphology will support adduction sounds, which would require amended neural programming. In this light it is interesting to note that the blue catfish, which produces only abduction sounds, first adducts its spine silently before stridulating (Ghahramani et al. 2014; Mohajer et al. 2015), suggesting the possibility of an existing step that may have occurred in those species that produce sound by stridulating in both directions.

The enlarged dorsal process, the rough surface of the channel in the cleithrum, and the fused pectoral girdle required to bind and lock the spines were co-opted to make sounds in most catfishes. Sound production was likely added secondarily to the defense function, and in many species sounds were further co-opted as agonistic and courtship signals.

2.6 Conclusion

In addition to reviewing the topic generally this chapter proposes a novel conceptual path to explore evolution of sound-producing mechanisms and propose the concept of exaptation as an entry to understand the myriad forms and solutions employed by fishes. Existing anatomical structures are first used in nonvoluntary sound production, which provides advantages and results in further selection and refinement of more sophisticated sonic organs. The examples have focused on fishes using stridulatory mechanisms to produce sounds but make it clear that a similar evolutionary history likely applies to mechanisms based on swim bladder and sound-producing muscles.

References

Alexander, R. M. (1966). Physical aspects of swimbladder function. *Biological Reviews, 41*(1), 141–176.

Amorim, M. C. P., Fonseca, P. J., & Almada, V. C. (2003). Sound production during courtship and spawning of *Oreochromis mossambicus*: Male–female and male–male interactions. *Journal of Fish Biology, 62*(3), 658–672.

Amorim, M. C. P., Knight, M. E., Stratoudakis, Y., & Turner, G. F. (2004). Differences in sounds made by courting males of three closely related Lake Malawi cichlid species. *Journal of Fish Biology, 65*, 1358–1371.

Amorim, M. C. P., Simoes, J. M., Mendonca, N., Bandarra, N. M., Almada, V. C., & Fonseca, P. J. (2010). Lusitanian toadfish song reflects male quality. *Journal of Experimental Biology, 213*(17), 2997–3004.

Amorim, M. C. P., & Vasconcelos, R. O. (2008). Variability in the mating calls of the Lusitanian toadfish *Halobatrachus didactylus*: Cues for potential individual recognition. *Journal of Fish Biology, 73*(6), 1267–1283.

Appelt, D., Shen, V., & Franzini-Armstrong, C. (1991). Quantitation of Ca ATPase, feet and mitochondria in superfast muscle fibres from the toadfish, Opsanus tau. *Journal of Muscle Research and Cell Motility, 12*(6), 543–552.

Barimo, J. F., & Fine, M. L. (1998). Relationship of swim-bladder shape to the directionality pattern of underwater sound in the oyster toadfish. *Canadian Journal of Zoology, 76*(1), 134–143.

Bass, A. H., & Baker, R. (1990). Sexual dimorphisms in the vocal control system of a teleost fish: Morphology of physiologically identified neurons. *Journal of Neurobiology, 21*, 1155–1168.

Bass, A. H., & Marchaterre, M. A. (1989). Sound-generating (sonic) motor system in a teleost fish (*Porichthys notatus*): Sexual polymorphisms and general synaptology of sonic motor nucleus. *Journal of Comparative Neurology, 286*(2), 154–169.

Bertucci, F., Attia, J., Beauchaud, M., & Mathevon, N. (2012). Sounds produced by the cichlid fish *Metriaclima zebra* allow reliable estimation of size and provide information on individual identity. *Journal of Fish Biology, 80*, 752–766.

Bertucci, F., Ruppé, L., Van Wassenbergh, S., Compère, P., & Parmentier, E. (2014). New insights into the role of the pharyngeal jaw apparatus in the sound-producing mechanism of *Haemulon flavolineatum* (Haemulidae). *The Journal of Experimental Biology, 217*(21), 3862–3869.

Bosher, B., Newton, S., & Fine, M. (2006). The spine of the channel catfish, *Ictalurus punctatus*, as an anti-predator adaptation: An experimental study. *Ethology, 112*, 188–195.

Boyle, K. S., Colleye, O., & Parmentier, E. (2014). Sound production to electric discharge: sonic muscle evolution in progress in *Synodontis* spp. catfishes (Mochokidae). *Proceedings of the Royal Society B: Biological Sciences,* 281(1791).

Boyle, K. S., Dewan, A. K., & Tricas, T. C. (2013). Fast drum strokes: Novel and convergent features of sonic muscle ultrastructure, innervation, and motor neuron organization in the pyramid butterflyfish (*Hemitaurichthys polylepis*). *Journal of Morphology, 274*(4), 377–394.

Boyle, K. S., & Tricas, T. C. (2010). Pulse sound generation, anterior swim bladder buckling and associated muscle activity in the pyramid butterflyfish, *Hemitaurichthys polylepis. The Journal of Experimental Biology, 213*(22), 3881–3893.

Boyle, K. S., & Tricas, T. C. (2011). Sound production in the longnose butterflyfishes (genus *Forcipiger*): Cranial kinematics, muscle activity and honest signals. *The Journal of Experimental Biology, 214*(22), 3829–3842.

Bradbury, J., & Vehrencamp, S. (1998). *Principles of animal communication.* Sunderland, MA: Sinauer Associates.

Brousseau, R. A. (1978). The pectoral anatomy of selected Ostariophysi 2. The Cypriniformes and Siluriformes. *Journal of Morphology, 150*, 79–116.

Burkenroad, M. D. (1930). Sound production in the Haemulidae. *Copeia, 1930*(1), 17–18.

Burkhead, W. S. (1972). Toxicity of stings of Ariid and Ictalurid catfishes. *Copeia, 1972*, 790–807.
Chardon, M. (1968). Anatomie comparée de l'appareil de Weber et des structures connexes chez les Siluriformes. *Annales du Musée Royal d'Afrique Centrale, 169*, 1–273.
Colleye, O., Frederich, B., Vandewalle, P., Casadevall, M., & Parmentier, E. (2009). Agonistic sounds in the skunk clownfish *Amphiprion akallopisos*: Size-related variation in acoustic features. *Journal of Fish Biology, 75*(4), 908–916.
Colleye, O., Nakamura, M., Frédérich, B., & Parmentier, E. (2012). Further insight into the sound-producing mechanism of clownfishes: What structure is involved in sound radiation? *The Journal of Experimental Biology, 215*(13), 2192–2202.
Colleye, O., Ovidio, M., Salmon, A., & Parmentier, E. (2013). Contribution to the study of acoustic communication in two Belgian river bullheads (*Cottus rhenanus* and *C. perifretum*) with further insight into the sound-producing mechanism. *Frontiers in Zoology, 10*(1), 71.
Colleye, O., & Parmentier, E. (2012). Overview on the diversity of sounds produced by clownfishes (Pomacentridae): Importance of acoustic signals in their peculiar way of life. *PLoS One, 7*(11), e49179.
Colleye, O., Vandewalle, P., Lanterbecq, D., Lecchini, D., & Parmentier, E. (2011). Interspecific variation of calls in clownfishes: degree of similarity in closely related species. *BMC Evolutionary Biology, 11*(1), 365.
Colson, D., Patek, S., Brainerd, E., & Lewis, S. (1998). Sound production during feeding in *Hippocampus* seahorses (Syngnathidae). *Environmental Biology of Fishes, 51*(2), 221–229.
Connaughton, M. A. (2004). Sound generation in the searobin (*Prionotus carolinus*), a fish with alternate sonic muscle contraction. *Journal of Experimental Biology, 207*(10), 1643–1654.
Connaughton, M., Taylor, M., & Fine, M. (2000). Effects of fish size and temperature on weakfish disturbance calls: Implications for the mechanism of sound generation. *Journal of Experimental Biology, 203*, 1503–1512.
Courtenay, W. (1971). Sexual dimorphism of the sound producing mechanism of the striped cusk eel, *Rissola marginata* (Pisces: Ophidiidae). *Copeia, 2*, 259–268.
Crawford, J. (1997). Hearing and acoustic communication in mormyrid electric fishes. *Marine and Freshwater Behaviour and Physiology, 29*, 65–86.
Cullen, J. A., Maie, T., Schoenfuss, H. L., & Blob, R. W. (2013). Evolutionary novelty versus exaptation: Oral kinematics in feeding versus climbing in the waterfall-climbing hawaiian goby *Sicyopterus stimpsoni. PLoS One, 8*(1), e53274.
De Jong, K., Bouton, N., & Slabbekoorn, H. (2007). Azorean rock-pool blennies produce size-dependent calls in a courtship context. *Animal Behaviour, 74*, 1285–1292.
de Pinna, M. C. C. (1996). A phylogenetic analysis of the Asian catfish families Sisoridae, Akysidae, and Amblycipitidae, with a hypothesis on the relationships of the neotropical Aspredinidae (Teleostei, Ostariophysi). *Fieldiana Zoology, 1996*, 1–83.
Demski, L. S., Gerald, J. W., & Popper, A. N. (1973). Central and peripheral mechanisms of teleost sound production. *American Zoologist, 13*, 1141–1167.
Diogo, R., Oliveira, C., & Chardon, M. (2001). On the osteology and myology of catfish pectoral girdle, with a reflection on catfish (Teleostei: Siluriformes) plesiomorphies. *Journal of Morphology, 249*, 100–125.
Egner, S. A., & Mann, D. A. (2005). Auditory sensitivity of sergeant major damselfish *Abudefduf saxatilis* from post-settlement juvenile to adult. *Marine Ecology Progress Series, 285*, 213–222.
Eichelberg, H. (1977). Fine structure of the drum muscles of the piranha (Serrasalminae, Characidae). *Cell and Tissue Research, 185*(4), 547–555.
Evans, R. R. (1973). The swimbladder and associated structures in western Atlantic sea robins (Triglidae). *Copeia, 1973*(2), 315–321.
Fawcett, D. W., & Revel, J. P. (1961). The sarcoplasmic reticulum of a fast-acting fish muscle. *Journal of Biophysical and Biochemical Cytology, 10*, 89–109.
Fay, R. R. (1988). *Hearing in vertebrates: A psychophysics databook*. Winnetka, IL: Hill-Fay Associates.

Feher, J., Waybright, T., & Fine, M. (1998). Comparison of sarcoplasmic reticulum capabilities in toadfish (*Opsanus tau*) sonic muscle and rat fast twitch muscle. *Journal of Muscle Research and Cell Motility, 19*(6), 661–674.

Ferry-Graham, L. A. (1999). Mechanics of ventilation in swellsharks, *Cephaloscyllium ventriosum* (Scyliorhinidae). *The Journal of Experimental Biology, 202*(11), 1501–1510.

Fine, M. L. (1978). Seasonal and geographical variation of the mating call of the oyster toadfish *Opsanus tau* L. *Oecologia, 36*, 45–57.

Fine, M. L. (2012). Swimbladder sound production: The forced response versus the resonant bubble. *Bioacoustics, 21*(1), 5–7.

Fine, M. L., Bernard, B., & Harris, T. M. (1993). Functional morphology of toadfish sonic muscle fibers: Relationship to possible fiber division. *Canadian Journal of Zoology, 71*(11), 2262–2274.

Fine, M., Friel, J., McElroy, D., King, C., Loesser, K., & Newton, S. (1997). Pectoral spine locking and sound production in the channel catfish *Ictalurus punctatus*. *Copeia, 1997*, 777790.

Fine, M., King, C., Friel, J., Loesser, K., & Newton, S. (1999). Sound production and locking of the pectoral spine of the channel catfish. *American Fisheries Society Symposium, 24*, 105–114.

Fine, M. L., King, C. B., & Cameron, T. M. (2009). Acoustical properties of the swimbladder in the oyster toadfish *Opsanus tau*. *Journal of Experimental Biology, 212*(21), 3542–3552.

Fine, M. L., & Ladich, F. (2003). Sound production, spine locking and related adaptations. In B. G. Kapoor, G. Arratia, M. Chardon, & R. Diogo (Eds.), *Catfishes* (pp. 248–290). Enfield, NH: Science Publishers.

Fine, M. L., Lahiri, S., Sullivan, A. D. H., Mayo, M., Newton, S. H., & Sismour, E. N. (2014). Reduction of the pectoral spine and girdle in domesticated channel catfish is likely caused by changes in selection pressure. *Evolution, 68*, 2102–2107. doi:10.1111/evo.12379.

Fine, M. L., Malloy, K. L., King, C. B., Mitchell, S. L., & Cameron, T. M. (2001). Movement and soung generation by toadfish swimbladder. *Journal of Comparative Physiology A, 187*, 371–379.

Fine, M. L., McElroy, D., Rafi, J., King, C. B., Loesser, K. E., & Newton, S. (1996). Lateralization of pectoral stridulation sound production in the channel catfish. *Physiology and Behavior, 60*, 753–757.

Fine, M. L., & Parmentier, E. (2015). Mechanisms of sound production. In F. Ladich (Ed.), *Sound communication in fishes* (pp. 77–126). Vienna: Springer.

Fine, M. L., & Waybright, T. D. (2015). Grunt variation in the oyster toadfish Opsanus tau: effect of size and sex. PeerJ 3, ei1330.

Fish, M. P. (1953). The production of underwater sounds by the northern searhorse, *Hippocampus hudsonius*. *Copeia, 1953*, 98–99.

Fitch, W. M. (1971). Toward defining the course of evolution: Minimum change for a specific tree topology. *Systematic Biology, 20*(4), 406–416.

Frédérich, B., Olivier, D., Litsios, G., Alfaro, M. E., & Parmentier, E. (2014). Trait decoupling promotes evolutionary diversification of the trophic and acoustic system of damselfishes. *Proceedings of the Royal Society B: Biological Sciences, 281*(1789), 20141047.

Frédérich, B., Sorenson, L., Santini, F., Slater, G. J., & Alfaro, M. E. (2013). Iterative ecological radiation and convergence during the evolutionary history of damselfishes (Pomacentridae). *The American Naturalist, 181*(1), 94–113.

Futuyma, D. J. (Ed.). (1986). *Evolutionary biology*. Sunderland, MA: Sinauer Associates.

Gainer, H. (1969). Multiple innervation of fish skeletal muscle. In G. A. Kerkut (Ed.), *Experiments in physiology and biochemistry* (Vol. 2, pp. 191–208). New York: Academic Press.

Gainer, H., Kusano, K., & Mathewson, R. F. (1965). Electrophysiological and mechanical properties of squirrelfish sound-producing muscle. *Comparative Biochemistry and Physiology, 14*(4), 661–671.

Gayet, M., & Meunier, F. J. (2003). Palaeontology and palaeobiogeography of catfishes. In G. Arratia, B. G. Kapoor, M. Chardon, & R. Diogo (Eds.), *Catfish* (Vol. 2, pp. 491–522). Enfield, NH: Science Publishers.

Gayet, M., & Van Neer, W. (1990). Caractères diagnostiques des épines de quelques silures africains. *Journal of African Zoology, 104*, 241–252.
Ghahramani, Z. N., Mohajer, Y., & Fine, M. L. (2014). Developmental variation in sound production in water and air in the blue catfish Ictalurus furcatus. *The Journal of Experimental Biology, 217*(23), 4244–4251.
Gould, S. J., & Vrba, E. S. (1982). Exaptation-a missing term in the science of form. *Paleobiology, 8*(1), 4–15.
Hallacher, L. E. (1974). The comparative morphology of extrinsic gasbladder musculature in the scorpionfish genus *Sabastes* (Pisces: Scorpaenidae). *Proceedings of the California Academy of Sciences, 40*, 59–86.
Hamoir, G., & Focant, B. (1981). Proteinic differences between the sarcoplasmic reticulums of the superfast swimbladder and the fast skeletal muscles of the toadfish *Opsanus tau. Molecular Physiology, 1*, 353–359.
Hamoir, G., Gerardin-Otthiers, N., & Focant, B. (1980). Protein differentiation of the superfast swimbladder muscle of the toadfish *Opsanus tau. Journal of Molecular Biology, 143*(1), 155–160.
Harris, G. G. (1964). Considerations on the physics of sound production by fishes. In W. N. Tavolga (Ed.), *Marine bio-acoustics* (Vol. 1, pp. 233–247). New York: Pergamon Press.
Hawkins, A. D. (1993). Underwater sound and fish behaviour. In T. J. Pitcher (Ed.), *Behaviour of Teleost Fishes* (2nd ed., pp. 129–169). London: Chapman & Hall.
Heard, S. B., & Hauser, D. L. (1995). Key evolutionary innovations and their ecological mechanisms. *Historical Biology, 10*(2), 151–173.
Hein, J. (1990). Reconstructing evolution of sequences subject to recombination using parsimony. *Mathematical Biosciences, 98*(2), 185–200.
Henglmuller, S., & Ladich, F. (1999). Development of agonistic behaviour and vocalization in croaking gourami. *Journal of Fish Biology, 54*, 380–395.
Heyd, A., & Pfeiffer, W. (2000). Über die Lauterzeugung der Welse (Siluroidei, Ostariophysi, Teleostei) und ihren Zusammenhang mit der Phylogenese und der Schreckreaktion. *Revue Suisse de Zoologie, 107*, 165–211.
Hill, G., Fine, M., & Musick, J. (1987). Ontogeny of the sexually dimorphic sonic muscle in three sciaenid species. *Copeia, 3*, 708–713.
Hirsch, J. E., Bigbee, J. W., & Fine, M. L. (1998). Continuous adult development of multiple innervation in toadfish sonic muscle. *Journal of Neurobiology, 36*, 348–356.
Holt, D. E., & Johnston, C. E. (2014). Sound production and associated behaviours in blacktail shiner *Cyprinella venusta*: A comparison between field and lab. *Environmental Biology of Fishes, 97*(11), 1207–1219.
Hubbs, C. L., & Hibbard, C. W. (1951). *Ictalurus lambda*, a new catfish, based on a pectoral spine from the lower Pliocene of Kansas. *Copeia, 1951*, 8–14.
Hughes, G. M. (1960). A comparative study of gill ventilation in marine teleosts. *Journal of Experimental Biology, 37*(1), 28–45.
Johnston, C. E., & Johnson, D. L. (2000). Sound production in *Pimephales notatus* (Rafinesque) (Cyprinidae). *Copeia, 2000*, 567–571.
Kaatz, I. M., & Stewart, D. J. (2012). Bioacoustic variation of swimbladder disturbance sounds in Neotropical doradoid catfishes (Siluriformes: Doradidae, Auchenipteridae): Potential morphological correlates. *Current Zoology, 58*(1), 171–188.
Kaatz, I. M., Stewart, D. J., Rice, A. N., & Lobel, P. S. (2010). Differences in pectoral fin spine morphology between vocal and silent clades of catfish (order Siluriformes): Ecomorphological implications. *Current Zoology, 56*, 73–89.
Kastberger, G. (1981a). Economy of sound production in piranhas (Serrasalminae, Characidae): I. Functional properties of sonic muscles. *Zoologische Jahrbücher Physiologie, 85*, 113–125.
Kastberger, G. (1981b). Economy of sound production in piranhas (Serrasalminae, Characidae): II. Functional properties of sound emitter. *Zoologische Jahrbücher Physiologie, 85*, 393–411.
Kéver, L., Boyle, K. S., Bolen, G., Dragicevic, B., Dulcic, J., & Parmentier, E. (2014a). Modifications in call characteristics and sonic apparatus morphology during puberty in *Ophidion rochei* (Actinopterygii: Ophidiidae). *Journal of Morphology, 275*(6), 650–660.

Kéver, L., Boyle, K., Dragicevic, B., Dulcic, J., Casadevall, M., & Parmentier, E. (2012). Sexual dimorphism of sonic apparatus and extreme intersexual variation of sounds in *Ophidion rochei* (Ophidiidae): First evidence of a tight relationship between morphology and sound characteristics in Ophidiidae. *Frontiers in Zoology, 9*(1), 34.
Kéver, L., Boyle, K. S., Dragičević, B., Dulčić, J., & Parmentier, E. (2014b). A superfast muscle in the complex sonic apparatus of *Ophidion rochei* (Ophidiiformes): Histological and physiological approaches. *The Journal of Experimental Biology, 217*(19), 3432–3440.
Klingenberg, C. P. (2008). Morphological integration and developmental modularity. *Annual Review of Ecology, Evolution, and Systematics, 39*(1), 115–132.
Kocher, T. D., Conroy, J. A., McKaye, K. R., & Stauffer, J. R. (1993). Similar morphologies of cichlid fish in Lakes Tanganyika and Malawi are due to convergence. *Molecular Phylogenetics and Evolution, 2*(2), 158–165.
Kratochvil, H. (1978). Der Bau des Lautapparates vom Knurrenden Gurami (*Trichopsis vittatus* Cuvier & Valenciennes) (Anabantidae, Belontiidae). *Zoomorphologie, 91*, 91–99.
Kratochvil, H. (1985). Beiträge zur Lautbiologie der Anabantoidei—Bau, Funktion und Entwicklung von lauterzeugenden Systeme. *Zoologische Jahrbücher Physiologie, 89*, 203–255.
Ladich, F., & Bass, A. H. (2005). Sonic motor pathways in piranhas with a reassessment of phylogenetic patterns of sonic mechanisms among teleosts. *Brain, Behavior and Evolution, 66*, 167–176.
Ladich, F., Bischof, C., Schleinzer, G., & Fuchs, A. (1992). Intra- and interspecific differences in agonistic vocalization in croaking gouramis (Genus: *Trichopsis*, Anabantoidei, Teleostei). *Bioacoustics, 4*, 131–141.
Ladich, F., & Fine, M. (2006). Sound-generating mechanisms in fishes: A unique diversity in vertebrates. In F. Ladich, S. P. Collin, P. Moller, & B. G. Kapoor (Eds.), *Communication in fishes* (Vol. 1, pp. 3–34). Enfield, NH: Science Publishers.
Lagardère, J. P., & Ernande, B. (2004). Émissions sonores enregistrées en marais salé et attribuées à l'anguille européenne. *Comptes Rendus Biologies, 327*(4), 353–359.
Larson, G., Stephens, P. A., Tehrani, J. J., & Layton, R. H. (2013). Exapting exaptation. *Trends in Ecology & Evolution, 28*(9), 497–498.
Lewis, M. K., Nahirney, P. C., Chen, V., Adhikari, B. B., Wright, J., Reedy, M. K., et al. (2003). Concentric intermediate filament lattice links to specialized Z-band junctional complexes in sonic muscle fibers of the type I male midshipman fish. *Journal of Structural Biology, 143*(1), 56–71.
Liem, K. F. (1985). Ventilation. In M. Hildebrand, D. M. Bramble, K. F. Liem, & D. B. Wake (Eds.), *Functional vertebrate morphology* (pp. 186–209). Cambridge, MA: Harvard University Press.
Lim, A. C. O., Chong, V. C., Chew, W. X., Muniandy, S. V., & Wong, C. S. (2015). Sound production in the tiger-tail seahorse *Hippocampus cones*: insights into the sound producing mechanisms. *Journal of the Acoustical Society of America, 138*, 404–412.
Lindstedt, S. L., McGlothlin, T., Percy, E., & Pifer, J. (1998). Task-specific design of skeletal muscle: Balancing muscle structural composition. *Comparative Biochemistry and Physiology Part B: Biochemistry and Molecular Biology, 120*(1), 35–40.
Longrie, N., Van Wassenbergh, S., Vandewalle, P., Mauguit, Q., & Parmentier, E. (2009). Potential mechanism of sound production in *Oreochromis niloticus* (Cichlidae). *Journal of Experimental Biology, 212*(21), 3395–3402.
Lundberg, J. G. (1997). Fishes of the Miocene La Venta fauna: Additional taxa, biotic and paleoenvironmental implications. In F. R. Kay, R. H. Maddern, R. L. Cifelli, & J. J. Flynn (Eds.), *Vertebrate paleontology in the Neotropics: The Miocene fauna of La Venta, Colombia* (pp. 67–91). Washington, DC: Smithsonian Institution Press.
Mahajan, C. L. (1963). Sound producing apparatus in an Indian catfish *Sisor rhabdophorus* Hamilton. *Journal of the Linnean Society of London, Zoology, 43*, 721–724.
Malavasi, S., Torricelli, P., Lugli, M., Pravoni, F., & Mainardi, D. (2003). Male courtship sounds in a teleost with alternative reproductive tactics, the grass goby, *Zosterisessor ophiocephalus*. *Environmental Biology of Fishes, 66*, 231–236.

Mann, D. A. (2006). Propagation of fish sounds. In F. Ladich, S. P. Collin, P. Moller, & B. G. Kapoor (Eds.), *Communication in fishes* (Vol. 1, pp. 107–120). Endfield, NH: Science Publishers.

Mann, D., Bowers-Altman, J., & Rountree, R. (1997). Sounds produced by the striped cusk-eel *Ophidion marginatum* (Ophidiidae) during courtship and spawning. *Copeia, 3*, 610–612.

Mann, D., & Lobel, P. S. (1998). Acoustic behaviour of the damselfish *Dascyllus albisella*: Behavioural and geographic variation. *Environmental Biology of Fishes, 51*, 421–428.

Markl, H. (1971). Schallerzeugung bei Piranhas (Serrasalminae, Characidae). *Journal of Comparative Physiology A, 74*(1), 39–56.

Mayr, E. (1989). *Histoire de la biologie. Diversité, évolution et hérédité*. Paris: Fayard.

McLennan, D. (2008). The Concept of Co-option: Why Evolution Often Looks Miraculous. *Evolution: Education and Outreach, 1*(3), 247–258.

Miano, J. P., Loesser-Casey, K. E., & Fine, M. L. (2013). Description and scaling of pectoral muscles in ictalurid catfishes. *Journal of Morphology, 274*, 467–477.

Millot, S., Vandewalle, P., & Parmentier, E. (2011). Sound production in red-bellied piranhas (*Pygocentrus nattereri*, Kner): An acoustical, behavioural and morphofunctional study. *Journal of Experimental Biology, 214*(21), 3613–3618.

Mohajer, Y. J., Ghahramani, Z. N., & Fine, M. L. (2015). Pectoral sound generation in the blue catfish *Ictalurus furcatus*. *Journal of Comparative Physiology A: Sensory Neural and Behavioral Physiology, 201*, 305–315.

Mok, H.-K., Parmentier, E., Chiu, K.-H., Tsai, K.-E., Chiu, P.-H., & Fine, M. (2011). An Intermediate in the evolution of superfast sonic muscles. *Frontiers in Zoology, 8*(1), 31.

Moulton, J. M. (1958). The acoustical behavior of some fishes in the Bimini area. *Biology Bulletin, 114*(3), 357–374.

Moulton, J. M. (1960). Swimming sounds and the schooling of fishes. *Biology Bulletin, 119*(2), 210–223.

Myrberg, A. A., Jr. (1981). Sound communication and interception in fishes. In W. N. Tavolga, A. N. Popper, & R. R. Fay (Eds.), *Hearing and sound communication in fishes* (pp. 395–425). New York: Springer.

Myrberg, A. A., Jr. (1997). Underwater sound: Its relevance to behavioural functions among fishes and marine mammals. *Marine and Freshwater Behavioural Physiology, 29*, 3–21.

Myrberg, A. A., Jr., Ha, S. J., & Shamblott, M. J. (1993). The sounds of bicolor damselfish (*Pomacentrus partitus*): Predictors of body size and a spectral basis for individual recognition and assessment. *Journal of the Acoustical Society of America, 94*, 3067–3070.

Myrberg, A. A., Jr., Mohler, M., & Catala, J. (1986). Sound production by males of a coral reef fish (*Pomacentrus partitus*): Its significance to females. *Animal Behaviour, 34*, 913–923.

Myrberg, A., Spanier, E., & Ha, S. (1978). Temporal patterning in acoustic communication. In E. Reese & F. J. Lighter (Eds.), *Contrasts in behaviour* (pp. 137–179). New York: Wiley.

Oliveira, T. P. R., Ladich, F., Abed-Navandi, D., Souto, A. S., & Rosa, I. L. (2014). Sounds produced by the longsnout seahorse: A study of their structure and functions. *Journal of Zoology, 294*(2), 114–121.

Oliver, S., & Lobel, P. (2013). Direct mate choice for simultaneous acoustic and visual courtship displays in the damselfish, *Dascyllus albisella* (Pomacentridae). *Environmental Biology of Fishes, 96*(4), 447–457.

Olivier, D., Frédérich, B., Herrel, A., & Parmentier, E. (2015). A morphological novelty for feeding and sound production in the yellowtail clownfish. *Journal of Experimental Zoology Part A: Ecological Genetics and Physiology*. doi:10.1002/jez.1907.

Olivier, D., Frederich, B., Spanopoulos-Zarco, M., Balart, E., & Parmentier, E. (2014). The cerato-mandibular ligament: A key functional trait for grazing in damselfishes (Pomacentridae). *Frontiers in Zoology, 11*(1), 63.

Ono, R. D., & Poss, S. G. (1982). Structure and innervations of the swimbladder musculature in the weakfish, *Cynoscion regalis* (Teleostei: Sciaenidae). *Canadian Journal of Zoology, 60*, 1955–1967.

Onuki, A., & Somiya, H. (2004). Two types of sounds and additional spinal nerve innervation to the sonic muscle in John Dory, *Zeus faber* (Zeiformes: Teleostei). *Journal of the Marine Biological Association of the United Kingdom, 84*(04), 843–850.

Ostrom, J. H. (1979). Bird flight: How did it begin? *American Scientist, 67*, 46–56.

Parmentier, E., Bouillac, G., Dragičević, B., Dulčić, J., & Fine, M. (2010a). Call properties and morphology of the sound-producing organ in *Ophidion rochei* (Ophidiidae). *Journal of Experimental Biology, 213*(18), 3230–3236.

Parmentier, E., Boyle, K. S., Berten, L., Brie, C., & Lecchini, D. (2011a). Sound production and mechanism in *Heniochus chrysostomus* (Chaetodontidae). *Journal of Experimental Biology, 214*(Pt 16), 2702–2708.

Parmentier, E., Chardon, M., & Vandewalle, P. (2002). Preliminary study on the ecomorphological signification of the sound-producing complex in Carapidae. In P. Aerts, K. D'Août, A. Herrel, & R. Van Damme (Eds.), *Topics in functional and ecological vertebrate morphology* (pp. 139–151). Maastricht, The Netherlands: Shaker Publishers.

Parmentier, E., Colleye, O., Fine, M., Frederich, B., Vandewalle, P., & Herrel, A. (2007). Sound production in the clownfish *Amphiprion clarkii. Science, 316*, 1006.

Parmentier, E., Compere, P., Casadevall, M., Fontenelle, N., Cloots, R., & Henrist, C. (2008). The rocker bone: A new kind of mineralised tissue? *Cell and Tissue Research, 334*, 67–79.

Parmentier, E., & Diogo, R. (2006). Evolutionary trends of swimbladder sound mechanisms in some teleost fishes. In F. Ladich, S. P. Collin, P. Moller, & B. G. Kapoor (Eds.), *Communication in fishes* (Vol. 1, pp. 45–70). Enfield, NH: Science Publishers.

Parmentier, E., Fabri, G., Kaatz, I., Decloux, N., Planes, S., & Vandewalle, P. (2010b). Functional study of the pectoral spine stridulation mechanism in different mochokid catfishes. *Journal of Experimental Biology, 213*(7), 1107–1114.

Parmentier, E., Fontenelle, N., Fine, M. L., Vandewalle, P., & Henrist, C. (2006a). Functional morphology of the sonic apparatus in *Ophidion barbatum* (Teleostei, Ophidiidae). *Journal of Morphology, 267*, 1461–1468.

Parmentier, E., Genotte, V., Focant, B., Goffinet, G., & Vandewalle, P. (2003a). Characterization of the primary sonic muscles in *Carapus acus* (Caparidae): A multidisciplinary approach. *Proceedings of the Royal Society Biological Sciences Series B, 270*, 2301–2308.

Parmentier, E., Kéver, L., Boyle, K., Corbisier, Y.-E., Sawelew, L., & Malavasi, S. (2013). Sound production mechanism in *Gobius paganellus* (Gobiidae). *The Journal of Experimental Biology, 216*(17), 3189–3199.

Parmentier, E., Kéver, L., Casadevall, M., & Lecchini, D. (2010c). Diversity and complexity in the acoustic behaviour of *Dacyllus flavicaudus* (Pomacentridae). *Marine Biology, 157*(10), 2317–2327.

Parmentier, E., Lagardere, J.-P., Braquegnier, J.-B., Vandewalle, P., & Fine, M. L. (2006b). Sound production mechanism in carapid fish: First example with a slow sonic muscle. *Journal of Experimental Biology, 209*, 2952–2960.

Parmentier, E., Tock, J., Falguière, J.-C., & Beauchaud, M. (2014). Sound production in *Sciaenops ocellatus*: Preliminary study for the development of acoustic cues in aquaculture. *Aquaculture, 432*, 204–211.

Parmentier, E., Vandewalle, P., Brie, C., Dinraths, L., & Lecchini, D. (2011b). Comparative study on sound production in different Holocentridae species. *Frontiers in Zoology, 8*(1), 12.

Parmentier, E., Vandewalle, P., & Lagardère, J. P. (2003b). Sound-producing mechanisms and recordings in *Carapini* species (Teleostei, Pisces). *Journal of Comparative Physiology. A, Neuroethology, Sensory, Neural, and Behavioral Physiology, 189*(4), 283–292.

Patek, S. N., Baio, J. E., Fisher, B. L., & Suarez, A. V. (2013). Mutlifunctionality and mechanical origins: Ballistic jaw protrusion in trap-jaw ants. *Proceeding of National Academy of Science of the U S A, 103*(34), 12787–12792.

Popper, A. N., & Fay, R. R. (2011). Rethinking sound detection by fishes. *Hearing Research, 273*(1–2), 25–36.

Ramcharitar, J., Gannon, D., & Popper, A. (2006). Bioacoustics of fishes of the family Scianidae (croackers and drums). *Transactions of the American Fisheries Society, 135*, 1409–1431.

Rice, A. N., & Bass, A. H. (2009). Novel vocal repertoire and paired swimbladders of the three-spined toadfish, *Batrachomoeus trispinosus*: Insights into the diversity of the Batrachoididae. *Journal of Experimental Biology, 212*(9), 1377–1391.

Rice, A. N., & Lobel, P. S. (2003). The pharyngeal jaw apparatus of the Cichlidae and Pomacentridae: Function in feeding and sound production. *Reviews in Fish Biology and Fisheries, 13*(4), 433–444.

Ripley, J. L., & Foran, C. M. (2007). Influence of estuarine hypoxia on feeding and sound production by two sympatric pipefish species (Syngnathidae). *Marine Environmental Research, 63*(4), 350–367.

Rome, L. C., Cook, C., Syme, D. A., Connaughton, M. A., Ashley-Ross, M., Klimov, A., et al. (1999). Trading force for speed: Why superfast crossbridge kinetics leads to superlow forces. *Proceedings of the National Academy of Sciences of the U S A, 95*, 5826–5831.

Rome, L. C., & Lindstedt, S. L. (1998). The quest for speed: Muscles built for high frequency contractions. *News in the Physiological Sciences, 13*, 261–268.

Rome, L. C., Syme, D. A., Hollingworth, S., Lindstedt, S., & Maylor, S. M. (1996). The whistle and the rattle: The design of sound producing muscles. *Proceedings of the National Academy of Sciences of the U S A, 93*, 8095–8100.

Rüber, L., & Adams, D. C. (2001). Evolutionary convergence of body shape and trophic morphology in cichlids from Lake Tanganyika. *Journal of Evolutionary Biology, 14*(2), 325–332.

Sand, O., & Hawkins, A. D. (1973). Acoustic properties of the cod swimbladder. *Journal of Experimental Biology, 58*(3), 797–820.

Schachner, G., & Schaller, F. (1981). Schallerzeugung and schallreaktionen beim antennenwels (Mandim) *Rhambdia sebae sebae. Zoologische Beitraege, 27*, 375–392.

Schaefer, S. A. (1984). Mechanical strength of the pectoral spine/girdle complex in *Pterygoplichthys* (Loricariidae: Siluroidei). *Copeia, 1984*, 1005–1008.

Schaeffer, P., Conley, K., & Lindstedt, S. (1996). Structural correlates of speed and endurance in skeletal muscle: The rattlesnake tailshaker muscle. *Journal of Experimental Biology, 199*(2), 351–358.

Schneider, H. (1967). Morphology and physiology of sound-producing mechanisms in teleost fishes. In W. N. Tavolga (Ed.), *Marine bioacoustics* (Vol. 2, pp. 135–158). Oxford, England: Pergamon Press.

Schulz-Mirbach, T., Hess, M., Metscher, B., & Ladich, F. (2013). A unique swim bladder-inner ear connection in a teleost fish revealed by a combined high-resolution microtomographic and three-dimensional histological study. *BMC Biology, 11*(1), 75.

Simpson, G. G. (1953). *The major features of evolution*. New York: Columbia University Press.

Sismour, E. N., Nellis, S. C., Newton, S. H., Mays, D. a., & Fine, M. L. (2013). An Experimental study of consumption of channel catfish *Ictalurus punctatus* by largemouth bass *Micropterus salmoides* when alternative prey are available. *Copeia, 2013*(2), 277–283.

Skoglund, C. (1961). Functional analysis of swimbladder muscles engaged in sound production of the toadfish. *Journal of Biophysical and Biochemical Cytology, 10*, 187–200.

Smith, F. M., & Croll, R. P. (2011). Autonomic control of the swimbladder. *Autonomic Neuroscience, 165*(1), 140–148.

Somlyo, A. V., Shurman, H., & Somlyo, A. P. (1977). Composition of sarcoplasmic reticulum in situ by electron probe x-ray microanalysis. *Nature, 268*, 556–558.

Sörensen, W. (1895). Are the extrinsic muscles of the air-bladder in some Siluroidae and the "elastic spring" apparatus of others subordinate to the voluntary production of sounds? What is, according to our present knowledge, the function of the Weberian ossicles? *Journal of Anatomy and Physiology, 29*, 205–229, 399–423, 518–552.

Sprague, M. W., & Luczkovich, J. J. (2001). Do striped cusk-eels *Ophidion marginatum* (Ophidiidae) produce the "chatter" sound attribuated to weakfish *Cynoscion regalis* (Scianidae)? *Copeia, 3*, 854–859.

Stadler, J. M. (2002). Evidence for a hydrodynamic mechanism of sound production by courting males of the notchtongue goby, *Bathygobius curacao* (Metzlar). *Bioacoustics, 13*, 145–152.

Stiassny, M. L. J. (1981). The phyletic status of the family Cichlidae (Pisces, Perciformes): A comparative anatomical investigation. *Netherlands Journal of Zoology, 31*, 275–314.

Tavolga, W. (1964a). Sonic characteristics and mechanisms in marine fishes. In W. N. Tavolga (Ed.), *Marine bio-acoustics* (pp. 195–211). New York: Pergamon Press.

Tavolga, W. N. (1964b). Sonic characteristics and mechanisms in marine fishes. In W. N. Tavolga (Ed.), *Marine Bio-acoustics* (pp. 195–211). Oxford, England: Pergamon Press.

Tavolga, W. N. (1971). Sound production and detection. In W. S. Hoar & D. J. Randall (Eds.), *Fish physiology* (Vol. 5, pp. 135–205). New York: Academic Press.

Tavolga, W. N. (1977). Mechanisms for directional hearing in the sea catfish (*Arius felis*). *Journal of Experimental Biology, 67*, 97–115.

Thorson, R. F., & Fine, M. L. (2002). Acoustic competition in the gulf toadfish Opsanus beta: acoustic tagging. *J. Acoust. Soc. Amer. 111*, 2302–2307.

Tower, R. W. (1908). The production of sound in the drumfishes, the searobin and the toadfish. *Annals of the New York Academy of Sciences, 18*, 149–180.

van Bergeijk, W. A. (1964). Directional and nondirectional hearing in fish. In W. N. Tavolga (Ed.), *Marine Bio-acoustics* (pp. 281–299). New York: Pergamon Press.

Vance, T. (2000). Variability in stridulatory sound production in the channel catfish, *Ictalurus punctatus*. *BIOS, 71*, 79–84.

Vandewalle, P., Havard, M., Claes, G., & Vree, F. D. (1992). Mouvements des mâchoires pharyngiennes pendant la prise de nourriture chez le *Serranus scriba* (Linné, 1758) (Pisces, Serranidae). *Canadian Journal of Zoology, 70*(1), 145–160.

Vandewalle, P., Parmentier, É., & Chardon, M. (2000). The branchial basket in teleost feeding. *Cybium, 24*(4), 319–342.

Vandewalle, P., Saintin, P., & Chardon, M. (1995). Structures and movements of the buccal and pharyngeal jaws in relation to feeding in *Diplodus sargus*. *Journal of Fish Biology, 46*, 623–656.

Vasconcelos, R. O., Carriço, R., Ramos, A., Modesto, T., Fonseca, P. J., & Amorim, M. C. P. (2012). Vocal behavior predicts reproductive success in a teleost fish. *Behavioral Ecology, 23*(2), 375–383.

Vasconcelos, R. O., & Ladich, F. (2008). Development of vocalization, auditory sensitivity and acoustic communication in the Lusitanian toadfish *Halobatrachus didactylus*. *Journal of Experimental Biology, 211*, 502–509.

Wagner, G. P. (1996). Homologues, natural kinds and the evolution of modularity. *American Zoologist, 36*(1), 36–43.

Wahlberg, M., & Westerberg, H. (2003). Sounds produced by herring (*Clupea harengus*) bubble release. *Aquatic Living Resources, 16*(3), 271–275.

Wainwright, P. (1989a). Functional morphology of the pharyngeal jaw apparatus in perciform fishes: An experimental analysis of the haemulidae. *Journal of Morphology, 200*(3), 231–245.

Wainwright, P. (1989b). Prey processing in haemulid fishes: Patterns of variation in pharyngeal jaw muscle activity. *Journal of Experimental Biology, 141*(1), 359–375.

Waybright, T. D., Kollenkirchen, U., & Fine, M. L. (1990). Effect of size and sex on grunt production in the oyster toadfish. *Abstracts of the Society for Neuroscience, 16*, 578.

Wright, J. (2009). Diversity, phylogenetic distribution, and origins of venomous catfishes. *BMC Evolutionary Biology, 9*(1), 282.

Young, I. S., & Rome, L. C. (2001). Mutually exclusive muscle designs: The power output of the locomotory and sonic muscles of the oyster toadfish (*Opsanus tau*). *Proceedings of the Royal Society of London Series B: Biological Sciences, 268*(1480), 1965–1970.

Zeldicht, M. L., & Fink, W. L. (1996). Heterochrony: Stability and innovation in the evolution of form. *Paleobiology, 22*, 241–254.

Chapter 3
Vocal Sound Production and Acoustic Communication in Amphibians and Reptiles

Kaitlen C. Colafrancesco and Marcos Gridi-Papp

Abstract Most amphibians and reptiles produce sounds with a larynx containing a pair of vocal cords. Clicking and hissing are common in both groups whereas tonal sounds are found most frequently in anurans and geckos. Calls can exceed 90 dB SPL at a distance of 1 m and they can have fundamental frequencies above 20 kHz. Calling is used mostly by males for courtship and territorial displays. Offspring and females call to synchronize hatching and to mediate maternal care. Adults and juveniles in many groups produce hissing when threatened. Amphibians and reptiles include more than 17,000 species. As a result of this diversity, major advances in the field of vocalization are made through exploratory research but also through careful experimentation and the use of novel technologies. Combining the study of vocal and auditory systems is important to explain issues such as the diversity of frequency tuning in the group. Many questions can also be answered through comparative studies in amphibians and reptiles because these groups have evolved independent solutions to common communication problems.

Keywords Amplitude modulation • Bellow • Call • Crocodile • Frequency modulation • Frog • Gecko • Hiss • Larynx • Nonlinearity • Toad • Turtle • Ultrasound • Vocal cord • Vocal sac

3.1 Introduction

Amphibians and reptiles (ectotherm tetrapods) are highly diverse classes of vertebrates, represented by 17,174 species (Frost 2014; Uetz 2014). Amphibians have an exclusive common ancestor, but reptiles share their common ancestor with birds. Avian vocalizations are discussed separately (Düring and Elemans, Chap. 5), as

---

K. Colafrancesco • M. Gridi-Papp (✉)
Department of Biological Sciences, University of the Pacific,
3601 Pacific Ave., Stockton, CA 95211, USA
e-mail: k_colafrancesco@u.pacific.edu; mgridipapp@pacific.edu

R.A. Suthers et al. (eds.), *Vertebrate Sound Production and Acoustic Communication*, Springer Handbook of Auditory Research 53,
DOI 10.1007/978-3-319-27721-9_3

birds are a highly diverse group with many specializations for acoustic communication. Amphibians and reptiles have colonized most environments and adapted to the majority of climates on our planet (Vitt and Caldwell 2013). All of the major groups, including caecilians (limbless amphibians), newts, salamanders, frogs, turtles, tuataras (order Rhynchocephalia), snakes, lizards, and crocodiles, can use the larynx to emit sound. This richness and diversity makes ectotherm tetrapods a rewarding subject for the study of vocalization mechanisms and evolution.

Vocalizations and acoustic communication are best known in anurans (frogs, toads, and treefrogs), possibly because most species produce conspicuous choruses that attract our attention (Rand 2001). The calls of thousands of species have been documented and many studies have explored the behavior and evolution of calling (Ryan 2001; Gerhardt and Huber 2002). The general morphology and physiology of vocalization has been described (Duellman and Trueb 1986; Feder and Burggren 1992), as has the basic neural and endocrine control of calling (Walkowiak 2006; Yang and Kelley 2008). Other amphibians such as salamanders and newts seem to lack conspicuous advertisement calls but they do emit vocalizations as part of their defense displays against predators (Brodie 1978).

Among reptiles, calls are used in a variety of contexts. Calls mediate interactions between mother and offspring in crocodilians (Vergne et al. 2009). Geckos vocalize in courtship and territorial interactions (Frankenberg 1982). Tortoises vocalize during mating (Galeotti et al. 2005).

This chapter summarizes the mechanisms used to produce vocalizations and the behavioral roles of vocalizations in the main groups of amphibians and reptiles. The best studied case, anurans, is addressed first to provide a context for interpretation of the remaining groups. Emphasis is placed on morphological structure and its relationship to frequency tuning. The neural and endocrine control of calling (Kelley et al. 2001; Wilczynski and Chu 2001; Walkowiak 2006) are not covered and readers are referred to other reviews for discussions of the temporal features of the calls (Greenfield 1994; Littlejohn and Ryan 2001; Wells 2010).

3.2 Vocal Apparatus

3.2.1 Hyoid and Larynx

The larynx is the source of vocal sound in anurans. It is supported by the hyoid, which is a cartilaginous plate that forms the floor of the mouth and has several processes (Fig. 3.1a; Duellman and Trueb 1986). The posteromedial processes are usually the only bony portion of the hyoid and they form a rigid V-shaped support for the larynx. The hyoid itself is supported mostly by muscles, and variation in muscular tension allows for extensive displacement of the hyolaryngeal apparatus along the longitudinal axis of the body. The hyoid articulates with the rest of the skeleton only at the cranium. The long and flexible anterior process of the hyoid attaches to the otic capsule and forms the posterior wall of the Eustachian tube.

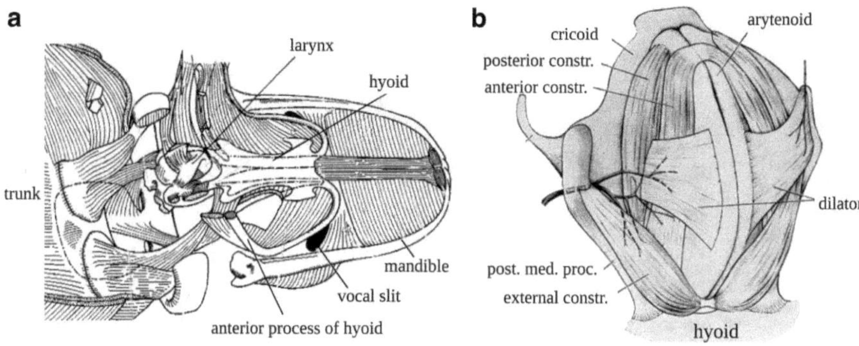

Fig. 3.1 Structure of the floor of the mouth and larynx in anurans. (**a**) Position of the larynx, supported by the hyoid, at the posterior side of the mouth in toads of the genus *Bufo*. Notice the vocal slits, lateral to the hyoid, connecting to the vocal sac. Next to the vocal slits, the anterior process of the hyoid extends to the posterior side of the mouth and attaches to the skull, forming the posterior wall of the Eustachian tube. (**b**) Major laryngeal structures in the European treefrog (*Hyla arborea*). The cricoid cartilage ring is positioned between the posteromedial processes of the hyoid (post. med. proc.) and contains the two arytenoid cartilages. The long laryngeal nerve innervates the anterior constrictor muscle (anterior constr.), the posterior constrictor muscle (posterior constr.), the external constrictor muscle (external constr.), and the dilator muscle. The short laryngeal nerve innervates only the dilator muscle [*Sources*: **a**. Martin and Gans (1972). **b**. Walkowiak (2006)]

The cricoid cartilage provides a structural frame for the larynx (Fig. 3.1b; Trewavas 1932; Schneider 1970). It forms a complete ring of cartilage positioned in the same plane as the posteromedial processes of the hyoid. A pair of highly concave arytenoid cartilages lies within the cricoid ring with which they articulate. The arytenoid cartilages in amphibians are relatively larger than in mammals. Together they occlude the passage of air through the larynx at rest. This passage opens when the arytenoids hinge apart. This opening can be caused by contraction of the laryngeal dilator muscles, or by increased subglottal air pressure (Paulsen 1965; Martin 1971).

Sound is produced by a pair of vocal cords that are mounted within the concavities of the arytenoid cartilages and attached to the cricoid ring (Fig. 3.2a; Trewavas 1932). Each vocal cord blocks the passage of air through the concavity of its arytenoid cartilage. As a consequence, air can cross the larynx only by passing through the medial separation between the two vocal cords. Laryngeal airflow sets the vocal cords into a passive vibration that produces sound. In contrast to the mammalian vocal folds, the vocal cords of anurans lack any muscular layers. The vocal cords are usually attached to the lateral edge of the arytenoid cartilage, from which they bulge anteriorly to expand into the concavity of the arytenoid cartilage. Each vocal cord has a straight, reinforced medial edge, where it contacts its symmetric counterpart during phonation. Some groups of anurans have this medial line of contact between the vocal cords expanded anteriorly, posteriorly, or in both directions (Schmid 1978). The functional significance of this variation has not yet been clarified experimentally.

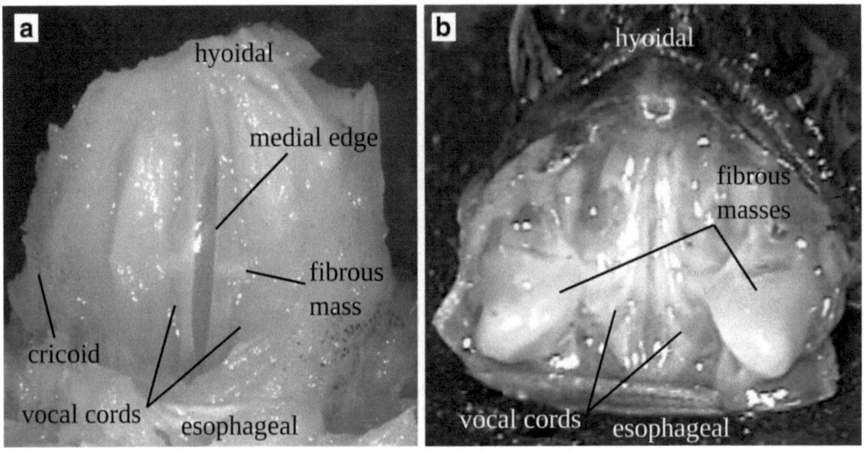

Fig. 3.2 Vocal cords and fibrous masses in two anurans. Larynges are shown from a postero-ventral perspective, with the posterior and ventral tissues of the larynx removed to expose the entire length of the vocal cords. (**a**) *Gray* treefrog (*H. versicolor*) from Austin, TX. Notice the nonmuscular vocal cords with small fibrous masses embedded near the middle of their lengths. The inner surfaces of the closed arytenoid cartilages are visible between the medial edges of the two vocal cords. (**b**) Túngara frog (*E. pustulosus*) from Panama. Notice how large and lateralized the fibrous masses are compared to those of the gray treefrog

Vocal cords frequently contain a thickened portion called fibrous mass near the center of the membrane. It lowers the fundamental frequency of vocal cord vibration by adding mass at the point where it has maximal effect on the tuning of the vibrating membrane (Martin 1971). In túngara frogs (*Engystomops pustulosus*), the fibrous mass is greatly enlarged and displaced posterolaterally (Fig. 3.2b; Drewry et al. 1982; Ryan and Drewes 1990). Males of this species can optionally overlay "chuck" notes to their regular "whine" call. Chucks have half of the fundamental frequency of the whine and sound distinctive. The addition of chucks increases the complexity of the advertisement call and its attractiveness to females (Ryan 1980, 1985). The partial disconnection of the fibrous mass from the vocal cord likely allows these frogs to call with or without fibrous mass vibration. Although the control mechanism of fibrous mass vibration has not yet been elucidated, ablation of this mass in live animals removes the complexity (odd harmonics) from the chucks without altering the whines (Gridi-Papp et al. 2006).

3.2.2 Laryngeal Muscles

Four pairs of laryngeal muscles have been identified in anurans. All of them lie external to the laryngeal cavity and have attachments to the arytenoid and cricoid cartilages, or to the hyoid (Fig. 3.1b; Trewavas 1932; Schneider 1970). As the vocal cords are not muscular, any control of their position or stretching has to be produced by muscles that are external to the laryngeal cavity. These muscles are innervated by the long and the short laryngeal branches of the vagus nerve.

The laryngeal dilator muscle is the sole opener of the passage through the larynx. This muscle originates near the distal end of the posteromedial process of the hyoid and inserts along the medial edge of the arytenoid cartilage. Simultaneous bilateral shortening of the muscle pulls the medial edges of the two arytenoid cartilages apart, pivoting them on their attachments to the cricoid cartilage.

The arytenoid cartilages can be moved to close the larynx by contraction of constrictor muscles that form a sphincter (ring of muscle) around the arytenoid cartilages. These muscle pairs form the external laryngeal constrictor muscle (also called hyolaryngeus muscle) and anterior laryngeal constrictor muscle (also called sphincter anterior muscle). Both of these muscles lie deep to the laryngeal dilator muscle. The external laryngeal constrictor muscle originates at the posteromedial process of the hyoid. The anterior laryngeal constrictor muscle originates from the cricoid cartilage.

A fourth pair of laryngeal muscles, the posterior laryngeal constrictor, is present in most anurans but lacking in toads of the genus *Bufo* (Trewavas 1932; Martin 1971). The posterior laryngeal constrictor muscle is also deep to the laryngeal dilator muscle. It runs parallel but posterior to the anterior laryngeal constrictor muscle and inserts on the arytenoid cartilages at the same point where the posterior medial ends of the vocal cords attach internally. Some fibers of this muscle extend into the larynx and attach to the vocal cords, raising the possibility that muscle contraction could position or tense the vocal cords and modify the sound (Gaupp and Ecker 1904; Trewavas 1932).

Besides vocalization, the laryngeal muscles are also crucial to respiration. The arytenoid cartilages lie adducted (closed) at rest and they form a unidirectional valve that only allows airflow from the lungs into the mouth. To pass air from the mouth into the lungs, the animal contracts the laryngeal dilator muscle, which opens the arytenoids (De Jongh and Gans 1969). The floor of the mouth is then lifted to push the air into the lungs and the laryngeal constrictor muscles contract immediately after, preventing back flow by closing the arytenoid cartilages.

3.2.3 Variation in Laryngeal Structure

The vocal apparatus is highly variable among anurans and the preceding description only represents the general case. Comparisons of representatives from different families of anurans revealed variation in the shapes of the vocal cartilages and of the vocal cords. The posterior laryngeal constrictor muscle is absent in some toads (Bufonidae) and it is small or lacks its ventral portion in some other groups (Trewavas 1932). Toads in the genus *Bufo* have a pair of free-hanging posterior membranes with unknown function attached to the cricoid upstream of the true vocal cords (Martin 1971). The anterior laryngeal constrictor muscle does not attach to the hyoid in fire-bellied toads (*Bombina*). In European spadefoot toads

(Pelobatidae), the cricoid ring is incomplete, having a gap in the mid-dorsal region, but in many other species it receives a secondary attachment of the dilator muscle (Trewavas 1932).

Sexual dimorphism is present in most species, with adult males developing larynges that are relatively larger than those of females. Females of many species occasionally emit courtship or release calls but only in contact with or at a short distance from the male. These calls are produced at greatly reduced intensities when compared to male advertisement calls (Emerson et al. 1999).

The most strikingly specialized larynges discovered in anurans are probably those found in tropical tongueless frogs (Pipidae) in which both the tadpoles and the adults live underwater. Male advertisement calls are sexually selected in the African members of this group and both male and female release calls have been recorded in many species (Tobias et al. 2011, 2014). Adult Marsabit clawed frogs (*Xenopus borealis*) of both sexes produce sound underwater without moving air through the larynx (Yager 1992, 1996). Vocal cords are absent and the cricoid is expanded, forming a cartilaginous box. The laryngeal cavity contains a pair of modified arytenoid cartilages to which the single pair of laryngeal muscles attaches. The two arytenoids have an elongated, ossified portion and they articulate with each other at a small cartilaginous disc. At rest, the discs of the two arytenoids are in contact with each other and contraction of the laryngeal muscle separates the discs. This separation produces a click that is the basis of the call. Males and females produce various call types distinguished by the temporal arrangement of the clicks and also by frequency.

In the African clawed frog (*Xenopus laevis*) males produce at least four types of calls with distinctive temporal structure and social role (Kelley and Tobias 1999). Females produce two call types in response to male calling or clasping. Ticking is a call given by unresponsive females and it suppresses male calling. Rapping, on the other hand, is given by receptive females and it stimulates male calling. The laryngeal structure is similar to that of the Marsabit clawed frog, and the male larynx is significantly larger than the female larynx (Fischer and Kelley 1991).

Male African clawed frogs also have larger nuclei in the brain circuitry dedicated to the control of calling than females. These vocal brain regions comprise the motor neurons of the cranial nerve nuclei IX and X in the caudal medulla and of the pretrigeminal nucleus of the dorsal tegmental area of the medulla (DTAM). Sex differences in these areas and in the larynx are triggered by sex hormones via androgen receptors in the vocal brain nuclei and laryngeal tissues (Zornik and Yamaguchi 2008). The timing of potentials produced by the laryngeal motor neurons differ for each call type during natural calling and match the timing of the clicks produced at the larynx (Yamaguchi and Kelley 2000). This simple relationship validates the recording of motor neuron output in isolated brains as a proxy for calling (Zornik and Kelley 2008). This experimental approach facilitated the study of the brain's response to different hormonal environments and it showed that the temporal patterns produced by the motor neurons can be masculinized by exposing the brain to androgens (Zornik and Yamaguchi 2008).

3.2.4 *Lungs*

The paired lungs of anurans can contain impressive volumes of air relative to body size, accounting for more than 30% of the body volume (De Jongh and Gans 1969). They tend to have a simple structure with internal septa increasing the respiratory surface near the outer walls and they are lined with surfactant (Hughes and Vergara 1978; Romer and Parsons 1986). They attach directly to the larynx without any intervening cartilaginous or muscular bronchi.

3.2.5 *Muscles of the Body Wall*

In the majority of anurans, the muscles of the flanks produce most of the power needed for vocalization by compressing the air in the lungs and forcing it out through the larynx (Gaupp and Ecker 1904; Martin and Gans 1972). These are the transverse, external oblique, and internal oblique muscles. Several species of anurans produce advertisement calls that require exhalations to be produced at rates considerably higher (50 Hz) than other types of voluntary movements. The trunk muscles are physiologically adapted to the task (Eichelberg and Obert 1976; Girgenrath and Marsh 1997, 1999). In addition, their fast contraction properties are sexually dimorphic, testosterone dependent, and they are maintained only during the breeding season (Girgenrath and Marsh 2003).

3.2.6 *Mouth and Vocal Sac*

The buccal cavity in anurans is connected to the vocal sac that is present in males only. The sac is formed by a fold of the buccal lining that extends through a pair of round or elongated passages (vocal slits) into the space between the hyoidal musculature and the intermandibular musculature (Fig. 3.1a; Liu 1935). Most commonly, a single vocal sac is centered under the throat (median subgular position), but it can also be split medially (paired subgular position) or it can be caudal to the head on each side of the body (paired lateral position). The wall of the vocal sac is very rich in elastic fibers between the thin layer of interhyoidal muscle and the skin (Jaramillo et al. 1997).

3.3 Vocalizations

3.3.1 *Relationship of Sound Production with Breathing*

Both breathing and vocalization involve passing air through the larynx. Most of the key structures and movements employed in these two processes are the same. The anuran breathing mechanism has been described in detail elsewhere

(De Jongh and Gans 1969; Vitalis and Shelton 1990). Abdominal (diaphragmatic) inspiration is lacking because it requires a rigid rib cage for the pulmonary air pressure to drop when the diaphragm contracts. The ribs of anurans are short and do not form a rigid enclosure around the lungs, as occurs in reptiles or mammals. To inflate the lungs, anurans perform a series of buccal pumping movements, each involving (1) opening the nares, (2) lowering the floor of the mouth to suck in air, (3) closing the nares and opening the larynx, and (4) elevating the floor of the mouth to push the air into the lungs.

Anurans also engage in oscillatory cycles of respiration without opening the larynx or closing the nares (De Jongh and Gans 1969). Each oscillatory cycle renews the small volume of air contained in the mouth without replacing the air in the lungs, thereby enabling gas exchanges in the highly vascularized mouth.

3.3.2 Basic Respiratory Mechanisms for Phonation

Most anurans produce sound during expiration when air is transferred from the lungs through the larynx and mouth into the vocal sac. Fire-bellied toads (*Bombina*), however, phonate during inspiration and painted frogs (*Discoglossus*) can produce sound during inspiration or expiration (Weber 1974; Walkowiak 1992). Inspiratory phonation is based on the same movements produced during respiratory inspiration. Sound is produced during or after contraction of the buccal pump muscles that lift the floor of the mouth (Strake et al. 1994; Walkowiak 2006). Inspiratory vocalizations have also been described in the Neotropical and distantly related Peters' four-eyed frog (*Pleurodema diplolister*; Hödl 1992) and Muller's termite frog (*Dermatonotus muelleri*; Giaretta et al. 2015). These findings indicate that inspiratory vocalizations may have evolved multiple times in anurans.

Before each bout of calling, the animal utilizes buccal pumping to inflate its lungs. The larynx is then locked in a closed position by contraction of the laryngeal constrictor muscles (Martin and Gans 1972). This behavior of overinflation and laryngeal locking is also exhibited in defensive behaviors to hinder ingestion by predators such as snakes (Duellman and Trueb 1986).

Contraction of the trunk muscles then elevates the subglottal pressure until the laryngeal constrictor muscles relax and allow for passive opening of the arytenoid cartilages. This initiates laryngeal airflow with passive vibration of the vocal cords and marks the onset of sound (Martin and Gans 1972). The larynx produces sound over a wide range of subglottal air pressures. Within this range, a nearly linear relationship is observed among subglottal pressure, laryngeal airflow, sound pressure, and sound intensity (Martin 1971; Gridi-Papp 2014). Above this pressure range, phonation tends to become noisy, respond nonlinearly to pressure or fail completely. Kime and collaborators proposed a quantitative framework for modeling the physical properties and acoustic output of the anuran larynx (Kime et al. 2013).

The mouth and nares are kept closed during phonation except in distress calls (Hödl and Gollmann 1986), and the vocal sac thus inflates as sound is produced.

At the end of phonation, the trunk muscles relax and the laryngeal dilator muscle contracts to prevent the arytenoid cartilages from closing. Air is then returned from the vocal sac into the lungs by elastic recoil and by contraction of the thin musculature of the vocal sac (Dudley and Rand 1991; Jaramillo et al. 1997).

3.3.3 Control over Call Frequency

Compared to the vocalizations of birds and mammals, anuran calls tend to be very simple and most commonly restricted to a single note without extensive frequency modulation. This raises the question of whether such simplicity stems from the neural control of vocalization or from limitations in laryngeal mechanics (Walkowiak 2006). Experiments in which air was passed at various pressures through the larynges of euthanized anurans have revealed that the fundamental frequency of the sound output is greatly influenced by the driving air pressure (in direct correlation). These experiments have also shown that anuran larynges can produce a wide range of fundamental frequencies, even in species that do not exhibit significant frequency modulation in their calls (Martin 1971; Gridi-Papp 2014). The structure of the larynx, therefore, does not restrict the fundamental frequency of the call to a fixed frequency.

Other mechanisms of frequency control based on the morphology of the larynx have been hypothesized, but these have not been verified experimentally. Based on anatomical evidence, the posterior laryngeal constrictor muscle has been suggested to stretch the vocal cords and allow the animal to control the frequency of the laryngeal output (see Sect. 3.2.2; Gaupp and Ecker 1904; Trewavas 1932). This hypothesis was refuted through experimental stimulation of this muscle that resulted in adduction of the two vocal cords but no visible tensioning (Schmidt 1972). This muscle was therefore suggested to control the onset of sound, but not its frequency.

It is possible that most anurans lack a mechanism for voluntary control of call frequency other than varying the driving pulmonary air pressure. Electrophysiological recordings show that all of the laryngeal muscles remain relaxed while the vocal cords produce sound during release calling in the gulf coast toad (*Bufo valliceps*; Martin and Gans 1972), and the northern leopard frog (*Lithobates pipiens*; Schmidt 1972). The generality of these results is unknown because similar measurements are not available during advertisement calling or from other species. Release calls were used in these experiments because they are readily produced by the experimental species when the individual is held by the axils. This mimics the natural condition in which males and unreceptive females produce release calls when clasped by a male, which promptly releases them on hearing the call (Sullivan and Wagner 1988; Tobias et al. 2014).

Males in many species of anurans are favored by natural selection to produce advertisement calls at high repetition rates and intensity (Gerhardt and Huber 2002). Active manipulation of vocal cord vibrations after phonation is initiated may, therefore, be avoided in favor of maximizing the intensity of the sound output. Experiments flowing humid air through the larynges of euthanized North American treefrogs showed that the acoustic intensity of the laryngeal output is directly correlated with the driving respiratory air pressure and with the fundamental frequency (Gridi-Papp 2008, 2014). When call frequency is altered by varying the driving air pressure, call intensity should change proportionally. Frequency modulation in the calls of several species of anurans can be explained based on this mechanism (e.g., *Bufo quercicus*, *E. pustulosus*; Dudley and Rand 1991). Cricket frogs (*Acris crepitans*) have been observed to lower the frequency and the intensity of their calls during aggressive interactions (Wagner 1989, 1992). In addition, Strecker's chorus frogs (*Pseudacris streckeri*) gradually increase both the intensity and frequency of the very first calls that they produce in the night (Gridi-Papp, pers. obs.).

3.3.4 Ultrasound, Advertisement Call Complexity, and Variability

The majority of anurans produce calls with fundamental frequencies between 100 Hz and 6 kHz, and have stereotyped advertisement calls with up to four types of notes. The Madagascar bright-eyed frog (*Boophis madagascariensis*), however, has been reported to produce highly variable calls (Narins et al. 2000). From 9 to 28 call types can be distinguished, depending on how they are classified. Most of the variability is based on call duration, amplitude envelope, or number and rate of notes. The call types also differ in tonality but not so much in fundamental frequency or in frequency modulation.

Extensive call variability in the frequency domain has been found in some South Asian torrent frogs of the genera *Odorrana* and *Huia*. They are distantly related within the family Ranidae (Cai et al. 2007; Stuart 2008) but share high-frequency communication (34 kHz in *O. tormota*, 38 kHz for *H. cavitympanum*), which they may have evolved to cope with the low-frequency broadband noise (<20 kHz) that dominates their breeding sites (Feng et al. 2006; Arch et al. 2009). Males of the concave-eared torrent frog (*O. tormota*) produce fundamental frequencies that vary between 5 and 10 kHz with high levels of energy in its harmonics up to 30 kHz and also around 60 kHz (Fig. 3.3a; Feng et al. 2002). Playback experiments with sequences containing only the ultrasonic harmonics of the call elicited increased calling, indicating that the animals can detect and respond to ultrasound. Their auditory sensitivity to ultrasound has also been confirmed through electrophysiological recordings from the midbrain and through laser vibrometry of the eardrum (Feng et al. 2006; Gridi-Papp et al. 2008). The thin eardrums of these anurans are recessed and allow middle ear transmission through a short ossicle to the inner ear.

In addition to communicating at high frequencies, the concave-eared torrent frog produces advertisement calls that are highly variable in duration and frequency modulation (Feng et al. 2002, 2009). The categorization of these calls is not trivial, and the behavioral role of this variation is unclear. Frequency modulation of the call

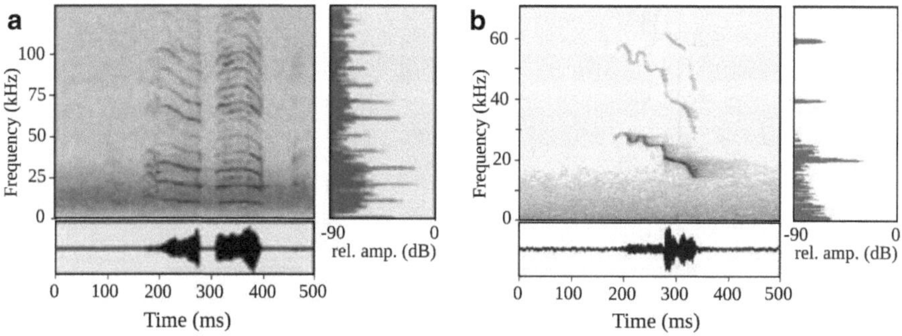

Fig. 3.3 High-frequency vocalizations in Asian frogs that breed along rocky streams. (**a**) Concave-eared torrent frog (*O. tormota*) from China. The third harmonic can exceed 30 kHz having equivalent energy content to those of the first two harmonics. (**b**) Hole-in-the-head frog (*H. cavitympanum*) from Borneo. The fundamental frequency can reach nearly 30 kHz. Extensive and highly variable frequency modulation is present in the calls. Notice the natural background noise peaking at low frequencies and gradually rolling off at about 20 kHz [Sources: **a**. Feng and Narins (2008). **b**. Arch et al. (2008)]

is not necessarily accompanied by amplitude modulation in the same direction, suggesting that these animals might be able to control fundamental frequency independently of pulmonary air pressure.

The larynx of male concave-eared torrent frogs is relatively small, but not qualitatively different from that described for anurans in general (Suthers et al. 2006). The medial edges of the vocal cords have a thin portion that is the most likely source of the high frequencies. Radiation of the ultrasound is mediated by two pairs of vocal sacs: one subgular pair and one lateral pair (Feng et al. 2002).

An even more extreme case of high-frequency communication is found in the hole-in-the-head frog (*Huia cavitympanum*; Fig. 3.3b). The advertisement calls of this species have a fundamental frequency usually between 10 and 20 kHz, but as in the concave-eared torrent frog, the calls are highly variable in frequency (Arch et al. 2008). Frequency modulation within calls can raise the fundamental frequency above 28 kHz, and some calls are produced entirely above 20 kHz. This species also breeds near noisy streams, exhibits recessed, thin eardrums and is sensitive to ultrasound (Arch et al. 2009). The genus *Huia* contains three more species. They all have recessed eardrums and call along fast flowing streams. Advertisement calls have been recorded for one of them, the Javan torrent frog (*H. mansonii*). These calls have extensive frequency modulation with the fundamental frequency ranging between 5 and 15 kHz (Boonman and Kurniati 2011).

3.3.5 *Nonlinear Vocal Cord Vibration Behavior*

Rapid, unpredictable transitions in acoustic structure have been described in the natural vocalizations of some mammals and birds (Fee et al. 1998; Fitch et al. 2002). Such fast changes have been shown experimentally to be based on nonlinearities in

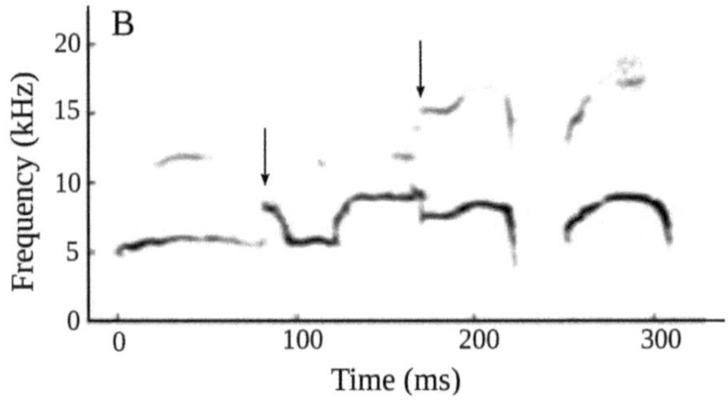

Fig. 3.4 Rapid transitions in the advertisement call of the concave-eared torrent frog (*O. tormota*). Unpredictable rapid transitions in frequency structure (*arrows*) were observed both in field recordings and in sound produced artificially by passive larynges in the laboratory (From Feng et al. 2002)

the vibratory response of vocal cords to subglottal pressure. This allows the animals to introduce sharp transitions into the acoustic structure of their calls without requiring fast or complex muscular coordination. These sharp transitions are not known from anuran advertisement calls, except in the concave-eared torrent frog (Fig. 3.4; Feng et al. 2006, 2009). They have been observed in experiments in which sound was produced by airflow through the larynges of euthanized frogs. These experiments revealed that sharp transitions in the sound structure can occur during gradual changes in the driving air pressure (Suthers et al. 2006). In North American treefrogs, experimental sound production showed nonlinear behavior near the low and high extremes of subglottal air pressure at which the larynx can phonate. Linear behavior was observed throughout most of the respiratory pressure range in which sound can be produced (Gridi-Papp 2014).

3.3.6 Control over Call Intensity

The simplest mechanism for control of call intensity could be based on control over the contraction strength of the trunk muscles. In many anurans, the larynx passively resists airflow and starts making sound when subglottal air pressure exceeds a threshold (Paulsen 1967; Martin 1971). Experiments in which humid air was passed through the larynges of euthanized North American treefrogs have revealed a range of nearly 40 dB of sound pressure that can be obtained by simply varying the subglottal air pressure above the threshold of phonation (Gridi-Papp 2014).

Other mechanisms for controlling call intensity involve muscular control of laryngeal resistance. Contraction of the laryngeal constrictor muscles can increase the laryngeal resistance to air flow by adducting the vocal cords and closing the arytenoid cartilages. Contraction of the laryngeal dilator muscle, on the other hand, opens the arytenoid cartilages, which also causes abduction of the vocal cords (Schmidt 1965, 1972; Martin 1971; Martin and Gans 1972). Whereas the dilator and constrictor muscles are known to contract immediately before and after a call, or between notes, all laryngeal muscles appear to remain relaxed while the vocal cords vibrate during release calling by the gulf coast toad (*Bufo valliceps*; Martin and Gans 1972).

Amplitude modulation within the call is common in amphibians. It can be generated in different ways that are not evident from the acoustic structure. This has created some confusion in the literature and the issue has been reviewed elsewhere (Littlejohn and Ryan 2001). One type of amplitude modulation is produced as a series of expiration–inspiration cycles. Air is returned to the lungs during a short interval between successive notes. As an example, Cope's gray treefrogs (*Hyla chrysoscelis*) produce notes at a rate of about 50 Hz (Girgenrath and Marsh 1997). Many toads in the genera *Bufo* and *Rhinella* produce long (>5 s) and pulsed advertisement calls through this process (Martin 1971; Martin and Gans 1972). Cycles of rapid inspirations between expiratory sounds are also found in songbirds. The short inspirations are called minibreaths and they facilitate the production of long songs (Hartley and Suthers 1989). In zebra finches, some minibreaths are used to produce inspiratory sound and make the song even less interrupted (Goller and Daley 2001). This same phenomenon also occurs in Muller's termite frog (Giaretta et al. 2015).

A second mechanism of amplitude modulation has been described in the gulf coast toad (*Bufo valliceps*), with pulses being produced within a single expiration (Martin 1971). The arytenoid cartilages vibrate passively and modulate the amplitude of the sound produced by the vocal cords. In a third proposed mode of amplitude modulation, the vocal cords themselves produce amplitude-modulated sound through complex vibration patterns. However, the precise mechanism has not been described (McAlister 1961; Schmidt 1972). Several toads in the genus *Bufo* and also hylids such as the cricket frog (*Acris crepitans*) produce amplitude modulation in their advertisement calls through both methods described in the preceding paragraphs: through expiration–inspiration and also within a single expiration.

3.3.7 *Resonance, Filtering, and Radiation of the Call*

Anurans lack an elongated neck and the larynx is positioned at the posterior end of the oral cavity. Vocal sounds radiate from the larynx into the mouth and through the vocal slits into the vocal sac. Amplification by cavity resonance was originally assumed to be the role of the vocal sac but it was later ruled out by experimental placement of calling anurans in an atmosphere of 20 % oxygen and 80 % helium (heliox; Capranica and Moffat 1983; Rand and Dudley 1993). No significant frequency shift was found in the spectra of calls recorded in heliox when compared to calls recorded in air. This showed that the vocal sac does not act as a resonance cavity in the species tested.

Several nonexclusive alternative hypotheses could explain why anurans call into a vocal sac with their mouths shut. It might allow them to call at a faster rate or at a lower energetic cost (Pauly et al. 2006). Reinflation of the lungs by recoil of the

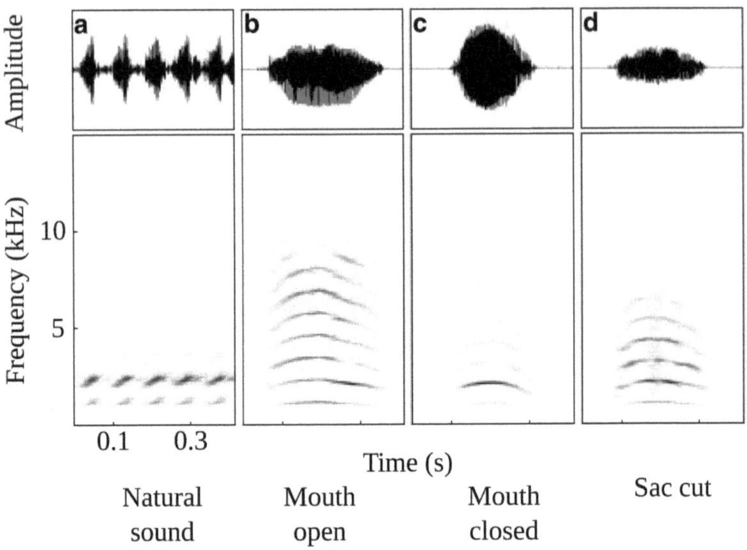

Fig. 3.5 Acoustic filtering through the tissues of the gray treefrog (*H. versicolor*) in a field recording (**a**) or in artificial sounds produced in the laboratory with controlled airflow through the larynx of a euthanized specimen (**b–d**). (**b**) Specimen with mouth wide open and vocal sac deflated. (**c**) Mouth closed and vocal sac inflating at each call. (**d**) Mouth closed and vocal sac partially ablated with scissors (From Gridi-Papp 2008)

vocal sac is much faster and potentially more energetically efficient than buccal pumping. Furthermore, recycling of the air at each call can also prevent dehydration. With extended calling at high rates, even minute losses of water could add up and hinder the performance of the animal. As an additional hypothesis, the vocal sac can act as a visual signal. Female túngara frogs are preferentially attracted to calls matched by the synchronized inflation of a vocal sac (Rosenthal et al. 2004; Taylor and Ryan 2013). In brilliant-thighed poison frogs (*Allobates femoralis*), male territorial aggression is dependent on simultaneous stimulation by sound and by an inflating vocal sac (Narins et al. 2005). In addition, the vocal sac may serve as an acoustic filter or as a sound radiator.

As anurans call with their mouths and nares shut, once sound enters the mouth and the vocal sac, it has to cross a tissue boundary to radiate into the environment and this produces an opportunity for filtering. Capranica and Moffat (1983) recorded the calls of anurans before and after damaging the vocal sac by cutting. Damage to the sac rendered the calls detuned, with the energy distributed in higher harmonics. This observation is supported by experimental flow of humid air through the larynges of euthanized gray treefrogs (*H. versicolor*; Fig. 3.5; Gridi-Papp 2008). With the mouth of the frog open, or the vocal sac cut, an extensive series of harmonics was recorded. When the mouth of the frog was held closed and the vocal sac was allowed to inflate, most of the energy in the higher harmonics was filtered out and the recordings were tuned to the second harmonic, as in natural calls. Additional evidence for extensive filtering comes from recordings of distress calls, which are commonly produced by Neotropical treefrogs when they are handled (Hödl and Gollmann 1986). These calls are always produced with the mouth open, and their energy is spread over a much greater range of harmonics than in advertisement calls.

Finally, vocal sacs also act as sound radiators. If the mouth of the anuran is open, sound is radiated out to the environment by the vocal cords. With the mouth closed and the thin vocal sac inflated, sound can be radiated from its entire surface, which is much greater than the combined surface of the vocal cords. Purgue (1995) measured the vibration response of the body surface in euthanized anurans stimulated by sound. He studied three species of anurans and found that the vocal sacs, but not other body parts, were tuned to the frequency of the advertisement calls. Most of the energy in the higher frequencies of the calls radiated through the surface of the vocal sac. He also found that the sound was up to 12 dB more intense when recorded across the body with the mouth closed and the vocal sac inflated than when recorded directly. His findings of increased tuning and amplitude were corroborated by laryngeal activation experiments comparing the output radiated with the mouth open or closed (Gridi-Papp 2008).

Sound that enters the buccal cavity does not radiate well through the bony skull. It does radiate through the middle ears because these are normally connected to the mouth by short and wide Eustachian tubes that are not collapsed at rest. Body surface measurements with a vibration probe showed that the eardrums radiate a modest proportion of the call's energy in treefrogs, but that in bullfrogs the eardrums radiate most of the energy in the call (Purgue 1995, 1997). Male bullfrogs (*Lithobates catesbeianus*) vocalize floating in the water and while they are considerably smaller than the females, they have much larger eardrums. Male bullfrogs also have relatively small vocal sacs when compared to other species that vocalize while floating, so the eardrums, which are above the water surface when the frog is calling, serve as the main radiators to broadcast mating calls.

Several species of anurans have been shown to select or modify the microenvironment of their calling sites to maximize radiation of the calls. Rosy bug-eyed frogs (*Eupsophus calcaratus*) and several species of giant burrowing frogs (*Heleioporus*) construct burrows that resonate and enhance the transmission distance of their calls (Bailey and Roberts 1981; Penna 2004). Bornean treehole frogs (*Metaphrynella sundana*) call from treeholes filled with water and adjust the frequency of their calls to match the resonance frequencies of the selected treeholes (Lardner and bin Lakim 2002).

3.4 Energy and Advertisement Strategies

Reproductive behaviors can involve costs such as increased risk of predation and large investment of energy. Male anurans substantially increase their consumption of oxygen when calling and they can experience even higher metabolic rates when spawning or building an egg nest (Bucher et al. 1982; Wells 2010). In the most vocal anuran species, a male may produce several thousand calls per hour at intensities greater than 90 dB SPL (at 1 m distance from the mouth) for several hours per night (Gerhardt 1975; Taigen and Wells 1985; Grafe and Thein 2001). Males typically lose body mass and lipid content during the breeding season because calling and territory defense reduce their foraging time and increase their energy expenditure (Jenssen 1972; Nally 1981; Woolbright and Stewart 1987). Females tend to make even larger investments in reproduction. In túngara frogs, a female invests more energy in the eggs of one breeding event than a male invests in calling during the entire breeding season (Ryan et al. 1983a).

Social interactions in the form of choruses and competition between males can alter energy budgets by influencing male calling behavior (Wells and Taigen 1984; Wells 1988). Females tend to prefer males that produce calls with high intensity and complexity, fast rate, and long duration (Gerhardt 1988). These preferences lead males to produce vigorous calls that also make them conspicuous and increase their risk of predation or parasitism (Ryan et al. 1983b; Bernal et al. 2006, 2007). Males can therefore reduce their costs when in small choruses by producing calls with few notes and at a low rate. In dense choruses with high competition, however, they tend to emit calls at high rates and with many notes (Schwartz and Wells 1985).

The sound of a chorus can stimulate males to call and it elicits positive phonotaxis (attraction) in both males and in gravid females (Bernal et al. 2009). Exposure of offspring to adult advertisement calls ranges from none in desert explosive breeders to daily exposure in species that inhabit tropical climates with constant weather and form choruses throughout the year. Males raised in isolation produce normal calls as adults and there is no evidence of vocal learning in anurans, either in calling males or in female preferences (Dawson and Ryan 2009, 2012).

Chorusing in males frequently leads to alternated calling, but there are also cases of synchronous calling (Greenfield 1994). Males compete by overlapping calls, increasing call complexity, calling more vigorously, producing aggressive calls, or engaging in physical combat. They can also adopt alternative tactics such as a satellite strategy, active searching for females, or polyandrous amplexus (Wells 2010). Features of the advertisement calls produced by males in a chorus are generally sufficient for species recognition and sexual selection. Anuran calls form important mechanisms of prezygotic isolation that can be the basis of reinforcement in the process of speciation (Blair 1974; Loftus-Hills and Littlejohn 1992). When these preferences fail to form a reproductive barrier, natural hybridization can take place (Castellano and Giacoma 1998; Pfennig and Simovich 2002; Pfennig 2007). In túngara frogs, males are as selective as females in responding with phonotaxis to male calls but males are less selective when responding with calling (Bernal et al. 2009). Female preferences can be a function of past evolutionary history (Ryan and Rand 1995), good genes (Welch et al. 1998), sensory exploitation (Ryan et al. 1990), or runaway sexual selection (Kirkpatrick and Ryan 1991).

3.5 Interactions Between Vocalization and Hearing

Although hearing and sound production are mostly mediated by different neural circuits and peripheral structures, these sensory and motor systems are linked by shared structures such as the oral cavity, by their proximity in the head, and by the functional need for matched tuning. Some recently raised issues reinforce the need for integrating the study of vocalization with that of hearing to gain a better understanding of the evolution of communication systems.

3.5.1 The Role of the Environment

The discovery of ultrasound in the calls of anurans has advanced the current understanding of the mechanisms and potential limitations of the vocal and auditory apparatuses of these animals (Feng et al. 2006). More generally, it has led researchers to ask why communication at very high frequencies has evolved in some South Asian frogs but apparently not in other anurans inhabiting the rest of the world.

High-frequency vocal and auditory capabilities are shared features of the concave-eared torrent frog, the hole-in-the-head frog, and their close relatives that call along noisy streams (Arch et al. 2009; Boonman and Kurniati 2011; Shen et al. 2011). It has therefore been proposed that high-frequency communication evolved in these animals to escape masking from the low-frequency environmental noise characteristic of rocky streams. If this is the driving factor in the evolution of high-frequency communication in anurans, then ultrasound is only one of a few solutions to a common problem.

In the New World, anurans of the genus *Hylodes* inhabit rocky stream environments and tend to produce long calls (2–4 s) with short notes (50 ms) separated by regular intervals that should facilitate signal extraction from noise (Haddad and Giaretta 1999; Wogel et al. 2004). In addition, their calls have high dominant frequencies (5 kHz at the third harmonic) and they produce visual signals, as several other diurnal Neotropical anurans do, by flagging their colored limbs (Hartmann et al. 2005; Preininger et al. 2013; Starnberger et al. 2014).

Among Australian anurans, species that vocalize along streams actually produce lower frequencies than species that vocalize at ponds (Hoskin et al. 2009). Background noise produced by other signaling animals in tropical forest has also been suggested to promote the usage of alternative signaling modes in anurans. Such modes could include seismic signals produced by the impact of the vocal sac hitting the ground during calling (Lewis and Narins 1985; Lewis et al. 2001).

Although anurans can use other modes of communication as alternative channels, it is important to consider that they can also use them as additional channels. The movements of the vocal sac are synchronized with sound production during calling. Experimental manipulation of the relative timings of the call and vocal sac inflation revealed that the animals integrate stimuli across modes such that the combined stimuli can produce different behavioral responses than each individual stimulus (Narins et al. 2005; Taylor and Ryan 2013). Furthermore, the receivers of the stimuli may differ in sex or species—predators or parasites—and have different access to additional cues. Male túngara frogs always call floating and produce ripples on the surface of the water. Females are preferentially attracted toward the multimodal stimulus over the acoustic stimulus only. Males respond with increased calling to the multimodal stimulus, and predatory bats are preferentially attracted to the multimodal stimulus (Halfwerk et al. 2014a, b). When the frog hears a bat approaching, it becomes immediately silent but its position is still marked for a few seconds by the slowly propagating concentric ripples on the surface of the water.

3.5.2 Closure of the Eustachian Tube

The middle ears of vertebrates are connected to the mouth by Eustachian tubes. It was long believed that in anurans the Eustachian tubes remain permanently open, providing a connection between the two ears, which function as pressure gradient receivers (Chung et al. 1978; Mason and Narins 2002a). In addition to employing ultrasound in its communication, however, the concave-eared torrent frog actively closes its Eustachian tubes during calling (Fig. 3.6; Gridi-Papp et al. 2008). Muscular contraction causes the anterior process of the hyoid to bend, deforming the wall of the Eustachian tube and blocking it. This mechanism was confirmed through electric

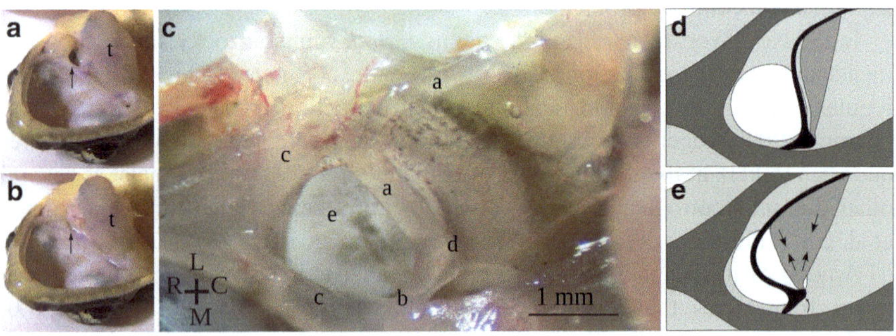

Fig. 3.6 Active Eustachian tube closure in the concave-eared torrent frog (*O. tormota*). (**a**) Ventral view of the open mouth of an awake male, showing the tongue (t) and the left Eustachian tube open (arrow). (**b**) Same as **a**, a moment later, showing the Eustachian tube closed. (**c**) Ventral view of the left Eustachian tube with the buccal skin removed. a = anterior process of the hyoid; b = hinged attachment of the anterior hyoid process to the skull; c = skull; d = submaxillary muscle; e = internal surface of the eardrum, with the extrastapes attached at the center. L = lateral; M = medial; R = rostral; and C = caudal. (**d**) Schematic representation of the structures shown in **c**, with the Eustachian tube open. (**e**) Closure of the Eustachian tube is produced by contraction (arrows) of the submaxillary muscle, which pivots the anterior hyoid process at its attachment on the skull (From Gridi-Papp et al. 2008)

stimulation in concave-eared torrent frogs, but it was not observed in northern leopard frogs subjected to the same procedure, showing that active Eustachian tube closure is not general to anurans.

Behavioral observation of the Eustachian tube closure is possible in the concave-eared torrent frog due to the transparent nature of the eardrums (Gridi-Papp et al. 2008). Males always close the Eustachian tubes during the phonatory phase of vocalization but this is the only context in which closure has been observed in the field. Eustachian tube closure is therefore likely to serve the role of protecting the ears from the intense acoustic output of the larynx or from increased air pressures in the mouth. As an alternative possibility, Eustachian tube closure could allow the concave-eared torrent frogs to shift the tuning of their middle ears to higher frequencies. The eardrum in these animals is recessed (Feng et al. 2006). With closure of the Eustachian tube, the volume of the middle ear is drastically reduced, thereby stiffening the eardrum and shifting the tuning of the middle ear to higher frequencies. Field observations, however, failed to reveal closure in response to playbacks of frog calls or stream noise.

3.5.3 *Protection of the Inner Ears*

In addition to Eustachian tube closure, the inner ear may be protected by muscle contractions that could restrict the movements of the auditory ossicles in two ways. First, anurans exhibit a stapedial (columellar) muscle extending between the suprascapula and the stapes (columella), and an opercular muscle extending from the suprascapula to the operculum. As both of these bones are lodged in in the oval window, they could interfere with the transmission of acoustic signals. Second, these bones do not move in and out like a piston, but instead they hinge against the edge of the oval window (Jørgensen and Kanneworff 1998; Mason and Narins 2002a, b). Gentle pulling on the opercular muscle with forceps produces movement of both bones, indicating that they are directly linked in their movements. The opercular muscle has also been suggested to transmit seismic vibrations from the ground and limbs to the operculum. Experimental stimulation or ablation of the opercular muscle did not alter the vibration response of the eardrum to sound, but ablation diminished the animal's sensitivity to seismic vibrations (Hetherington 1985, 1987, 1994).

An alternative possibility for the role of the operculum is protection from excessive bulging of the eardrum when the animal pushes the floor of the mouth up in order to inflate the lungs through buccal pumping. Such increases in buccal air pressure produce much greater bulging of the eardrum than does sound, and they push the stapes footplate into the oval window. Contraction of the opercular muscle could oppose this movement of the stapes, protecting the hair cells in the inner ear (Mason and Narins 2002a). This hypothesis is supported by electrophysiological recordings that have shown that the opercular muscle contracts in synchrony with buccal pumping (Hetherington and Lombard 1983).

3.6 Vocalizations by Other Amphibians

In addition to anurans, the class Amphibia includes caudates (newts and salamanders) and caecilians (amphibians without legs). Little is known about communication in caecilians, possibly because of their fossorial (subterranean) habit and restricted geographic distribution compared to other amphibians. A few accounts of sound production have reported clicks and two other sounds produced by caecilians of four different genera (Fig. 3.7a; Largen et al. 1972; Thurow and Gould 1977; Duellman and Trueb 1986). Clicks of the Mexican burrowing caecilian (*Dermophis mexicanus*) were recorded from a captive individual originating from Colombia or Peru. The clicks were 50 ms in duration and had a dominant frequency of about 2400 Hz. This frequency is apparently the sixth harmonic but the series is smeared by the noisy structure of the short sound. The animal emitted clicks as single notes, pairs of notes or bursts, but the sound was only audible to the researcher within a 1–3 m distance. The mechanism of production is unknown, but the animal would maintain its mouth closed and the sounds were suggested to be radiated through the nares. The larynges of caecilians exhibit arytenoid cartilages, laryngeal dilator muscles, and two pairs of constrictor muscles that form a sphincter around the insertions of the laryngeal dilator muscle, in an arrangement equivalent to that described above (see Sect. 3.2.2) for anurans (Duellman and Trueb 1986).

Salamanders and newts of various genera emit soft sounds, hissing or clicks, produced by snapping the mouth shut. The sounds are accompanied by biting or defensive displays against predators and no role in communication with conspecifics

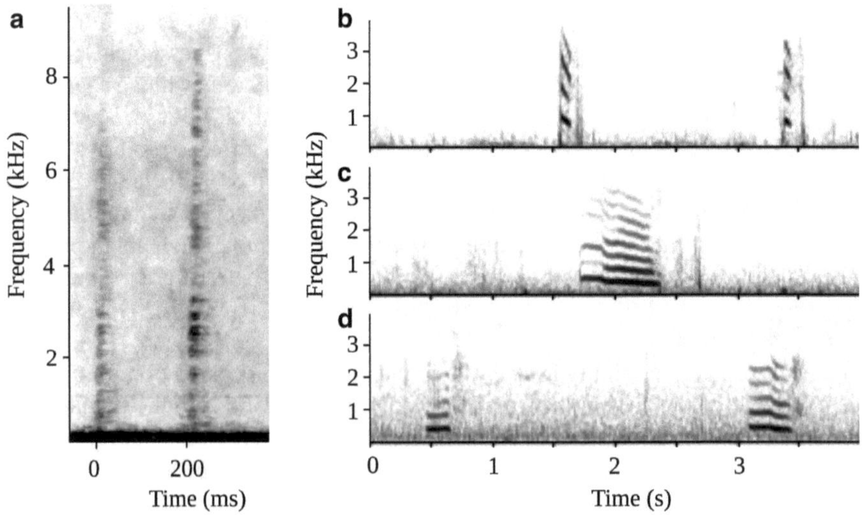

Fig. 3.7 (**a**) Pair of clicks produced by a caecilian (*D. mexicanus*) in captivity. (**b–d**) Mounting calls produced by males of three species of tortoise during copulation (FFT window = 256 samples, sampling rate = 8 kHz). (**b**) *Testudo hermanni*. (**c**) *Testudo graeca*. (**d**) *Testudo marginata* [*Sources*: **a**. Thurow and Gould (1977). **b–d**. Sacchi et al. (2004)]

has been proposed (Brodie 1978). These amphibians have a larynx with arytenoid cartilages, dilator and constrictor muscles, although some species may lack a laryngeal constrictor muscle. The lungless salamanders *Plethodon glutinosus* and *Desmognathus fusca* have been reported to emit short sounds when handled (Neill 1952), even though the entire family of lungless salamanders (Plethodontidae) lacks lungs as well as the entire larynx and associated muscles (Wilder 1896; Duellman and Trueb 1986).

3.7 Vocalization by Reptiles

3.7.1 Turtles

Several types of vocalizations are produced by chelonians (turtles) in air and underwater. A third of the species with described courtship behavior are known to vocalize. The most vocal species belong to the families of tortoises (Testudinidae), softshell turtles (Trionychidae), pond turtles (Emydidae), and batagurid turtles (Bataguridae; Gans and Maderson 1973; Galeotti et al. 2005). In tortoises, most calling is done by males and it is associated with courtship and mounting (Fig. 3.7b). The signals vary among species from noisy to highly tonal, with a frequency of 100–700 Hz, a duration of 80–1000 ms, and intercall intervals of 0.3–2.9 s. The dominant frequency of the calls is correlated with body mass, while call duration and rate are correlated with male mounting success. Nocturnal chorusing has been described in the Travancore tortoise (*Geochelone travancorica*), but its behavioral role is unknown (Campbell and Evans 1972).

The larynx in tortoises is formed by two arytenoid and one cricoid cartilages supported by the hyoid (Sacchi et al. 2004). It is controlled by two pairs of muscles: one dilator and one constrictor, which pivot the arytenoid cartilages over the cricoid, opening and closing the glottis. The vocal cords are rich in elastic fibers, but they differ from those of other vertebrates because they connect the arytenoid cartilages to the hyoid, and therefore they are not located inside the larynx. Another peculiarity is the presence of two diverticula in the ventral wall of the cricoid, which are suggested to act as cavity resonators.

The freshwater northern snake-necked turtle (*Chelodina oblonga*) from Western Australia exhibits a diverse acoustic repertoire with 17 types of calls produced underwater, including harmonic series with frequency modulation, noisy calls, clicks in isolation or arranged in trains, and transitions between these types (Giles et al. 2009). Females of the Arrau turtle (*Podocnemis expansa*) vocalize with six types of calls in various contexts involving nesting, aggregation of the hatchlings, and migration with the offspring to the flooded forest (Ferrara et al. 2014a, b). Juveniles vocalize inside the eggs, in the nest, and in the water, apparently eliciting vocal responses from the mother (Ferrara et al. 2013). The existence of acoustic communication in these freshwater turtle species and its use in parental care suggest that turtle vocalization might be more widespread than previously suspected.

3.7.2 *Tuatara*

Only two extant species of tuatara exist (*Sphenodon puctatus* and *S. guntheri*) and they can be found in islands around New Zealand. These are the only representatives of one of the four orders of reptiles that used to be much more species-rich in the Mesozoic. Contact with humans and mammalian predators extinguished tuatara from New Zealand's main island (Towns and Daugherty 1994).

Tuatara have an unusual biology. They are nocturnal, have a long life span, and females lay eggs every 2–5 years (Bogert 1953; Cree 1994; Nelson et al. 2002). Incubation of the offspring takes about 1 year to complete and in their odd system of temperature sex determination, high temperatures tend to induce the development of males (Nelson et al. 2004; Mitchell et al. 2008). Females guard their nests and this reduces damage from excavation by other females (Refsnider et al. 2009). Tuatara produce grunts as part of an elaborate courtship ritual and also during feeding or when producing aggressive displays (Gans et al. 1984).

3.7.3 *Snakes*

Snakes utilize a variety of mechanisms for sound production, including scale rubbing, tail rattling, and hissing (Gans and Maderson 1973; Young 2003). Hissing involves inflating the lungs and then emitting a jet of air that can be continuous, intermittent, and sometimes explosive. These sounds do not involve vocal cords. They have a noisy frequency spectrum ranging from 3 to 13 kHz (Young 1991). The larynx in snakes is small and it lacks vocal cords, but as in other vertebrates, it is formed by two arytenoid cartilages that articulate with a cricoid cartilage. The intrinsic musculature includes three pairs of muscles: the dilator laryngis muscle, the sphincter laryngis muscle, and the arytenocricoideus muscle (Young 2000). Hissing sounds are produced in conjunction with aposematic (warning) behaviors and might serve to deter predators. The king cobra (*Ophiophagus hannah*) and the mangrove ratsnake (*Gonyosoma oxycephalum*) produce a hiss tuned to low frequencies (600 Hz in the king cobra) which is made possible by tracheal diverticula that act as cavity resonators (confirmed with heliox in the ratsnake (Young 1991). The pine snake (*Pituophis melanoleucus*) produces two simultaneous sounds at up to 90 dB SPL through vibration of a laryngeal septum (Young et al. 1995). Hissing is also present in lizards, turtles, and crocodilians, where it is always associated with defensive behaviors and frequently followed by lunges or biting.

3.7.4 *Lizards*

Among lizards, geckos are best known for producing complex tonal calls with harmonic structure and for an elaborate calling apparatus with elastic vocal cords (Fig. 3.8a; Moore et al. 1991; Yu et al. 2011). The vocal cords vary significantly in

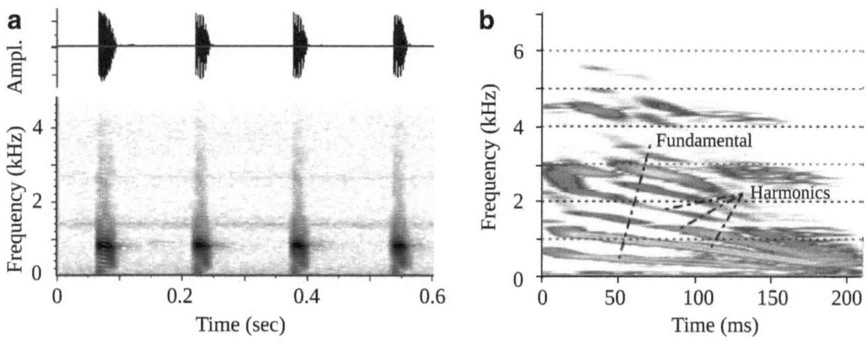

Fig. 3.8 (**a**) Vocalization of a barking gecko (*P. garrulus*). (**b**) Prehatching call of the Nile crocodile (*C. niloticus*) recorded a few hours before hatching [*Sources*: **a**. Hibbitts et al. (2007). **b**. Vergne et al. (2012)]

position and orientation (Russell et al. 2000). Geckos can produce intense sounds and have a variety of call types that are mostly involved in territorial or courtship behaviors (Hibbitts et al. 2007). Juveniles and adults of both sexes are vocal (Frankenberg 1982). The highest frequency vocalizations are from Australian pygopod lizards, which produce a very fast series of clicks that sound like tones. They have a dominant frequency of about 8 kHz, with harmonics containing significant amounts of energy above 15 kHz. This spectral range is matched by hearing sensitivity up to 4 kHz with a drop of about 20 dB at 8–12 kHz (Weber and Werner 1977; Manley and Kraus 2010).

3.7.5 Crocodiles

Crocodilians have an elaborate vocal system (Vergne et al. 2009). Adults of both sexes produce an intense "bellow" during courtship or defense. In American alligators (*Alligator mississippiensis*), this call reaches 91–94 dB SPL at 1 m from the source in air with the dominant frequency at 125–250 Hz. The call is also coupled to the water, where it reaches 121–125 dB at 1 m with a lower dominant frequency of 20–100 Hz (Todd 2007). In Chinese alligators (*A. sinensis*) the bellow reaches nearly 91 dB at 5 m from the source in air (Wang et al. 2007).

Noisy hisses are emitted by females during nest defense and by males during territorial interaction in various crocodilians (Garrick et al. 1978; Britton 2001; Wang et al. 2007). These sounds have a lower intensity (60 dB SPL at 2 m) and a longer duration (several seconds) than the bellow. Females also produce soft grunts to attract their offspring during several weeks of posthatching maternal care. Grunts have a short duration (0.1 s) and low fundamental frequency (100 Hz).

Offspring produce prehatching calls in the egg, with linearly descending frequency modulation, most energy at the fundamental frequency (500 Hz), harmonics up to 3 kHz, and a duration of about 200 ms (Fig. 3.8b; Britton 2001). These calls synchronize the hatching of young in the nest and stimulate the female to uncover the eggs (Vergne and Mathevon 2008). After hatching, the calls become longer, with a higher frequency, broader bandwidth, and increased intensity. When feeding or moving in a group, juveniles also produce contact calls with an acoustic structure similar to that of posthatching calls. If they are handled or in the presence of a predator, the intensity, dominant frequency, and frequency modulation of the calls increase. Like adults, juveniles can hiss when threatened. Through contact calls, juveniles and adults of various species communicate using the same parameters of frequency modulation (Vergne et al. 2012). They respond similarly to contact calls of their own species or of other species. Experiments have shown that this is because similar contact calls have common roles across species and not because of lack of discrimination ability.

3.8 Summary

All major groups of amphibians and reptiles produce sounds with the glottis, formed by a cricoid cartilage ring and a pair of arytenoid cartilages. The glottis can be opened by a pair of dilator muscles that open the arytenoids and closed by constrictor muscles that produce the opposite movement. Vocal cords and resonance chambers are present in some groups but lacking in others. Clicking and hissing are present in most groups, whereas tonal sounds with harmonic structure are most common in anurans and geckos, but also observed in some of the other groups. Crocodilian bellows are the most intense vocalizations, followed closely by anuran advertisement calls. The highest fundamental frequencies seen in ectotherm vertebrates are produced by an anuran (>20 kHz), followed by limbless lizards. Vocalizations of anurans and tortoises are mostly produced by males and mediate mating, whereas in crocodilians and some freshwater turtles vocalizations are also produced by offspring and females to synchronize hatching and to mediate maternal care. In geckos, adults and juveniles of both sexes employ specific calls in advertisement and territorial behavior. All major reptile groups and some caudates produce hissing when threatened.

Major advances in our understanding of communication in ectotherm tetrapods, such as the discovery of ultrasound in frogs, or parenting vocalizations in freshwater turtles, continue to be produced through exploratory research. Equally important advancements such as the elucidation of the mechanisms of phonation or their behavioral and evolutionary significance result from careful experimentation and from the application of novel technologies. This diversity of rewarding experimental approaches is not surprising, given that this group of vertebrates includes more than 17,000 species. A holistic approach to the study of vocal and auditory systems is necessary to explain the matched evolution of frequency tuning and issues involving morphology shared by the two systems such as the Eustachian tubes. Integrating studies across groups of amphibians and reptiles can also be rewarding, as these groups exhibit independent solutions to common communication problems and employ vocalization in complementary aspects of their natural history.

Acknowledgments The authors thank Cristina O. Gridi-Papp and the editors for reviewing the manuscript.

References

Arch, V. S., Grafe, T. U., Gridi-Papp, M., & Narins, P. M. (2009). Pure ultrasonic communication in an endemic Bornean frog. *PLoS One, 4*(4), e5413.

Arch, V. S., Grafe, T. U., & Narins, P. M. (2008). Ultrasonic signaling by a Bornean frog. *Biology Letters, 4*(1), 19–22.

Bailey, W. J., & Roberts, J. D. (1981). The bioacoustics of the burrowing frog *Heleioporus* (Leptodactylidae). *Journal of Natural History, 15*(4), 693–702.

Bernal, X. E., Page, R. A., Rand, A. S., & Ryan, M. J. (2007). Cues for eavesdroppers: Do frog calls indicate prey density and quality? *The American Naturalist, 169*(3), 409–415.

Bernal, X. E., Rand, A. S., & Ryan, M. J. (2006). Acoustic preferences and localization performance of blood-sucking flies (*Corethrella* Coquillett) to túngara frog calls. *Behavioral Ecology, 17*(5), 709–715.

Bernal, X. E., Rand, A. S., & Ryan, M. J. (2009). Task differences confound sex differences in receiver permissiveness in túngara frogs. *Proceedings of the Royal Society B: Biological Sciences, 276*(1660), 1323–1329.

Blair, W. F. (1974). Character displacement in frogs. *American Zoologist, 14*(4), 1119–1125.

Bogert, C. M. (1953). Body temperatures of the tuatara under natural conditions. *Zoologica, 38*(4), 63–64.

Boonman, A., & Kurniati, H. (2011). Evolution of high-frequency communication in frogs. *Evolutionary Ecology Research, 13*(2), 197–207.

Britton, A. R. C. (2001). Review and classification of call types of juvenile crocodilians and factors affecting distress calls. In G. C. Grigg, F. Seebacher, & C. E. Franklin (Eds.), *Crocodilian biology and evolution* (pp. 364–377). Chipping Norton, England: Surrey Beatty & Sons.

Brodie, E. D. (1978). Biting and vocalization as antipredator mechanisms in terrestrial salamanders. *Copeia, 1978*(1), 127–129.

Bucher, T. L., Ryan, M. J., & Bartholomew, G. A. (1982). Oxygen consumption during resting, calling, and nest building in the frog *Physalaemus pustulosus*. *Physiological Zoology, 55*(1), 10–22.

Cai, H.-X., Che, J., Pang, J.-F., Zhao, E.-M., & Zhang, Y.-P. (2007). Paraphyly of Chinese *Amolops* (Anura, Ranidae) and phylogenetic position of the rare Chinese frog, *Amolops tormotus*. *Zootaxa, 1531*, 49–55.

Campbell, H. W., & Evans, W. E. (1972). Observations on the vocal behavior of chelonians. *Herpetologica, 28*, 277–280.

Capranica, R. R., & Moffat, A. J. M. (1983). Neurobehavioral correlates of sound communication in anurans. In J.-P. Ewert, R. R. Capranica, & D. J. Ingle (Eds.), *Advances in vertebrate neuroethology* (pp. 710–730). New York: Plenum Press.

Castellano, S., & Giacoma, C. (1998). Stabilizing and directional female choice for male calls in the European green toad. *Animal Behaviour, 56*(2), 275–287.

Chung, S. H., Pettigrew, A., & Anson, M. (1978). Dynamics of the amphibian middle ear. *Nature, 272*(5649), 142–147.

Cree, A. (1994). Low annual reproductive output in female reptiles from New Zealand. *New Zealand Journal of Zoology, 21*(4), 351–372.

Dawson, B., & Ryan, M. J. (2009). Early experience leads to changes in the advertisement calls of male *Physalaemus pustulosus*. *Copeia, 2009*(2), 221–226.

Dawson, B., & Ryan, M. J. (2012). Female preferences are not altered by early acoustic experience in the neotropical frog *Physalaemus pustulosus*. *Journal of Herpetology, 46*(4), 535–538.

De Jongh, H. J., & Gans, C. (1969). On the mechanism of respiration in the bullfrog, *Rana catesbeiana*: A reassessment. *Journal of Morphology, 127*(3), 259–289.

Drewry, G. E., Heyer, W. R., & Rand, A. S. (1982). A functional analysis of the complex call of the frog *Physalaemus pustulosus. Copeia, 1982*(3), 636–645.

Dudley, R., & Rand, A. S. (1991). Sound production and vocal sac inflation in the túngara frog, *Physalaemus pustulosus* (Leptodactylidae). *Copeia, 1991*(2), 460–470.

Duellman, W. E., & Trueb, L. (1986). *Biology of amphibians*. Baltimore: Johns Hopkins University Press.

Eichelberg, H., & Obert, H.-J. (1976). Fat and glycogen utilization in the larynx muscles of fire-bellied toads (*Bombina bombina* L.) during calling activity. *Cell and Tissue Research, 167*(1), 1–10.

Emerson, S. B., Greig, A., Carroll, L., & Prins, G. S. (1999). Androgen receptors in two androgen-mediated, sexually dimorphic characters of frogs. *General and Comparative Endocrinology, 114*(2), 173–180.

Feder, M. E., & Burggren, W. W. (1992). *Environmental physiology of the amphibians*. Chicago: University of Chicago Press.

Fee, M. S., Shraiman, B., Pesaran, B., & Mitra, P. P. (1998). The role of nonlinear dynamics of the syrinx in the vocalizations of a songbird. *Nature, 395*(6697), 67–71.

Feng, A., & Narins, P. (2008). Ultrasonic communication in concave-eared torrent frogs (*Amolops tormotus*). *Journal of Comparative Physiology. A, Neuroethology, Sensory, Neural, and Behavioral Physiology, 194*(2), 159–167.

Feng, A., Narins, P., & Xu, C.-H. (2002). Vocal acrobatics in a Chinese frog, *Amolops tormotus. Naturwissenschaften, 89*(8), 352–356.

Feng, A. S., Narins, P. M., Xu, C.-H., Lin, W.-Y., Yu, Z.-L., Qiu, Q., et al. (2006). Ultrasonic communication in frogs. *Nature, 440*(7082), 333–336.

Feng, A. S., Riede, T., Arch, V. S., Yu, Z., Xu, Z.-M., Yu, X.-J., et al. (2009). Diversity of the vocal signals of concave-eared torrent frogs (*Odorrana tormota*): Evidence for individual signatures. *Ethology, 115*(11), 1015–1028.

Ferrara, C. R., Vogt, R. C., Giles, J. C., & Kuchling, G. (2014a). Chelonian vocal communication. In G. Witzany (Ed.), *Biocommunication of animals* (pp. 261–274). New York: Springer.

Ferrara, C. R., Vogt, R. C., & Sousa-Lima, R. S. (2013). Turtle vocalizations as the first evidence of posthatching parental care in chelonians. *Journal of Comparative Psychology, 127*(1), 24.

Ferrara, C. R., Vogt, R. C., Sousa-Lima, R. S., Tardio, B. M., & Bernardes, V. C. D. (2014b). Sound communication and social behavior in an Amazonian river turtle (*Podocnemis expansa*). *Herpetologica, 70*(2), 149–156.

Fischer, L. M., & Kelley, D. B. (1991). Androgen receptor expression and sexual differentiation of effectors for courtship song in *Xenopus laevis. Seminars in Neuroscience, 3*(6), 469–480.

Fitch, W. T., Neubauer, J., & Herzel, H. (2002). Calls out of chaos: The adaptive significance of nonlinear phenomena in mammalian vocal production. *Animal Behaviour, 63*(3), 407–418.

Frankenberg, E. (1982). Vocal behavior of the Mediterranean house gecko, *Hemidactylus turcicus. Copeia, 1982*, 770–775.

Frost, D. R. (2014). *Amphibian species of the world: An online reference.* New York: American Museum of Natural History. Retrieved May 8, 2015, from http://research.amnh.org/herpetology/amphibia/index.html.

Galeotti, P., Sacchi, R., Fasola, M., & Ballasina, D. (2005). Do mounting vocalizations in tortoises have a communication function? A comparative analysis. *Herpetological Journal, 15*(2), 61–71.

Gans, C., Gillingham, J. C., & Clark, D. L. (1984). Courtship, mating and male combat in tuatara, *Sphenodon punctatus. Journal of Herpetology, 18*(2), 194–197.

Gans, C., & Maderson, P. F. A. (1973). Sound producing mechanisms in recent reptiles: Review and comment. *American Zoologist, 13*(4), 1195–1203.

Garrick, L. D., Lang, J. W., & Herzog, H. A. (1978). Social signals of adult American alligators. *Bulletin of the AMNH, 160*, 155–192.

Gaupp, E., & Ecker, A. (1904). *A. Ecker's und R. Wiedersheim's anatomie des frosches*. Braunschweig, Germany: Vieweg.
Gerhardt, H. C. (1975). Sound pressure levels and radiation patterns of the vocalizations of some North American frogs and toads. *Journal of Comparative Physiology. A, Neuroethology, Sensory, Neural, and Behavioral Physiology, 102*(1), 1–12.
Gerhardt, H. C. (1988). Acoustic properties used in call recognition by frogs and toads. In B. Fritzsch, M. J. Ryan, W. Wilczynski, T. E. Hetherington, & W. Walkowiak (Eds.), *The evolution of the amphibian auditory system* (pp. 455–483). New York: Wiley.
Gerhardt, H. C., & Huber, F. (2002). *Acoustic communication in insects and anurans*. Chicago: University of Chicago Press.
Giaretta, A. A., Vo, P., Herche, J., Tang, J., & Gridi-Papp, M. (2015). Reinterpreting features of the advertisement call of *Dermatonotus muelleri* (Boettger, 1885; Anura, Microhylidae). *Zootaxa, 3972*(4), 595–598.
Giles, J. C., Davis, J. A., McCauley, R. D., & Kuchling, G. (2009). Voice of the turtle: The underwater acoustic repertoire of the long-necked freshwater turtle, *Chelodina oblonga. Journal of the Acoustical Society of America, 126*(1), 434–443.
Girgenrath, M., & Marsh, R. L. (1997). In vivo performance of trunk muscles in tree frogs during calling. *Journal of Experimental Biology, 200*(24), 3101–3108.
Girgenrath, M., & Marsh, R. L. (1999). Power output of sound-producing muscles in the tree frogs *Hyla versicolor* and *Hyla chrysoscelis. Journal of Experimental Biology, 202*(22), 3225–3237.
Girgenrath, M., & Marsh, R. L. (2003). Season and testosterone affect contractile properties of fast calling muscles in the gray tree frog *Hyla chrysoscelis. American Journal of Physiology. Regulatory, Integrative and Comparative Physiology, 284*(6), 1513–1520.
Goller, F., & Daley, M. A. (2001). Novel motor gestures for phonation during inspiration enhance the acoustic complexity of birdsong. *Proceedings of the Royal Society of London, Series B: Biological Sciences, 268*(1483), 2301–2305.
Grafe, T. U., & Thein, J. (2001). Energetics of calling and metabolic substrate use during prolonged exercise in the European treefrog *Hyla arborea. Journal of Comparative Physiology B: Biochemical, Systemic, and Environmental Physiology, 171*(1), 69–76.
Greenfield, M. D. (1994). Synchronous and alternating choruses in insects and anurans: Common mechanisms and diverse functions. *American Zoologist, 34*(6), 605–615.
Gridi-Papp, M. (2008). The structure of vocal sounds produced with the mouth closed or with the mouth open in treefrogs. *Journal of the Acoustical Society of America, 123*(5), 2895–2902.
Gridi-Papp, M. (2014). Is the frequency content of the calls in North American treefrogs limited by their larynges? *International Journal of Evolutionary Biology, 2014*, e198069.
Gridi-Papp, M., Feng, A. S., Shen, J.-X., Yu, Z.-L., Rosowski, J. J., & Narins, P. M. (2008). Active control of ultrasonic hearing in frogs. *Proceedings of the National Academy of Sciences of the U S A, 105*(31), 11014–11019.
Gridi-Papp, M., Rand, A. S., & Ryan, M. J. (2006). Animal communication: Complex call production in the túngara frog. *Nature, 441*(7089), 38.
Haddad, C. F., & Giaretta, A. A. (1999). Visual and acoustic communication in the Brazilian torrent frog, *Hylodes asper* (Anura: Leptodactylidae). *Herpetologica, 55*(3), 324–333.
Halfwerk, W., Dixon, M. M., Ottens, K. J., Taylor, R. C., Ryan, M. J., Page, R. A., et al. (2014a). Risks of multimodal signaling: bat predators attend to dynamic motion in frog sexual displays. *The Journal of Experimental Biology, 217*(17), 3038–3044.
Halfwerk, W., Jones, P. L., Taylor, R. C., Ryan, M. J., & Page, R. A. (2014b). Risky ripples allow bats and frogs to eavesdrop on a multisensory sexual display. *Science, 343*(6169), 413–416.
Hartley, R. S., & Suthers, R. A. (1989). Airflow and pressure during canary song: Direct evidence for mini-breaths. *Journal of Comparative Physiology. A, Neuroethology, Sensory, Neural, and Behavioral Physiology, 165*(1), 15–26.
Hartmann, M. T., Giasson, L. O. M., Hartmann, P. A., & Haddad, C. F. B. (2005). Visual communication in Brazilian species of anurans from the Atlantic forest. *Journal of Natural History, 39*(19), 1675–1685.

Hetherington, T. E. (1985). Role of the opercularis muscle in seismic sensitivity in the bullfrog, *Rana catesbeiana. Journal of Experimental Zoology, 235*(1), 27–34.
Hetherington, T. E. (1987). Physiological features of the opercularis muscle and their effects on vibration sensitivity in the bullfrog *Rana catesbeiana. Journal of Experimental Biology, 131*(1), 189–204.
Hetherington, T. E. (1994). The middle ear muscle of frogs does not modulate tympanic responses to sound. *Journal of the Acoustical Society of America, 95*(4), 2122–2125.
Hetherington, T. E., & Lombard, R. E. (1983). Electromyography of the opercularis muscle of *Rana catesbeiana*: An amphibian tonic muscle. *Journal of Morphology, 175*(1), 17–26.
Hibbitts, T. J., Whiting, M. J., & Stuart-Fox, D. M. (2007). Shouting the odds: vocalization signals status in a lizard. *Behavioral Ecology and Sociobiology, 61*(8), 1169–1176.
Hödl, W. (1992). Reproductive behaviour in the neotropical foam-nesting frog *Pleurodema diplolistris* (Leptodactylidae). *Amphibia-Reptilia, 13*(3), 263–274.
Hödl, W., & Gollmann, G. (1986). Distress calls in neotropical frogs. *Amphibia-Reptilia, 7*(1), 11–21.
Hoskin, C. J., James, S., & Grigg, G. C. (2009). Ecology and taxonomy-driven deviations in the frog call–body size relationship across the diverse Australian frog fauna. *Journal of Zoology, 278*(1), 36–41.
Hughes, G. M., & Vergara, G. A. (1978). Static pressure-volume curves for the lung of the frog (*Rana pipiens*). *Journal of Experimental Biology, 76*(1), 149–165.
Jaramillo, C., Rand, A. S., Ibáñez, R., & Dudley, R. (1997). Elastic structures in the vocalization apparatus of the túngara frog *Physalaemus pustulosus* (Leptodactylidae). *Journal of Morphology, 233*(3), 287–295.
Jenssen, T. A. (1972). Seasonal organ weights of the green frog, *Rana clamitans* (Anura, Ranidae), under natural conditions. *Transactions of the Illinois State Academy of Science, 65*, 15–24.
Jørgensen, M. B., & Kanneworff, M. (1998). Middle ear transmission in the grass frog, *Rana temporaria. Journal of Comparative Physiology. A, Neuroethology, Sensory, Neural, and Behavioral Physiology, 182*(1), 59–64.
Kelley, D. B., & Tobias, M. L. (1999). Vocal communication in *Xenopus laevis*. In M. Hauser & M. Konishi (Eds.), *Neural mechanisms of communication* (pp. 9–35). Cambridge, MA: MIT Press.
Kelley, D. B., Tobias, M. L., Horng, S., & Ryan, M. J. (2001). Producing and perceiving frog songs: Dissecting the neural bases for vocal behaviors in *Xenopus laevis*. In M. J. Ryan (Ed.), *Anuran communication* (pp. 156–166). Washington, DC: Smithsonian Institution Press.
Kime, N. M., Ryan, M. J., & Wilson, P. S. (2013). A bond graph approach to modeling the anuran vocal production system. *The Journal of the Acoustical Society of America, 133*(6), 4133–4144.
Kirkpatrick, M., & Ryan, M. J. (1991). The evolution of mating preferences and the paradox of the lek. *Nature, 350*(6313), 33–38.
Lardner, B., & bin Lakim, M. (2002). Animal communication: Tree-hole frogs exploit resonance effects. *Nature, 420*(6915), 475.
Largen, M. J., Morris, P. A., & Yalden, D. W. (1972). Observations on the caecilian *Geotrypetes grandisonae* Taylor (Amphibia Gymnophiona) from Ethiopia. *Monitore Zoologico Italiano. Supplemento, 4*(1), 185–205.
Lewis, E. R., & Narins, P. M. (1985). Do frogs communicate with seismic signals? *Science, 227*(4683), 187–189.
Lewis, E. R., Narins, P. M., Cortopassi, K. A., Yamada, W. M., Poinar, E. H., Moore, S. W., et al. (2001). Do male white-lipped frogs use seismic signals for intraspecific communication? *American Zoologist, 41*(5), 1185–1199.
Littlejohn, M. J., & Ryan, M. J. (2001). Patterns of differentiation in temporal properties of acoustic signals of anurans. In M. J. Ryan (Ed.), *Anuran communication* (pp. 102–120). Washington, DC: Smithsonian Institution Press.
Liu, C. C. (1935). Types of vocal sac in the Salientia. *Proceedings of the Boston Society of Natural History, 41*, 19–40.

Loftus-Hills, J. J., & Littlejohn, M. J. (1992). Reinforcement and reproductive character displacement in *Gastrophryne carolinensis* and *G. olivacea* (Anura: Microhylidae): A reexamination. *Evolution, 46*(4), 896–906.
Manley, G. A., & Kraus, J. E. (2010). Exceptional high-frequency hearing and matched vocalizations in Australian pygopod geckos. *Journal of Experimental Biology, 213*(11), 1876–1885.
Martin, W. F. (1971). Mechanics of sound production in toads of the genus *Bufo*: Passive elements. *Journal of Experimental Zoology, 176*(3), 273–293.
Martin, W. F., & Gans, C. (1972). Muscular control of the vocal tract during release signaling in the toad *Bufo valliceps*. *Journal of Morphology, 137*(1), 1–27.
Mason, M. J., & Narins, P. M. (2002a). Vibrometric studies of the middle ear of the bullfrog *Rana catesbeiana* II. The operculum. *Journal of Experimental Biology, 205*(20), 3167–3176.
Mason, M. J., & Narins, P. M. (2002b). Vibrometric studies of the middle ear of the bullfrog *Rana catesbeiana* I. The extrastapes. *Journal of Experimental Biology, 205*(20), 3153–3165.
McAlister, W. H. (1961). The mechanics of sound production in North American *Bufo*. *Copeia, 1961*(1), 86–95.
Mitchell, N. J., Kearney, M. R., Nelson, N. J., & Porter, W. P. (2008). Predicting the fate of a living fossil: How will global warming affect sex determination and hatching phenology in tuatara? *Proceedings of the Royal Society of London B: Biological Sciences, 275*(1648), 2185–2193.
Moore, B. A., Russell, A. P., & Bauer, A. M. (1991). Structure of the larynx of the tokay gecko (*Gekko gecko*), with particular reference to the vocal cords and glottal lips. *Journal of Morphology, 210*(3), 227–238.
Nally, R. C. (1981). On the reproductive energetics of chorusing males: Energy depletion profiles, restoration and growth in two sympatric species of *Ranidella* (Anura). *Oecologia, 51*(2), 181–188.
Narins, P. M., Grabul, D. S., Soma, K. K., Gaucher, P., & Hödl, W. (2005). Cross-modal integration in a dart-poison frog. *Proceedings of the National Academy of Sciences of the U S A, 102*(7), 2425–2429.
Narins, P. M., Lewis, E. R., & McClelland, B. E. (2000). Hyperextended call note repertoire of the endemic Madagascar treefrog *Boophis madagascariensis* (Rhacophoridae). *Journal of Zoology, 250*(3), 283–298.
Neill, W. T. (1952). Remarks on salamander voices. *Copeia, 1952*, 195–196.
Nelson, N. J., Keall, S. N., Brown, D., & Daugherty, C. H. (2002). Establishing a new wild population of tuatara (*Sphenodon guntheri*). *Conservation Biology, 16*(4), 887–894.
Nelson, N. J., Thompson, M. B., Pledger, S., Keall, S. N., & Daugherty, C. H. (2004). Do TSD, sex ratios, and nest characteristics influence the vulnerability of tuatara to global warming? *International Congress Series, 1275*, 250–257.
Paulsen, K. (1965). Beitrage zur anatomie und physiologie des froschkehlkopfes (mit besonderer berucksichtigung der stimmbildung). *Zeitschrift für Wissenschaftliche Zoologie, 172*, 1–16.
Paulsen, K. (1967). *Das prinzip der stimmbildung in der wirbeltierreihe und beim menschen*. Frankfurt, Germany: Akademische Verlagsgesellschaft.
Pauly, G. B., Bernal, X. E., Rand, A. S., & Ryan, M. J. (2006). The vocal sac increases call rate in the túngara frog *Physalaemus pustulosus*. *Physiological and Biochemical Zoology, 79*(4), 708–719.
Penna, M. (2004). Amplification and spectral shifts of vocalizations inside burrows of the frog *Eupsophus calcaratus* (Leptodactylidae). *Journal of the Acoustical Society of America, 116*(2), 1254–1260.
Pfennig, K. S. (2007). Facultative mate choice drives adaptive hybridization. *Science, 318*(5852), 965–967.
Pfennig, K. S., & Simovich, M. A. (2002). Differential selection to avoid hybridization in two toad species. *Evolution, 56*(9), 1840–1848.
Preininger, D., Boeckle, M., Freudmann, A., Starnberger, I., Sztatecsny, M., & Hödl, W. (2013). Multimodal signaling in the small torrent frog (*Micrixalus saxicola*) in a complex acoustic environment. *Behavioral Ecology and Sociobiology, 67*(9), 1449–1456.

Purgue, A. P. (1995). *The sound broadcasting system of the bullfrog*. PhD thesis, University of Utah.

Purgue, A. P. (1997). Tympanic sound radiation in the bullfrog *Rana catesbeiana*. *Journal of Comparative Physiology. A, Neuroethology, Sensory, Neural, and Behavioral Physiology, 181*(5), 438–445.

Rand, A. S. (2001). A history of frog call studies 405 BC to 1980. In M. J. Ryan (Ed.), *Anuran communication* (pp. 8–19). Washington, DC: Smithsonian Institution Press.

Rand, A. S., & Dudley, R. (1993). Frogs in helium: The anuran vocal sac is not a cavity resonator. *Physiological Zoology, 66*(5), 793–806.

Refsnider, J. M., Keall, S. N., Daugherty, C. H., & Nelson, N. J. (2009). Does nest-guarding in female tuatara (*Sphenodon punctatus*) reduce nest destruction by conspecific females? *Journal of Herpetology, 43*(2), 294–299.

Romer, A. S., & Parsons, T. S. (1986). *The vertebrate body*. New York: Saunders College.

Rosenthal, G. G., Rand, A. S., & Ryan, M. J. (2004). The vocal sac as a visual cue in anuran communication: An experimental analysis using video playback. *Animal Behaviour, 68*(1), 55–58.

Russell, A. P., Rittenhouse, D. R., & Bauer, A. M. (2000). Laryngotracheal morphology of Afro-Madagascan geckos: A comparative survey. *Journal of Morphology, 245*(3), 241–268.

Ryan, M. J. (1980). Female mate choice in a neotropical frog. *Science, 209*(4455), 523–525.

Ryan, M. J. (1985). *The túngara frog: A study in sexual selection and communication*. Chicago: University of Chicago Press.

Ryan, M. J. (2001). *Anuran communication*. Washington, DC: Smithsonian Institution Press.

Ryan, M. J., Bartholomew, G. A., & Rand, A. S. (1983a). Energetics of reproduction in a neotropical frog, *Physalaemus pustulosus*. *Ecology, 64*(6), 1456–1462.

Ryan, M. J., & Drewes, R. C. (1990). Vocal morphology of the *Physalaemus pustulosus* species group (Leptodactylidae): Morphological response to sexual selection for complex calls. *Biological Journal of the Linnean Society, 40*(1), 37–52.

Ryan, M. J., Fox, J. H., Wilczynski, W., & Rand, A. S. (1990). Sexual selection for sensory exploitation in the frog *Physalaemus pustulosus*. *Nature, 343*, 66–67.

Ryan, M. J., & Rand, A. S. (1995). Female responses to ancestral advertisement calls in túngara frogs. *Science, 269*, 390–392.

Ryan, M. J., Tuttle, M. D., & Barclay, R. M. R. (1983b). Behavioral responses of the frog-eating bat, *Trachops cirrhosus*, to sonic frequencies. *Journal of Comparative Physiology. A, Neuroethology, Sensory, Neural, and Behavioral Physiology, 150*(4), 413–418.

Sacchi, R., Galeotti, P., Fasola, M., & Gerzeli, G. (2004). Larynx morphology and sound production in three species of Testudinidae. *Journal of Morphology, 261*(2), 175–183.

Schmid, E. (1978). Contribution to the morphology and histology of the vocal cords of central European anurans (Amphibia). *Zoologische Jahrbucher Anatomie, 5*, 133–150.

Schmidt, R. S. (1965). Larynx control and call production in frogs. *Copeia, 1965*(2), 143–147.

Schmidt, R. S. (1972). Action of intrinsic laryngeal muscles during release calling in leopard frog. *Journal of Experimental Zoology, 181*(2), 233–243.

Schneider, H. (1970). Morphologie des larynx von *Hyla a. arborea* (L.) und *Hyla meridionalis* Boettger (Amphibia, Anura). *Zoomorphology, 66*(4), 299–309.

Schwartz, J. J., & Wells, K. D. (1985). Intra- and interspecific vocal behavior of the neotropical treefrog *Hyla microcephala*. *Copeia, 1985*(1), 27–38.

Shen, J.-X., Xu, Z.-M., Feng, A. S., & Narins, P. M. (2011). Large odorous frogs (*Odorrana graminea*) produce ultrasonic calls. *Journal of Comparative Physiology. A, Neuroethology, Sensory, Neural, and Behavioral Physiology, 197*(10), 1027–1030.

Starnberger, I., Preininger, D., & Hödl, W. (2014). From uni-to multimodality: Towards an integrative view on anuran communication. *Journal of Comparative Physiology A, 200*(9), 777–787.

Strake, J., Luksch, H., & Walkowiak, W. (1994). Audio-motor interface in anurans. *European Journal of Morphology, 32*(2–4), 122–126.

Stuart, B. L. (2008). The phylogenetic problem of *Huia* (Amphibia: Ranidae). *Molecular Phylogenetics and Evolution, 46*, 49–60.

Sullivan, B. K., & Wagner, W. E., Jr. (1988). Variation in advertisement and release calls, and social influences on calling behavior in the Gulf Coast toad (*Bufo valliceps*). *Copeia, 1988*, 1014–1020.

Suthers, R. A., Narins, P. M., Lin, W.-Y., Schnitzler, H.-U., Denzinger, A., Xu, C.-H., et al. (2006). Voices of the dead: Complex nonlinear vocal signals from the larynx of an ultrasonic frog. *Journal of Experimental Biology, 209*(24), 4984–4993.

Taigen, T., & Wells, K. (1985). Energetics of vocalization by an anuran amphibian (*Hyla versicolor*). *Journal of Comparative Physiology B: Biochemical, Systemic, and Environmental Physiology, 155*(2), 163–170.

Taylor, R. C., & Ryan, M. J. (2013). Interactions of multisensory components perceptually rescue túngara frog mating signals. *Science, 341*(6143), 273–274.

Thurow, G. R., & Gould, H. J. (1977). Sound production in a caecilian. *Herpetologica, 33*(2), 234–237.

Tobias, M. L., Evans, B. J., & Kelley, D. B. (2011). Evolution of advertisement calls in African clawed frogs. *Behaviour, 148*(4), 519–549.

Tobias, M. L., Korsh, J., & Kelley, D. B. (2014). Evolution of male and female release calls in African clawed frogs. *Behaviour, 151*(9), 1313–1334.

Todd, N. P. M. (2007). Estimated source intensity and active space of the American alligator (*Alligator mississippiensis*) vocal display. *The Journal of the Acoustical Society of America, 122*(5), 2906–2915.

Towns, D. R., & Daugherty, C. H. (1994). Patterns of range contractions and extinctions in the New Zealand herpetofauna following human colonisation. *New Zealand Journal of Zoology, 21*(4), 325–339.

Trewavas, E. (1932). The hyoid and larynx of the Anura. *Royal Society of London Philosophical Transactions Series B: Biological Sciences, 222*, 401–527.

Uetz, P. (2014). *The reptile database.* Heidelberg, Germany: European Molecular Biology Laboratory. Retrieved May 8, 2015, from http://www.reptile-database.org.

Vergne, A. L., Aubin, T., Martin, S., & Mathevon, N. (2012). Acoustic communication in crocodilians: Information encoding and species specificity of juvenile calls. *Animal Cognition, 15*(6), 1095–1109.

Vergne, A. L., & Mathevon, N. (2008). Crocodile egg sounds signal hatching time. *Current Biology, 18*(12), 513–514.

Vergne, A. L., Pritz, M. B., & Mathevon, N. (2009). Acoustic communication in crocodilians: From behaviour to brain. *Biological Reviews, 84*(3), 391–411.

Vitalis, T. Z., & Shelton, G. (1990). Breathing in *Rana pipiens*: The mechanism of ventilation. *Journal of Experimental Biology, 154*(1), 537–556.

Vitt, L. J., & Caldwell, J. P. (2013). *Herpetology: An introductory biology of amphibians and reptiles* (4th ed.). Boston: Academic.

Wagner, W. E., Jr. (1989). Graded aggressive signals in Blanchard's cricket frog: vocal responses to opponent proximity and size. *Animal Behaviour, 38*(6), 1025–1038.

Wagner, W. E., Jr. (1992). Deceptive or honest signaling of fighting ability? A test of alternative hypotheses for the function of changes in call dominant frequency by male cricket frogs. *Animal Behaviour, 44*(3), 449–462.

Walkowiak, W. (1992). Acoustic communication in the fire-bellied toad: An integrative neurobiological approach. *Ethology Ecology and Evolution, 4*(1), 63–74.

Walkowiak, W. (2006). Call production and neural basis of vocalization. In P. M. Narins, A. S. Feng, R. R. Fay, & A. N. Popper (Eds.), *Hearing and sound communication in amphibians* (pp. 87–112). New York: Springer.

Wang, X., Wang, D., Wu, X., Wang, R., & Wang, C. (2007). Acoustic signals of Chinese alligators (*Alligator sinensis*): Social communication. *The Journal of the Acoustical Society of America, 121*(5), 2984–2989.

Weber, E. (1974). Comparative bio-acoustic investigations in the *Discoglossus pictus*, Otth 1837 and the *Discoglossus sardus* Tschudi 1837 (Discoglossidae, Anura). *Zoologische Jahrbucher Physiologie, 78*, 40–84.

Weber, E., & Werner, Y. L. (1977). Vocalizations of two snake-lizards (Reptilia: Sauria: Pygopodidae). *Herpetologica, 33*(3), 353–363.
Welch, A. M., Semlitsch, R. D., & Gerhardt, H. C. (1998). Call duration as an indicator of genetic quality in male gray tree frogs. *Science, 280*(5371), 1928–1930.
Wells, K. D. (1988). The effect of social interactions on anuran vocal behavior. In B. Fritzsch, W. Wilczynski, M. J. Ryan, T. Hetherington, & W. Walkowiak (Eds.), *The evolution of the amphibian auditory system* (pp. 9–18). New York: Wiley.
Wells, K. D. (2010). *The ecology and behavior of amphibians*. Chicago: University of Chicago Press.
Wells, K. D., & Taigen, T. L. (1984). Reproductive behavior and aerobic capacities of male American toads (*Bufo americanus*): Is behavior constrained by physiology? *Herpetologica, 40*(3), 292–298.
Wilczynski, W., & Chu, J. (2001). Acoustic communication, endocrine control, and the neurochemical systems of the brain. In M. J. Ryan (Ed.), *Anuran communication* (pp. 23–35). Washington, DC: Smithsonian Institution Press.
Wilder, H. H. (1896). Lungless salamanders. *Anatomischer Anzeiger, 12*, 182–192.
Wogel, H., Abrunhosa, P. A., & Weber, L. N. (2004). The tadpole, vocalizations and visual displays of *Hylodes nasus* (Anura: Leptodactylidae). *Amphibia-Reptilia, 25*(2), 219–226.
Woolbright, L. L., & Stewart, M. M. (1987). Foraging success of the tropical frog, *Eleutherodactylus coqui*: The cost of calling. *Copeia, 1987*, 69–75.
Yager, D. D. (1992). A unique sound production mechanism in the pipid anuran *Xenopus borealis. Zoological Journal of the Linnean Society, 104*(4), 351–375.
Yager, D. D. (1996). Sound production and acoustic communication in *Xenopus borealis*. In R. C. Tinsley & H. R. Kobel (Eds.), *The biology of Xenopus* (pp. 121–141). Oxford, England: Clarendon.
Yamaguchi, A., & Kelley, D. B. (2000). Generating sexually differentiated vocal patterns: Laryngeal nerve and EMG recordings from vocalizing male and female African clawed frogs (*Xenopus laevis*). *The Journal of Neuroscience, 20*(4), 1559–1567.
Yang, E. J., & Kelley, D. B. (2008). Hormones and the regulation of vocal patterns in amphibians: *Xenopus laevis* vocalizations as a model system. In D. Pfaff, A. Arnold, A. Etgen, S. Fahrbach, & R. Rubin (Eds.), *Hormones, brain, and behavior* (pp. 693–706). New York: Academic.
Young, B. A. (1991). Morphological basis of "growling" in the king cobra, *Ophiophagus hannah. Journal of Experimental Zoology, 260*(3), 275–287.
Young, B. A. (2000). The comparative morphology of the larynx in snakes. *Acta Zoologica, 81*(2), 177–193.
Young, B. A. (2003). Snake bioacoustics: Toward a richer understanding of the behavioral ecology of snakes. *The Quarterly Review of Biology, 78*(3), 303–325.
Young, B. A., Sheft, S., & Yost, W. (1995). Sound production in *Pituophis melanoleucus* (Serpentes: Colubridae) with the first description of a vocal cord in snakes. *Journal of Experimental Zoology, 273*(6), 472–481.
Yu, X., Peng, Y., Aowphol, A., Ding, L., Brauth, S. E., & Tang, Y.-Z. (2011). Geographic variation in the advertisement calls of *Gekko gecko* in relation to variations in morphological features: Implications for regional population differentiation. *Ethology Ecology and Evolution, 23*(3), 211–228.
Zornik, E., & Kelley, D. B. (2008). Regulation of respiratory and vocal motor pools in the isolated brain of *Xenopus laevis. The Journal of Neuroscience, 28*(3), 612–621.
Zornik, E., & Yamaguchi, A. (2008). Sexually differentiated central pattern generators in *Xenopus laevis. Trends in Neurosciences, 31*(6), 296–302.

Chapter 4
Locomotion-Induced Sounds and Sonations: Mechanisms, Communication Function, and Relationship with Behavior

Christopher James Clark

Abstract Motion-induced sound is an intrinsic byproduct of essentially all animal behavior. Locomotion-induced sounds that have evolved specialization for communication are termed sonations. The null hypothesis is that locomotion-induced sounds are noncommunicative (adventitious), produced by nonspecialized morphology, and are involuntary. A sound is a sonation if it is produced by specialized morphology, or is produced voluntarily. The production of locomotion-induced sound can be examined at two levels: the animal motions (kinematics) that lead to sound production and the physical acoustic mechanism(s) that generate(s) the sound itself. The physical acoustics of locomotion induced sound are diverse, with both aerodynamic and structural mechanisms, including aeroelastic flutter, percussion, stridulation, and presumably many other undescribed mechanisms. There is a direct sound–motion correspondence between aspects of an animal's motions and ensuing locomotion-induced sounds, especially in the time domain. This correspondence has two implications. One is experimental: sound recordings are a useful and perhaps underutilized source of data about animal locomotion. The second is behavioral: locomotion-induced sounds intrinsically contain information about an animal's motions (such as wingbeat frequencies) that may be of interest to other animals. Therefore, locomotion-induced sounds are intrinsically suited to be mechanisms by which animals directly evaluate the locomotor performance of other animals, such as during courtship. The sound–motion correspondence is also a constraint. Sonations seem less acoustically diverse than vocalizations. Because they require discrete behaviors to be produced, animals also have somewhat fewer opportunities to produce sonations strategically, and few sonations are frequency-modulated. Sound production mechanisms of sonations are external to the animal and therefore are easy to manipulate experimentally on wild animals, making sonations an ideal, underutilized system for testing hypotheses about acoustic function.

C.J. Clark (✉)
Department of Biology, University of California, Riverside,
Riverside, CA 92521, USA
e-mail: cclark@ucr.edu

R.A. Suthers et al. (eds.), *Vertebrate Sound Production and Acoustic Communication*, Springer Handbook of Auditory Research 53,
DOI 10.1007/978-3-319-27721-9_4

Keywords Adventitious • Communication • Flight • Flutter • Hummingbird • Kinematics • Manakin • Mechanical sound • Nonvocal • Voluntariness • Wing

4.1 Introduction

Many vertebrates produce nonvocal communication sounds, which are sometimes called mechanical sounds (Manson-Bahr and Pye 1985), particularly with respect to birds. Humans are no exception: we applaud a performance by clapping our hands or announce our presence by knocking on a door. Many of the best-known nonvocal signals are produced by birds, including the winnowing displays of snipe (*Gallinago* and *Coenocorypha* spp.), wing snaps of manakins (Pipridae; Fig. 4.1), or the diverse wing and tail sounds produced by displaying hummingbirds (Darwin 1871).

Examples apart from birds include ground thumping in rodents and other mammals (Randall 2001), tail rattling of rattlesnakes, or stridulation of modified spines in streaked tenrecs (*Hemicentetes semispinosus*) (Endo et al. 2010). Nonvocal acoustic communication has evolved hundreds of times, and as essentially any body

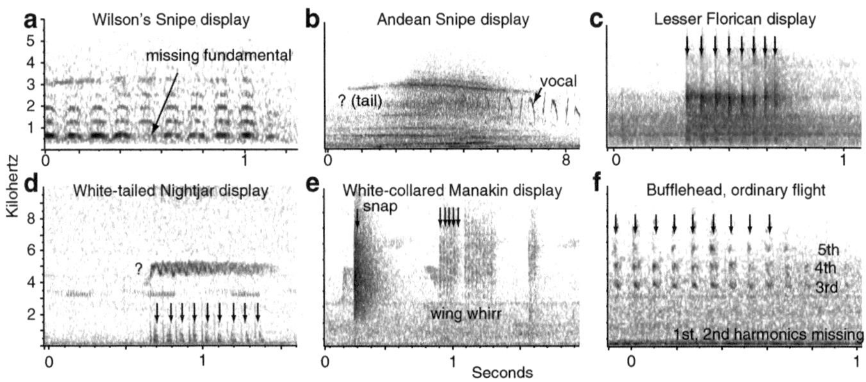

Fig. 4.1 Examples of the diversity of locomotion-induced sounds of birds. *Vertical arrows* indicate sound produced during wingbeats; question marks indicate physical mechanism is unresolved. (**a**) Male Wilson's snipe (*Gallinago delicata*) produces pulses of tonal sounds with its outer tail feathers during a dive display; *arrow* indicates a missing fundamental (*source*: XC #13808). (**b**) Male Andean snipe (*G. jamesoni*) produces multiple frequencies of sound, presumably with the tail, during a dive (XC #2014). (**c**) Male lesser florican (*Sypheotides indicus*) produces a knocking sound with the wings during a jump display (sound from Bhatt and Bardolia 2006). (**d**) Male white-tailed nightjar (*Hydropsalis cayennensis*) produces a tonal sound via an unknown mechanism during a flight display (ML #60664). (**e**) Male white-collared manakin (*Manacus candei*) produces two different types of sonations in display, first a snap and then a wing whirr (the latter was termed "snort" by Bostwick and Prum 2003; ML #72655). (**f**) Bufflehead (*Bucephala albeola*) of unknown sex produces tonal sound produced during ordinary flight. The third to fifth harmonics are present while the first and second harmonics of sound are missing (ML #57549). Note differing timescales of spectrograms; FFT: 512-sample Hann window, 50 % overlap. Sound source abbreviations: XC, www.Xeno-Canto.org; ML, Macaulay Library, macaulaylibrary.org

part can produce sound, nonvocal sounds are more diverse morphologically (though probably not acoustically) than vocal communication mechanisms. This chapter explores general features in connection with how these sounds are produced by terrestrial vertebrates, a brief overview of some of the physical acoustic mechanisms, their relationship with behavior, and what is known about how they evolve.

The evolutionary origin of communicative sounds is clear: They arose out of incidental byproducts of behaviors (adventitious sounds) that were converted to communication sounds after receivers attended to them (Darwin 1871; Ewing 1989). This occurs because virtually all motions generate sound, including inanimate motion, such as splashing of water or whooshing of wind in the trees. Sounds and vibrations are also an omnipresent byproduct of all animal motions, including the familiar rustling of a mouse or lizard scurrying through dry leaves, or the whoosh of the wings of a bat or bird that passes close by. Perhaps motion-induced sounds are underappreciated because they often do not have a particular communication function and can be unwanted noise that obscures a vocalization of interest. Everyday experience suggests that it is moving relatively silently that is difficult, such as a stalking cat or hunting owl, or as anyone trying to slip unnoticed out of a quiet room can attest.

These observations provide an axiom with two parts. First, all animal motions (i.e., nearly all animal behaviors) produce sound, audible or not. Second, the converse is also true: All significant animal sounds arise from a behavior. Therefore, study of animal bioacoustics can encompass nearly all of animal behavior. These two fields, animal behavior and bioacoustics, are partially divorced in practice because vocalizations of terrestrial vertebrates arise mostly from behaviors of the larynx or syrinx, structures specialized for the production of sound, whereas most other behaviors and morphology are irrelevant to vocalizations. These neat distinctions become blurred when nonvocal sounds are considered. Although all behaviors make sound, the issue is whether these sounds are audible. Further, every part of an animal is a sound-producing structure, even if not specialized for sound production. These features produce conceptual challenges: Simplifying assumptions that are usually reasonable for vocalizations, such as that they are voluntary, or function for communication, are often not reasonable for nonvocal sounds. This chapter explores these issues in more detail. Birds have been a focus of some recent work, and this chapter focuses preponderantly on them (Fig. 4.1).

4.1.1 Definitions

It is difficult to define a term by what it is not, so this chapter defines nonvocal sounds as locomotion-induced sounds. Broadly, all behaviors can be classified as locomotion, eating, or breathing (Barlow 1967). Of these three, nonvocal sounds physically originate from locomotion and are largely produced by interactions between the animal's integument and the environment. Eating and breathing, by contrast, are coupled pharyngeal functions and the associated morphology forms the vocal tract. Vocalizations, broadly defined, are the acoustic byproducts of the behaviors of eating or breathing. Vocalizations originate from motion inside the animal and include sounds made by fluid as it leaves the animal. Included under this broad definition of vocalization are the "voiced" sounds of the larynx or syrinx, and also "voiceless" sounds such as human whispering, sneezes, coughing, tongue clicks, bill snaps, chewing, and even passing gas. A narrower definition would consider vocalizations to include only "voiced" sounds of two nonhomologous structures, the larynx and syrinx (e.g., Au and Suthers 2014). This strict-sense definition of vocalizations would treat as nonvocal bill snaps and communication farts (Wilson et al. 2003), but by this count, also nonvocal are dolphin calls and human whispers (Au and Suthers 2014)! Such a broad definition of "nonvocal" is not useful, as whispering and communication farts have more in common with the "voiced" sounds generated by the larynx/syrinx than with locomotion-induced sounds. Therefore, this chapter employs the broad definition of vocalization, and nonvocal sounds are only those produced by behaviors related to locomotion.

Nonvocal sounds can be categorized either descriptively or functionally, and neither naming scheme is perfect. Considering function, adventitious sounds are incidental byproducts of motion that lack communication function. Darwin (1871) was the first to propose that these incidental sounds may be co-opted for communication, which he termed instrumental music. This poetic term did not enter popular usage, and the nonvocal sounds of birds instead came to be called mechanical sounds, without reference to function (Manson-Bahr and Pye 1985). Bostwick and Prum (2003) proposed the term "sonation" to mean nonvocal sounds modulated *and evolved* for communication function, where "sonate" is the corresponding verb. This term was to be analogous to the word phonation, which describes strict-sense vocalizations. One major problem arises from attempts to use the term sonation rigorously, however. Most syrinx/larynx sounds are likely to be functional and not adventitious, and so it is reasonable to assume they are phonations, even if the function is totally unknown. By contrast, it is often not clear whether a particular nonvocal sound has evolved for communication, and is therefore a sonation. Assuming that an unstudied phonation has a function will only rarely be controversial. By contrast, most nonvocal sounds are not communicative, and diagnosis of a sonation can be tricky and assumption laden. Therefore, the label sonation cannot be applied indiscriminately.

There are two principal criteria used to diagnose a sonation: specialized morphology and voluntariness. The easier criterion to apply is specialized morphology. If the morphology used to make the sound has evolved a functional form closely linked to the acoustic form of the sound, it is almost certainly a sonation. For instance, rattlesnakes have evolved a rattle, tenrecs have evolved modified spines, or many birds have evolved feathers with highly specialized shapes, and none of these morphologies have functions besides production of sound. These are reasonably assumed to be sonations, even if the sound is not voluntary, per the second criterion.

Morphology is often not obviously specialized for sound production. The second best criterion for a sonation is voluntariness, whether the sound is produced intentionally and production can be modulated by the animal. This complex topic is discussed further at the end of Sect. 4.2.1. Voluntariness is clearest when the kinematics (behaviors) that produce the sound are distinctive and specialized for sound production.

Even with these two criteria, ambiguous cases are not hard to find, even in human behavior, where intent can be deciphered. Human footsteps are normally adventitious and yet contain information used by others. One might recognize the approach of a particular colleague by his or her footsteps. Yet footsteps are also sometimes voluntarily modulated; for instance, that colleague might tiptoe to avoid alerting someone to his or her presence. Is tiptoeing communication? It fails the morphology criteria (feet/shoes do not seem adapted for acoustic communication) and the sounds produced are not voluntary either. The related behavior of tap dancing is a sonation, the product of cultural evolution. Thus, tap dancing is arguably a sonation, whereas distinctive, individual-specific footsteps are not.

Consider another example: If a person puts a bell on a cat or a horse that then jingles when the animal walks, is the resulting sound a sonation, or adventitious? It is modified morphology; it is not voluntarily modulated by the animal as it walks, but it is voluntary in the sense that the bell was placed voluntarily on the animal. It also *is* an attempt to communicate, by the human. However, if the purpose is to alert birds to the presence of the cat, it is arguably not a sonation, because birds have not evolved, or learned, to be aware that jingling is a signal of approaching danger. On the other hand, if the bell is to alert other people of an approaching horse, it is a sonation.

In neither of these examples is this logic iron clad: Both human footsteps and bells on animals are debatable sonations. The important role of learning has been ignored, and as the context is human behavior, intent is easier to assess than it is in animal behavior. It is often not clear *a priori* whether many types of nonvocal animal sounds are sonations or adventitious; function is a hypothesis to be tested, not assumed. The conservative approach is therefore to use the term sonation only when there is distinctive morphology or behavior obviously associated with the production of the nonvocal sound, or after experiments testing function. This means that many locomotion-induced sounds that might be sonations cannot be immediately recognized as such.

As an alternative, sounds can be named in the context of the underlying behavior, neutral to possible function. For instance, sounds produced in flight are "flight sounds," those produced during displays are "display sounds," and so forth. Some such sounds already have their own vernacular names, such as footsteps or clapping. In addition to functional neutrality, such a name should be descriptive and not imply a particular physical acoustic mechanism if the aptness of that mechanism is unclear. For instance, snapping, clapping, and drumming are related percussive mechanisms with distinctive acoustic forms and are often appropriate names, as in "wing clapping." But other widely used terms are misnomers. For instance, many sounds produced by bird wings have been called "wing whistles" as a description of their tonality (Miller and Inouye 1983; Barrera et al. 2011). However, a whistle is a specific aerodynamic mechanism that produces highly tonal sound that is often high pitched (Wilson et al. 1971). This aerodynamic mechanism, although common in wind musical instruments and certain mouth-generated human vocalizations (Fletcher 1992; Fletcher and Rossing 1998), has not been demonstrated for any nonvocal sound. The bird sounds called "wing whistles" are instead produced by aeroelastic flutter (Clark et al. 2013b). Whistle is also not a suitable synonym for tonal, because mechanisms involving mechanical (structural) resonance also produce tonal sound, such as the sound of a plucked guitar string, and one would not say "the plucked guitar string whistled pleasantly."

The disadvantage of a descriptive naming scheme is it can be cumbersome without pointing to whether or how the sound is biologically interesting.

4.2 Mechanisms

The physical mechanisms that produce locomotion-induced sounds derive from an interaction between morphology and behavior. The analogy of a musical instrument is apt: an animal's morphology is the instrument while its behavior is how the instrument is played. Section 4.2.1 explores the relationship between kinematics (motions) and sound, with emphasis on an experimental perspective. This topic, kinematics, is revisited in Sect. 4.3.1 from a functional perspective and the role of kinematics in communication. Later parts of this section provide a qualitative overview of the physical acoustics of locomotor-induced sound. This topic, and especially Sect. 4.2.5 (solid interactions), has been too poorly studied to allow a comprehensive overview. Entirely neglected are physical acoustics of locomotion-induced sounds that originate inside the animal, such as rattlesnake tail-shaking, clicking of tendons/ligaments across bony processes in joints, as in reindeer (*Rangifer tarandus*) and other ungulates (Bro-Jørgensen and Dabelsteen 2008), or "cracking" of joints, as in human knuckles, which is caused by cavitation of dissolved gasses in synovial fluid (Unsworth et al. 1971). Moreover, as Parmentier and Fine (Chap. 2) and Narins et al. (Chap. 7) address hydroacoustics and seismic communication respectively, only examples from airborne nonvocal sound are provided.

4.2.1 Animal Kinematics and Sound

Locomotion is a key component of animal natural history. Every motion an animal performs has an acoustic signature. This acoustic signature contains information about the animal's motions, potentially revealing its location in space, its velocity, and especially, discrete events of locomotion such as individual footfalls or wing flaps. The information contained in these sounds might be used by a potential receiver (Sect. 4.3), or experimentally useful to a scientist. In the context of describing the courtship dive of Anna's hummingbird (*Calypte anna*), Clark (2009) termed the relationship between sound and motion a "1:1 correspondence" because the

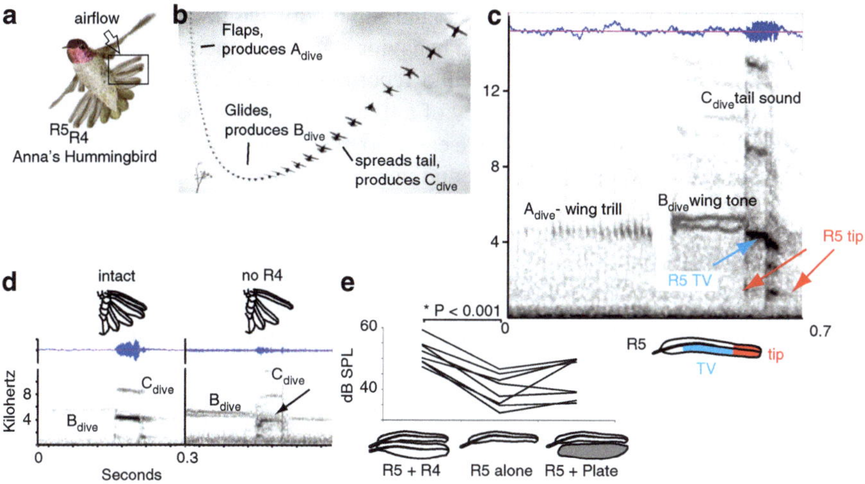

Fig. 4.2 Male Anna's hummingbird dive and produce sound with the wings and two tail-feathers, R5 and R4. (**a**) Male Anna's hummingbird (*Calypte anna*), with approximate direction of airflow over the tail during a dive. (**b**) Wing and tail kinematics during the dive correspond to sounds A_{dive}, B_{dive}, and C_{dive} produced during the dive. (**c**) Dive sound. Within C_{dive}, two sounds are produced: an approximately 4-kHz sound produced by the trailing vane (TV) of the feather R5 (teal), and a second, an approximately 1.2 kHz sound is produced by the tip of R5 (*red*) as the tail is spread and shut (*note*: the tip mode is particularly pronounced in this recording). (**d, e**) Feather R4 amplifies the dive sound through aerodynamic sympathetic vibration. (**d**) Removing R4 from a bird reduces amplitude of the dive sound. (**e**) Tested in a wind tunnel, presence of R4 increases SPL by 12 dB, because R4 vibrates in response to R5's flutter, whereas a flat plate as a control does not [**a, d, e** modified from Clark, Elias, & Prum (2011). Aeroelastic flutter produces hummingbird feather songs. *Science*, 333, 1430–1433. Reprinted with permission from AAAS. **b** modified from Clark (2009) under the author's copyright]

observed kinematics have a component frequency or timing that, after accounting for measurement error, exactly matches the frequencies or timings in the associated sound (Fig. 4.2b, c).

This correspondence does not imply that all animal motions are audible. Rather, given detection of locomotion-induced sound, temporal patterns within the sound must exactly match some aspect of the underlying motions of the animal, after correcting for the sound delay and other sources of error. The sound delay is the result of the difference between speed of light, which is nearly infinite, and the speed of sound in air (c), which is approximately 340 m s^{-1}, and varies slightly with temperature. For synchronized video and sound recordings, where the video is recording at a frame rate of n frames s^{-1} and the microphone is distance (d) meters from the animal, the lag (l) in video frames is

$$l = nd / c \tag{4.1}$$

The effect of sound delay is thus exacerbated when recording with high-speed video (high n) or at significant distances (d) from the subject.

The direct correspondence between sound and motion has broad utility in experimental design. A sound recording of a behavior can yield much of the same information as a video. For instance, a sound recording reveals stride frequency of a walking animal, or wingbeat frequency of a flying one (Ortiz-Crespo 1980; Hunter and Picman 2005). Some behaviors are easier to record with sound than with video, owing to the ability of sound to go around minor obstacles, lack of a need to focus, and the wider field of reception of microphones.

That said, the relationship between sound and motion is sometimes subtle, and connecting sounds with motions requires assumptions or prior knowledge about how the sound is produced. For example, in a flight display called the pendulum display (Fig. 4.3), a male Allen's hummingbird (*Selasphorus sasin*) rapidly flips his tail up and down in time with pulses of a "chirruping" sound (Aldrich 1939; Mitchell 2000). This exact match between tail motions and sound suggested to observers that these chirruping sounds were produced by the tail (Aldrich 1939; Mitchell 2000); the wings were flapped too fast to see with the naked eye. However, experimental evidence overturned this initial assessment. Birds missing their tail still produced these chirruping sounds (Clark 2014), and a high-speed video of the display revealed that when producing the chirruping sounds, the birds, in addition to moving their tail, also changed their wing kinematics, briefly flapping the wings with a contralateral asymmetry, so that a 1:1 match also existed between these asymmetrical wing motions and sound. The dorsoventral tail motions, though obvious to the naked eye,

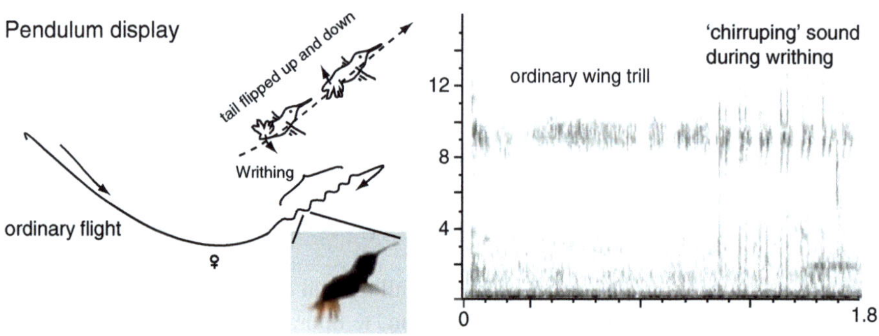

Fig. 4.3 Male Allen's hummingbird (*Selasphorus sasin*) pendulum display shows why the principle that sound and motion correspond directly must be applied carefully: Not all correlations are causal. Male Allen's hummingbirds produce an ordinary wing trill during all modes of flight. In the pendulum display, males fly in a shallow U over a female (*left*). On the downswing of the display, ordinary flight produces the ordinary wing trill, while on the upswing, the male "writhes." This writhing is apparent to the naked eye as a rapid flipping up and down of the widely spread tail (*left*); the wings, by contrast, are a blur. Produced in exact synchrony with writhing are pulses of "chirruping" sound (*right*). The exact synchrony of the tail-flipping with the pulses of "chirruping" led previous workers to suggest the tail produces the chirruping sound (Aldrich 1939; Mitchell 2000). However, this is incorrect: The chirruping sound is a modified version of the wing trill, produced by changes in wing kinematics that are visible only with a high-speed video. The tail's motions are correlated with sound production because the tail produces balancing forces against the atypical wing kinematics, not because the tail itself produces sound (Clark and Mitchell 2013)

were simply correlated with the asymmetrical wing motions, which actually produce the sound (Clark and Mitchell 2013).

This example demonstrates that inferring sound production from casual observation of kinematics alone may be misleading, and manipulative experiments (Sect. 4.2.2) that test the physical mechanism provide a stronger inference as to how the sound is produced.

4.2.1.1 Voluntariness

An important kinematic issue is sometimes not trivial to discern: To what degree is a locomotion-induced sound produced voluntarily? Voluntariness is the degree to which the animal can modulate the sound *independently* of the kinematics of ordinary motions, thereby permitting it to produce the sound strategically. For instance, many hummingbirds produce specialized wing sounds (wing trills) during ordinary flight, which seem muted or even silent during hovering, but are loud during sharp maneuvers and high-speed flight (Clark, pers. obs.). To what degree are these acoustic differences voluntary, allowing the animals to become stealthy or noisy if they wish?

The voluntariness of locomotion-induced sounds falls along a spectrum. Although the two ends of this spectrum, completely voluntary and entirely involuntary, are reasonably clear, intermediate examples are harder to diagnose and may require experimentation. Sounds easily identified as voluntary have two properties: (1) the sound requires specific, easily identifiable gross kinematics to produce and (2) these kinematics seem specialized for sound production. For instance, a hummingbird spreads its tail to make sound at only a specific point of its dive display, or a manakin snaps its wings together with kinematics not observed in ordinary flight (Bostwick and Prum 2003; Clark and Feo 2008). In these cases it is reasonable to assume the sound is voluntary, without detailed experimentation. Involuntary sounds instead have the following two properties: (1) the gross kinematics that produce them are general, not specific, and (2) the sound is always produced during general locomotor behavior. For instance, turbulence-induced whooshing sounds, although faint, are always present during bird flight. The only way an animal can modulate involuntary sounds is by changing the underlying behavior, such as by ceasing locomotion altogether.

In between these two extremes are cases where the voluntariness of the sound may be hard to assess. These are locomotion-induced sounds that do not have a distinct, obvious kinematic correlate, but they are also intermittent, observed in some types of seemingly ordinary locomotion, but not others. The tonal wing trills of many doves are an example; casual observation indicates these sounds are produced by only some individuals and/or in only some flight contexts (particularly during takeoff). Barrera et al. (2011), in a study of alarm function of the wing trill of zenaida dove (*Zenaida aurita*), implicitly assume the wing trill is voluntary because it is not produced in all modes of flight. However, production of a sound in only some contexts does not mean the sound is under voluntary control. Instead, the kinematic conditions necessary for involuntary production may be specific. Flutter of a wing feather (the likely physical mechanism underlying most tonal wing trills in birds, such as doves) requires local air velocity to exceed a threshold, U^*. The simplest explanation of a sound observed in some motions and not others is that U^* is exceeded in some cases but not others. Thus, a pigeon taking off normally (unalarmed) may be relatively quiet because it does not flap its wings quite fast enough to exceed U^* and produce sound, whereas a pigeon taking off in alarm, with only a small change in kinematics, does exceed U^* and produce tonal sound. This need not be voluntary.

Designing experiments to test whether a sound is voluntary requires an appropriate null model. As involuntary sounds abound in nature and this is clearly the ancestral character state if one goes far enough back in the phylogenetic tree, involuntary control is the null. Voluntary control of sound is often the derived character state, and is best regarded as the alternate hypothesis. Under this null, sound production is physically, strictly constrained and prescribed by gross locomotor kinematics. For sounds produced in apparently normal kinematics, the alternate hypothesis holds that the animal must alter something, an "invisible switch" (i.e., not easily observed within seemly ordinary motions), to turn sound on or off, without otherwise affecting locomotion. For instance, perhaps pigeons produce the wing trill voluntarily because they are able to alter a subtle component of their wing kinematics to turn wing trill production on or off, with little overall effect on flight. There are many ways this could occur. For instance, suppose they slightly change wing pronation at the beginning of the downstroke, and this affects bending or overlap of a crucial feather relative to its neighbor (nonoverlap is essential for flutter: Feo and Clark 2010), thereby activating sound production without otherwise affecting locomotion. Such a kinematic change is "invisible" because it would be hard to detect.

Unfortunately, obtaining the right data to clearly reject this null will sometimes be difficult, even when the null is in fact incorrect. The most intuitive approach is to record the animal's motions when producing and not producing the sound of interest, and search for a general kinematic correlate with sound production. However, this approach may be frustrating and inconclusive, for the switch itself may not be easily revealed. The simplest prediction the invisible switch hypothesis makes is that sound production does *not* correlate with gross kinematics, which is problematic because the absence of a correlation always has alternative explanations. By contrast, it will be easy to fail to reject the null; the null predicts some general kinematic variable will be different in sound-producing versus non-sound-producing kinematic contexts. For instance, the null predicts alarmed pigeons flap their wings faster on average than nonalarmed (notwithstanding spurious correlations; Sect. 4.2.1). Instead, to support the invisible switch hypothesis, an experimenter must find the switch itself, which constitutes the exact physical features that produce sound, and what kinematics the animal uses to turn sound production on and off. The experimenter must show (or argue) that changing these kinematics is not otherwise a crucial component for locomotion, and is therefore under voluntary control.

4.2.2 Experimental Approaches

Nonvocal sounds are, by definition, produced by an animal's external morphology. This affords easy access to the hypothesized sound source, facilitating experiments on the physical mechanisms of sound production with relatively benign effects on the animal, as has been done on hummingbirds (Miller and Inouye 1983; Clark and Feo 2008) and snipe (Tuck 1972). Whereas most manipulative experiments have sought to identify the physical origin of sound, Miller and Inouye (1983) tested function by using glue to silence the wing trill produced by the tips of the outer wing feathers of territorial male broad-tailed hummingbirds (*Selasphorus platycercus*). They found that these silenced males then tended to lose their territories, but regained them once the glue was removed; control birds with glue on non-sound-producing wing feathers were not as strongly affected. Miller and Inouye's (1983) study demonstrates the feasibility of performing field manipulations to study the ecological context and function of locomotion-induced sounds using simple experiments. Birds in particular are highly amenable: Feathers, if experimentally plucked, are regrown in a few weeks.

Feathers used for sonations often have a noteworthy shape, but not all unusually shaped feathers produce sound—and what constitutes an "unusual" feather shape may not be clear *a priori*. The most rigorous way to show that a given structure produces a particular sound is to demonstrate that the structure is both necessary and sufficient to produce the sound. Tests of both necessity and sufficiency can be feasible on feathers and are therefore an appropriate starting point for studies of function or evolution. More limited data, such as on sufficiency alone, can be ambiguous. For instance, essentially all feathers tested in a wind tunnel can flutter and are sufficient to generate sound, as discussed further in Sect. 4.2.4, and so this type of data by itself does not provide strong grounds for concluding that a feather actually does produce sound in flight (Clark et al. 2013a, b).

Paired experiments that demonstrated both necessity and sufficiency settled disagreement on the nonvocal origin of sonations in hummingbirds (Bostwick 2006). Male Anna's hummingbirds produce a loud *chirp* (the dive sound) during a display dive (Fig. 4.2). Rodgers (1940) demonstrated sufficiency by finding that whipping the outer tail feather through the air produced a sound similar to the *chirp* (see also Aldrich 1939). But Baptista and Matsui (1979) argued the dive sounds were vocal, owing to spectral similarity between the dive chirp and a portion of the species' obviously vocal song. Moreover in the sister species, Costa's hummingbird (*Calypte costae*), the dive sound and entire song were nearly identical to each other (Baptista and Matsui 1979; Baptista 2001). Neither side had tested whether tail feathers are necessary for the dive sound, so Clark and Feo (2008) resolved the disagreement by showing that removing the outer tail feathers from territorial males completely eliminated the bird's ability to produce the chirp. They also replicated Rodger's (1940) result, demonstrating the same feathers were sufficient to produce the sound. As the vocal versus nonvocal origin of other hummingbird sounds had been questioned (Baptista and Matsui 1979; Pytte and Ficken 1994), Clark and Feo performed similar experiments on Costa's (Clark and Feo 2010), black-chinned (*Archilochus alexandri*: Feo and Clark 2010), Calliope (*Selasphorus calliope*: Clark 2011), and Allen's hummingbird (*S. sasin*: Clark 2014). They showed that in every case, sounds of disputed origin were nonvocal.

To date, there seem to be no cases in which a bird sound claimed to be nonvocal was later demonstrated to be vocal. Every debated instance has turned out to be nonvocal, implying that if it does not sound vocal, it probably is not. On the other hand, there are many instances in which the hypothesized mechanism was incorrect (Clark 2008; Fig. 4.3). Therefore, a logical starting point for experiments on apparently nonvocal sounds is with simple tests intended to verify the mechanistic origin of sound. For instance, the streaked tenrec produces sound with modified hair (Endo et al. 2010), and videos seem to imply the mechanism is frictional (BBC 2011). A possible starting point for a study of function would be to test what happens to sound production when one or more of the modified hairs is removed or otherwise manipulated.

4.2.3 Physical Mechanisms

There are multiple physical mechanisms by which motion generates sound, in two categories: aerodynamic mechanisms and solid mechanisms. This is perhaps an oversimplification, and this section does not provide an extensive overview of physical acoustics. Major issues, such as of impedance and other mechanisms that modulate amplitude, are ignored. Instead this section provides a brief overview of the physical acoustic mechanisms known or hypothesized to contribute to locomotion-induced sounds.

Physically speaking, sound is vibration of fluid, a longitudinal oscillation in which both fluid pressure and velocity vary at a point in space (Ewing 1989; Fletcher 1992). The magnitude of the velocity component is high in the near field, close to the source, but diminishes and is nearly negligible in the far field, away from the source. Because sound is oscillating (changing) pressure and velocity, it therefore originates from any process that results in a *change* in local fluid pressure or velocity. All accelerations and structural vibrations therefore produce sound.

4.2.4 Aerodynamic Mechanisms

Aerodynamic origins of sound involve air flowing around a solid object. Sound is generated by any flow conditions that produce a change in pressure at a point on the surface of the object. Change in pressure of a point arises as a necessary consequence of unsteady or dynamic motions, meaning the motion has a significant (nonzero) acceleration. All animal motions have a dynamic component, and any temporal part of an animal accelerates, it displaces the surrounding fluid (air or water), resulting in a change in pressure. Some of this pressure change radiates away from the animal as longitudinal pressure waves. As most animal motions are low frequency, the resulting sound is infrasound, and will often be inaudible. This simple mechanism explains the humming sound of insect wings (Sueur et al. 2005; Bae and Moon 2008) and hummingbirds; these are audible because the wingbeat frequency itself is audible. As it is a physical consequence of production of aerodynamic force, Lentink et al. (2015) have shown that infrasound produced by a flying animal may be used to measure the underlying aerodynamic forces that were generated in flight. So although these sounds are largely biologically inconsequential, measurement of flight infrasound may be a technique with further applications for study of the biomechanics of flight.

More complex aerodynamic interactions arise from generation of turbulence, an aerodynamic mechanism that applies to all flying vertebrates (Vogel 1994). Turbulence, which is random fluid motion, can be modeled as a random or semirandom spectrum (in both space and time) of vortices with varying angular velocities (frequencies) and strengths, where a single vortex is a spinning packet of fluid (Blake 1986; Vogel 1994). Turbulently moving fluid is more or less the aerodynamic near field; turbulence flowing past an ear generates low-frequency, atonal sound as a near-field effect, such as the whooshing sound of waving a hand close by one's ear. Turbulence generated by air flowing past a microphone produces the same effect and is the reason microphones often require windscreens.

Vortices have low-pressure centers, meaning that each time a vortex changes strength (forming or dissipating), there is a change in this pressure, resulting in sound (Blake 1986). A vortex dissipating away from a solid structure, such as in an animal's wake, apparently produces relatively little sound, because an isolated vortex acts as a quadrupole sound source,[1] an inefficient radiator of sound (Blake 1986). By contrast, a vortex that forms adjacent to a solid such as an animal's wing or body will act as a dipole sound source, a more efficient radiator of sound. As vortex formation is often random with most energy at low frequency, the sound of turbulence forming is atonal and low frequency (Blake 1986). These vortex formation/dissipation mechanisms are the likely origin of the whooshing sounds animals make in flight, such as the flapping sounds passerine birds produce in ordinary flight (Fournier et al. 2013).

Owls that hunt by ear have multiple wing features that change how turbulence develops on the wing, shifting the vorticity power spectrum toward lower frequencies (Kroeger et al. 1972; Sarradj et al. 2011; Geyer et al. 2013). The amplitude of vortex-induced sound rises with a high power of velocity (Lighthill 1952), so owls also reduce their acoustic signature by flying slowly, and only in fast-flying birds, such as the stoop of a peregrine falcon (*Falco peregrinus*) or aerial dive of a marbled murrelet (*Brachyramphus marmoratus*, which produces a loud jet-like sound:

[1] Dipoles and quadrupoles are models of sound sources, where a dipole is two adjacent sources of opposite phase, and a quadrupole is four adjacent sources of alternating phase; see p. 171 of Fletcher and Rossing (1998).

Nelson and Hamer 1995) are turbulence-generated whooshes audible at distances of tens of meters.

Another aerodynamic mechanism is whistling, a term sometimes mistakenly applied to locomotion-induced sound on account of the sound's tonality. There are multiple types of whistles, the simplest of which produce tonal sound through vortex formation that is not random, but is driven at a particular frequency by an *aerodynamic* feedback mechanism. Such a mechanism is the result of an aerodynamic interaction with a solid structure. The sound of the wind whistling in one's ear or at the corner of a building are examples (Blake 1986; Fletcher 1992). Whistles are often coupled to (and the acoustic frequency driven by) air-filled cavities that act as Helmholtz resonators[2] (Fletcher 1992) such as the human mouth during whistling (Fletcher and Rossing 1998). Mechanical resonance (dynamic feedback from the solid structure) is unimportant to whistles. Although most human-designed whistles are associated with rigid structures, the structure could vibrate. If it does, under the whistle model it vibrates in forced response to fluid flow, as a vortex-induced vibration. No examples of whistled nonvocal sounds are yet known in animals. Sounds produced by flying birds are sometimes called "wing whistles," but this name appears to be a functional misnomer and no bird is known to actually whistle with its wings or tail (Clark et al. 2013b). Instead, the available evidence implicates wing stiffness and structural resonance as having an important physical role in these sounds, meaning they are instead produced by a different mechanism, aeroelastic flutter.

Aeroelastic flutter, or flutter for short, is a dynamic interaction that is the result of coupling between aerodynamic forces and the structural properties of a stiff, flat object such as a feather in flowing fluid. Flutter is the mechanism by which hummingbird feathers produce sound (Clark et al. 2011). Above a critical velocity, U^*, energy from the airflow overcomes damping and the feather spontaneously oscillates at a structural resonance frequency (Clark et al. 2013a, b). Nearly all flutter described thus far has been limit-cycle (periodic) flutter, in which the feather flutters at a discrete frequency, plus harmonics. Chaotic (non-limit cycle) flutter is possible (Alben and Shelley 2008), and occasionally individual feathers tested in a wind tunnel flutter this way, although no cases of chaotic flutter in actual bird flight are yet documented (Clark et al. 2013a). The frequency and mode shape[3] of flutter are set by local flow conditions (feather orientation relative to flow, air speed, presence of neighboring feathers) and also feather resonance properties (stiffness, size, shape). Changes in any of these independent variables can produce both linear and nonlinear responses in flutter frequency. The resulting sound is tonal with strong (sometimes dominant) harmonics, with frequencies ranging from a few hundred

[2] A Helmholtz resonator is a cavity with a characteristic aerodynamic resonance determined by its geometry. For example, blowing across the top of an empty beer bottle causes it to act as a Helmholtz resonator.

[3] Mode shape is the "shape" of a resonance frequency (normal mode) of a structure, that is, the distribution of motion of all points across a structure at a given frequency, when that structure is mechanically excited in an ideal way. Airflow is not an ideal source of excitation, so technically speaking, a fluttering feather exhibits not a mode shape but an *operating deflection shape* (Richardson 1997). This subtle distinction is ignored here, following Clark et al. (2013a).

hertz to as high as 10 kHz (Clark et al. 2013c). High-pitched sounds produced by flutter can superficially resemble whistles, but flutter can also produce low-pitched and buzzing sounds akin to the flight sound of a bee (Clark et al. 2011, 2013a).

Aeroelastic flutter lends itself to interspecific acoustic diversity, as small changes of feather shape produce differences in pitch, amplitude, and harmonic structure. Flutter-induced sounds are often not especially loud, but in some cases can carry for 100 m or more, apparently due to amplification mechanisms described later in this section. The shape of the sound field of fluttering feathers has not been measured, but the nature of the feather motion implies that it is a dipole, suggesting these sounds have strong directionality (Clark et al. 2013a).

Work thus far suggests that all flight feathers may spontaneously flutter under the right aerodynamic conditions, because flutter is a passive mechanism intrinsic to flat airfoils in fast flowing fluid (Clark et al. 2011, 2013a, b). Most individual feathers, tested in a wind tunnel, have many more possible modes of flutter than tend to be expressed in the flight of birds. That flutter occurs spontaneously demonstrates why this type of sound production could evolve easily. Experiments on living snipe and snipe feathers (*Gallinago* and *Coenocorypha* spp.) have demonstrated that flutter is responsible for the winnowing sounds they produce (Reddig 1978; Miskelly 1990).

Research on the mechanics of flutter has focused on single, isolated feathers mounted in a wind tunnel, because this is a tractable experimental paradigm. However, nearly all birds that produce sound with flutter have multiple neighboring feathers, making feather–feather interactions a possibility. In Anna's hummingbird, the outer tail feather (R5) produces an approximately 4 kHz sound during a courtship display (Clark and Feo 2008). Its proximal neighbor (R4) does not produce this sound on its own, but the presence of R4 amplifies the sound produced by R5 by roughly 12 dB (Fig. 4.2). This interaction is not structural, because the effect can be produced when the neighboring feathers do not physically touch, so it thus must be an aerodynamically driven sympathetic vibration (Fig. 4.2d, e; Clark et al. 2011). It seems possible that this type of interaction is widespread in bird sonations, because amplitude (loudness) is a variable of paramount importance for communication. Another type of feather–feather interaction occurs when two neighboring feathers flutter at different frequencies, $f1$ and $f2$. If the feathers are coupled, heterodyne (sideband) interactions appear at $f1 \pm f2$. This occurs in Allen's Hummingbird (Clark et al. 2011; Clark 2014). In this species the tail feathers R3 produces $f1$ (~2 kHz), while R4 produces $f2$ (e.g., 7 kHz). When the two feathers are together, interaction frequencies of $f1 \pm f2$ (5 kHz, 9 kHz) appear as well. A third type of feather–feather interaction was demonstrated in Calliope hummingbird (*Selasphorus calliope*), in which flutter-induced collisions between neighboring fluttering feathers produce an atonal, buzzing sound (Clark 2011), discussed further in Sect. 4.2.5.

In addition to flutter, there are additional unidentified aeroacoustic mechanisms that apply to animals. The sounds produced by ruffed grouse (*Bonasa umbellus*) are made as the male beats his wings against the air with no physical contact between wings and another structure. The resulting low-frequency atonal pulses of sound carry hundreds of meters (Archibald 1974). The aeroacoustic basis of this sound has not been established. It could be simply due to direct pressure changes caused by dynamic motion of the wings; whether this mechanism alone can account for the amplitude of the sound is unclear. As an alternative, the wings might force air out of the space between the wings and body to a degree sufficient to produce additional sound (see clapping, Sect. 4.2.5).

Even less clear is the physical basis of wing whirring wing sounds produced by birds such as toadies (Todidae) and manakins in the genera *Pipra* and *Manacus* (Fig. 4.1; Bostwick and Prum 2003), for which the acoustic mechanism remains unknown. One mechanism that obviously does not apply to living vertebrates is a sonic boom, as produces the crack of a bullwhip (Bostwick and Prum 2003). This mechanism requires local velocity of some part of the animal to exceed c (~340 m s^{-1} in air), a velocity many times higher than the fastest speeds of any animal or animal appendage, although it is plausible the tails of sauropod dinosaurs could reach it (Myhrvold and Currie 1997).

The aerodynamic mechanisms described in the preceding paragraphs produce significant levels of sound only at high speeds, mostly above 10 m s^{-1} in air. A hovering or slow-flying bird or bat flapping its wings has local wingtip velocities of approximately 10 m s^{-1}, which is why all flying vertebrates produce one or more of the previously mentioned acoustic signatures in flight. Most of these aerodynamic mechanisms are unimportant for terrestrial locomotion, owing to low velocity. For terrestrial animals it is instead interactions with a solid substrate, or at the air–water interface that tend to dominate their acoustic signatures.

4.2.5 Structural Mechanisms

Air is a nearly uniform medium, meaning that the aerodynamic sound production mechanisms described in Sect. 4.2.4 are dependent largely on the animal's morphology and kinematics. By contrast, many sounds of terrestrial locomotion vary substantially with the local substrate. Everyday experience shows that the same human foot actuated under similar kinematics produces different sounds when walking on snow, leaf litter, through dry grass, or through mud, on account of differences in the physical interactions of the foot with each of these substrates.

When two solid objects collide, rub, or otherwise move relative to each other, the result is local structural deformations and vibrations. Mechanisms generating vibrations include collision (percussion), rubbing (stick-and-slip), sudden material failure (as in a stick snapping underfoot), or other physical interactions between the two structures. Structural vibrations induced by physical contact produce airborne sound, because a surface vibration of a solid structure produces an equivalent vibration in the layer of air attached to the surface. This is due to the no-slip condition, in which the layer of air that is in contact with a solid at the solid–air interface moves along with the solid; the two do not slip relative to each other (Vogel 1994).

The proximity of two solid objects can also induce forced air movement. If air becomes trapped (restricted) between two moving objects, local pressures can briefly rise greatly, resulting in a significant amount of additional sound. For instance, clapping hands causes sound not by percussion-induced vibrations of the skin, but from shockwaves associated with air forced out from between the two hands. This is easy to demonstrate by observing how the sound of clapping changes with how the hands are cupped: Cupping affects the volume of air and local geometry of how the air is forced out (Fletcher 2013).

There appears to be high mechanistic diversity of how structural sounds may be produced, and no conceptual overview of all of the ways animal motions generate structural sounds. This may be a fruitful avenue for future research and synthesis. For the remainder of this section, attention is focused on recent research on how manakins and hummingbirds produce sounds via interacting solid structures.

Male manakins (Pipridae) are lekking birds that defend small courts and perform athletic displays for females. Many species produce sonations, particularly snapping sounds but also including other mechanisms (Fig. 4.1) (Prum 1998; Bostwick and Prum 2003, 2005). Snaps are percussive, produced by abrupt, impulsive physical contact between a wing and another structure, including the other wing, the body, or between individual wing feathers within a wing (*click* sounds of *Pipra mentalis*). The sounds produced are sudden, short, broadband impulses of sound, (Bostwick and Prum 2003). Male manakins have thickened wing feather shafts and enlarged, sexually dimorphic muscles associated with these displays (Schultz et al. 2001).

Instead of a strictly percussive mechanism of sound production, the club-winged manakin (*Machaeropterus deliciosus*) uses stridulation to produce pure tones with its secondary wing feathers (Bostwick and Prum 2005; Bostwick et al. 2009, 2012). Males elevate the wings over their back, and then rapidly, repeatedly collide the medial secondary feathers of the opposing wings together. The fifth secondary wing feather (S5) is a pick, S6 is a file, and the two together produce loud tonal sound with a fundamental frequency of 1.5 kHz (Fig. 4.4).

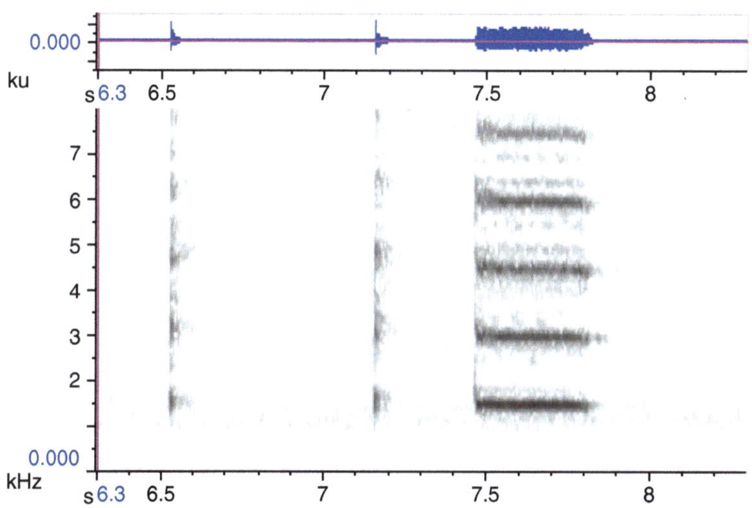

Fig. 4.4 Spectrogram and waveform of sonation of club-winged manakin (*Machaeropterus deliciosus*), produced by wing stridulation. Recording from Xeno-Canto #213391

This physical source of sound is augmented by two additional mechanisms that amplify amplitude. First, the wing feathers are stiff and, in comparison to wing feathers of other manakin species, have resonance frequencies tuned to 1.5 kHz. Neighboring feathers vibrate coherently and in phase, as indicated by their response to mechanical stimulation. This means the feathers of one wing are undamped and act as a functional unit, ringing collectively in response to the input frequency of the pick and file (Bostwick et al. 2009). Second, the ulna, the bone into which these secondary feathers insert, is robust and densely mineralized (Bostwick et al. 2012). This implies that its structural properties are also tuned for sound production. The precise role the ulna plays in sound production is unclear. It provides a solid, dense attachment point for the secondary feathers, stronger than in typical passerines. It may serve as a node (like the handle of a tuning fork), modulating the vibration of neighboring wing feathers while structurally isolating them from the rest of the bird, reducing damping of the feathers and thereby increasing the amplitude of the sound (Bostwick et al. 2012).

One final physical acoustic question not yet addressed is whether there is a phase-inverting mechanism between the wings, similar to stridulatory mechanisms in crickets (Bennet-Clark 1999). Without such a mechanism, when the two wings are struck together, they may vibrate in antiphase (a phase offset of 180°) relative to each other, causing destructive interference between the two wings, thereby reducing the amplitude of the sound. But that such a mechanism is needed is not entirely clear, as the two wings are held approximately coplanar when the bird sonates, and so might already have axes of vibration rotated 90° relative to the input impulses. One way to test this may be to examine the shape of the sound field around a sonating manakin.

Another mechanism that appears to be widespread in terrestrial vertebrates is rubbing or rustling, as in the sounds apparently produced by the modified hairs of streaked tenrec (Endo et al. 2010). Multiple physical mechanisms seem plausible explanations of such sounds, such as from stick-and-slip friction of two surfaces in contact. Feathers rubbing against feathers appear to produce *fanning* sonations in manakins in the genus *Manacus* (Bostwick and Prum 2003), and seem to be widespread in other birds, such as in adventitious sounds of wing and tail feathers rubbing against each other as a bird preens itself.

In addition to the wings and tail, the head, bill, and feet of animals also serve as percussive instruments. Woodpeckers drum with their bill adventitiously when foraging, but also seek out resonant surfaces to amplify loudness, and drum as a sonation to declare a territory (Stark et al. 1998). Ruddy ducks (*Oxyura jamaicensis*) thump their bill against their upper breast, producing an accelerating train of dull, quiet *thuds* in a close-range courtship display (Clark, pers. obs.). Many mammals drum, striking the ground with their feet or other parts of the body (Randall 2001), signals that may transmit vibrationally as well as acoustically.

A structural mechanism mentioned in Sect. 4.2.3 involves flutter-induced collisions. Male Calliope hummingbirds produce a strange buzzing sound with a dominant frequency of approximately 1 kHz, and modulated in pulses at 0.25 kHz,

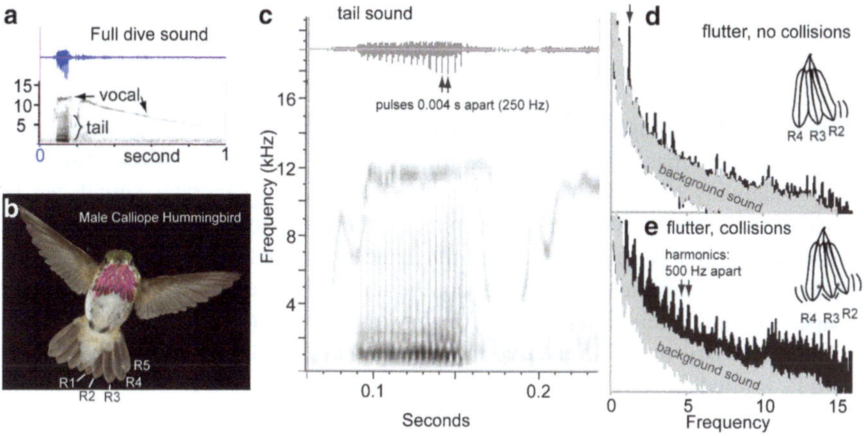

Fig. 4.5 Dive sound of male calliope hummingbird (*Selasphorus calliope*) includes sound produced by tail feathers fluttering and colliding with each other. (**a**) Dive sound includes both vocal and tail-generated elements. (**b**) Adult male calliope hummingbird with sound-producing tail feathers (rectrices, R1–R5) indicated. (**c**) Tail-generated portion of the sound recorded at 24 kHz, FFT: Hann, 50 % overlap, window size of 128 samples to emphasize the time domain. (**d, e**) Wind tunnel experiments to reproduce acoustic mechanism, in which three tail feathers were mounted adjacent to each other, in an orientation similar to that observed in the bird. *Black* is spectrum of interest, *gray* is the background noise of the wind tunnel with feathers present but not fluttering. (**d**) One or more feathers fluttered but did not collide, replicating only the approximately 1 kHz fundamental frequency of the dive sound (*arrow*), and not the acoustic energy greater than 1 kHz in the dive sound. (**e**) Feathers fluttered and collided at approximately 0.5 kHz, resulting in energy transfer to higher frequencies, replicating the acoustic structure observed in the dive sound. Power spectra of **d, e** produced in Raven 1.4 with a 65,536 FFT sample window [Modified from Clark (2011). Wing, tail, and vocal contributions to the complex acoustic signals of courting Calliope hummingbirds. *Current Zoology*, 57, 187–196. Reprinted with permission from the editorial office, *Current Zoology*]

during their courtship dive (Fig. 4.5). Unlike in relatives, removing single tail feathers had small effects on the buzzing sound, whereas removing the entire tail eliminated this sound. This implied that the source was the tail feathers as a group, rather than individual feathers (Clark 2011). Single tail feathers tested in a wind tunnel produced little sound, and had an easily elicited torsional mode at 0.5 kHz (Clark 2011; Clark et al. 2011). When sets of three tail feathers were tested together, this torsional mode seemed to facilitate collisions between neighboring feathers. These feather–feather collisions produced a buzzing sound rich in harmonics, similar to the dive sound (Fig. 4.5e) (Clark 2011).

Many other birds are reported to have multiple neighboring feathers with modified shape that may interact (Trail and Donahue 1991; Lane et al. 2007), especially guans (Delacour and Amadon 1973). Whether these other taxa also have similar, aerodynamically driven structural sound production mechanisms remains to be seen.

4.2.6 Source-Filter Theory

Closely related species may evolve acoustic differences in their sonations, as has happened in hummingbirds, snipe, or manakins (Prum 1998; Clark 2014). Because the physical mechanism that produces the sound is accessible to subtle manipulations, relatively nuanced hypotheses of how the source generates sound can be developed and tested. Many hummingbirds studied have multiple unusually shaped tail feathers, and yet experiments demonstrate only some of these feathers were necessary and sufficient for sound production. Other feathers with noteworthy shapes, generally neighbors of the necessary and sufficient feather, were not themselves necessary or sufficient. A vibrating feather may aerodynamically interact with its immediate neighbors, acting as coupled oscillators (Sect. 4.2.3), and complex interactions between individual feathers are physically plausible in most birds that produce sound via flutter. To provide a theoretical framework for understanding these complex interactions and how they evolve, Clark (2014) proposed a source-filter model of sonations that he applied to the evolution of sonations generated by the tail in hummingbirds in the genus *Selasphorus*.

The sound source is defined as the minimum set of structures both necessary and sufficient to produce quantifiable components of the sound of interest (see Sect. 4.2.2), while filters are adjacent structures to which the source is plausibly coupled, either structurally or aerodynamically. In the simplest cases, one single individual feather (or feather region) was both necessary and sufficient for sound production, making that feather the only source and a "lynchpin" for sound production (Table 1 in Clark 2014). This lynchpin model does not apply to all species: in Calliope hummingbirds, the tail feathers in aggregate are the source (Fig. 4.5).

Filters are by themselves unnecessary and/or insufficient for production of *quantifiable* components of sound, where amplitude and aspects of timbre are hard to quantify, particularly in the field. Because they by definition lack evidence of being the source, filters are always hypothetical in a particular species. As they may vibrate in forced response to a neighboring source feather to which they are coupled, and are presumably evolutionarily tuned to do so, this model predicts that filters are prone, over evolutionary time, to become sources. Moreover, filters could be spectral filters, for instance amplifying only some of the bandwidth of a source, such as a harmonic higher than the fundamental. As such, the model predicts that sonations could switch over evolutionary time from one frequency to another by hopping from one harmonic to another. This model explains patterns of evolution of mechanical sounds within the hummingbird genus *Selasphorus*. The ancestral character state is to produce a sound at approximately 1 kHz + integer harmonics by feather R2 source, and has evolved to a 2 kHz + integer harmonics sound from an R3 source in Allen's hummingbird. This could not have been the result of small, gradual changes from 1 to 2 kHz as the source shifted from R2 to R3, because Allen's R2 still has the ancestral character state, a "ghost" fundamental frequency of approximately 1 kHz, that is expressed when feather R3 is missing (Clark 2014). Therefore the simplest explanation is that the dominant frequency has "hopped" from the fundamental frequency of R2 to the second harmonic, which became the new fundamental frequency of R3, a process termed "harmonic hopping" (Kingston and Rossiter 2004; Robillard et al. 2013). This model provides a mechanism of how fundamental frequency of communication sounds can hop from one discrete frequency to another, a topic discussed further in Sect. 4.3.4.

This source-filter model should not be confused with the source-filter model of vocalizations (see Taylor et al., Chap. 8). As discussed in Sect. 4.4.1, animals seem to have greater control over the acoustic form of vocalizations than over nonvocal sounds. In the source-filter model of vocalizations, source and filter are also coupled, but animals control mechanical properties of the filter independently of the source, resulting in much of the diversity of the acoustic form of vocalizations. The source-filter model of sonations, by contrast, invokes no evidence of independent control of the filter. Rather, the model is a tool to understand how mechanical sounds evolve, in the context of experiments that show that a particular structure is either not necessary or insufficient for sound production in a particular species, yet phylogenetic evidence implies that such structures have played a role in sound production in the past and in sister taxa.

4.3 Function and Behavioral Significance

Both sonations and adventitious sounds play important roles in the biology of locomotion. Locomotion-induced sounds may alert individuals to the presence of an animal and its motions (Randall 2001). In addition to their role in communication, these sounds also play roles in predator–prey interactions. Owls can successfully capture prey in total darkness, using only the adventitious sounds of locomotion of prey (Konishi 1973). One widespread response of prey to a potential predator is to freeze, thereby ceasing production of locomotion-generated sound. Similarly, one hypothesis for the silent flight of owls is that reduction of self-noise masks from the prey the sound of the owl's approach (Konishi 1973). This section hereafter ignores predator–prey interactions and acoustic stealth and focuses on how sonations attain communication function.

Sonations appear to serve all of the same communication functions that vocalizations do, including as alarms for conspecifics, such as rodent thumping (Randall 2001) or dove flight sounds (Hingee and Magrath 2009); as aposomatic warnings (rattlesnake rattles), in territorial advertisement in place of undirected vocal song (ruffed grouse, broadbills in the genus *Smithornis*), and during displays directed toward conspecifics, as in manakins or hummingbirds. Vocal morphology seems to have arisen relatively few times in ancient lineages, providing a limited phylogenetic sample size with which to seek inferences as to why vocal acoustic communication originally evolved. By contrast, many sonations have evolved recently, yielding at least hundreds of independent phylogenetic origins of these behaviors. In bird sonations produced with the wings and tail, the majority of these types of sonations arise as sexual behaviors produced predominantly by one sex (usually males), either in place of vocal song, or during courtship displays directed at females (Clark, pers. obs.).

4.3.1 Kinematics Revisited: Animal Behavior

Section 4.2.1 explored the relationship between animal kinematics and sound from an experimental point of view, including discussion of whether these sounds are voluntary. This section turns to how the relationship between sound and motion influences the potential information content that nonvocal behaviors contain. The direct correspondence between motion and locomotion-induced sound (Sect. 4.2.1) has a key implication for animal behavior: Locomotion-induced sounds intrinsically contain information about the animal and its movement. The sound is a physical record of the behavior. Locomotion-induced sounds are not arbitrary in form, as vocal songs (especially learned songs) may be; they are physically constrained by the motions that produce them.

Timing variables seem likely to be especially effectively encoded acoustically. Perhaps they are easier for a receiver to evaluate accurately and precisely than even visual observation of the same behavior. Sound is an acoustic record of repeated motions, particularly for events measured in the time domain of a spectrogram. For instance, the wingbeat frequencies of displaying hummingbirds, manakins, or flappet larks (*Mirafra cinnamomea*) are nearly doubled during production of sonations (Norberg 1991; Bostwick and Prum 2003; Feo and Clark 2010). This change in frequency is easily heard or measured in a spectrogram (Fig. 4.6a, b). Although it has not yet been demonstrated that females pay attention to the wingbeat frequency of courting males in species such as these, this seems likely. Playback experiments show that hummingbirds respond to and use the wing sounds to identify the species/sex of other individuals (Hunter and Picman 2005; Hunter 2008).

In crested pigeons, Hingee and Magrath (2009) used playback experiments to demonstrate that the wing sound of a pigeon taking off normally did not elicit a response from a flock of pigeons (Fig. 4.6a), but playing back the wing sounds of a

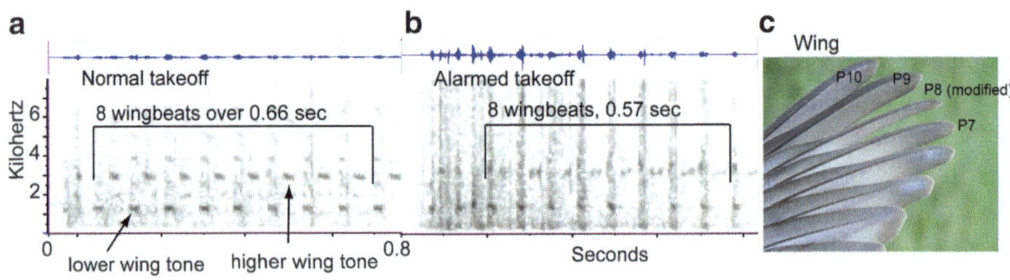

Fig. 4.6 Wing sounds produced by crested pigeon (*Ocyphaps lophotes*) during normal takeoff, alarmed takeoff, and wing showing modified primary feather P8. (**a**) During flight, this species produces two temporally offset sounds, lower and higher wing tone. The lower tone is likely produced during the downstroke. In normal takeoff this bird had a wingbeat frequency of 12.1 Hz (8/0.66), whereas in alarmed takeoff (**b**) the wingbeat frequency was 14 Hz (8/0.57). Hingee and Magrath (2009) demonstrated that pigeons flee in alarm to this higher wingbeat frequency, and not the lower frequency of ordinary takeoff. (**c**) The wing has a modified feather, P8, which is narrowed and stiffened, apparently for sound production of some or all of the wing sound. Sounds courtesy of Trevor Murray, photo of specimen 5653 from the Australian National Wildlife Collection, courtesy Robert Magrath. FFT: 512 samples; Hann, 50 % overlap (44 kHz)

bird taking off in alarm, with wingbeat frequency at a slightly higher tempo (Fig. 4.6b), caused the whole flock to take off in response. Further playback experiments demonstrated that this effect was not attributable to differences in the sound's amplitude. This result implies that the birds responded to the increase in wingbeat frequency. This simple experiment shows crested pigeons listen to each other and that the wing sounds serve as an alarm signal. Because the sound pulses are produced directly by the wingbeat, they are not susceptible to cheating, unlike vocalizations, although if they are under voluntary control (Sect. 4.2.1.1) the animal might chose not to produce the sound.

Beyond time-domain variables, additional information may be encoded by a sonation, such as sound features often measured in the frequency domain. What information frequency-domain variables contain depends on the particular physical mechanism producing the sound. In the case of flutter-induced sounds generated by feathers (e.g., hummingbird tail feathers), the pitch in many species encodes information about the signaler's flight velocity. In many feathers, amplitude and pitch of flutter-induced sound both change systematically with speed (Clark et al. 2011). But as the slopes of these relationships varied substantially among species tested (Clark et al. 2013b, c), data from one species cannot be readily extrapolated to another.

4.3.2 *Relationship with Displays*

Darwin (1871) was the first to observe that vigorous courtship displays may lend themselves to sonations, stating:

> We have seen that some birds during courtship flutter, shake or rattle their unmodified feathers together; and if the females were led to select the best performers, the males which possessed the strongest or thickest or most attenuated feathers, situated on any part of the body, would be the most successful; and thus by slow degrees the feathers might be modified to any extent. The females... would not notice the slight successive alternation in shape, but only the sounds thus produced. (Chap. XIII, p. 67)

Two factors may promote sonations arising during displays rather than other social contexts. First, displays are often vigorous, involving rapid appendage movements, and thereby tend to incidentally produce more adventitious sound than more placid movements. Second, if receivers are attuned to subtle differences in display performance, it may be relatively easy for female preferences to switch to male attributes that contain information about these differences. There is not yet evidence for which of these two factors may play a larger role in driving the evolution of new acoustic communication systems.

The coupling of sound to motion means that information about locomotion itself is intrinsically contained in locomotion-induced sound. This makes sonations a likely sensory modality for direct assessment of a locomotor performance of another individual. In courtship displays, females might evaluate subtle differences between males against poorly understood performance criteria. Aspects of these performance criteria, whatever they are, may be more efficiently transmitted or perceived acoustically than visually. For instance, suppose female auditory systems encode temporal frequency more precisely than visual systems (Sect. 4.3.1): If so, attention to acoustic characters of displays may allow finer discrimination between male performances than would visual observation of the same display. This is true regardless of the adaptive value of female preferences, that is, regardless of whether one believes females derive useful information from "honest" signaling on the part of the male, or whether female preferences are arbitrary (Prum 2010).

4.3.3 Relationship with Vocalizations

There is limited evidence in birds that vocalizations and sonations trade off, with sonations apparently replacing vocalizations. In *Pseudocolypteryx* flycatchers, Bostwick and Zyskowski (2001) document closely related taxa, some of which produce a regular vocalization, others of which have replaced the sound with a bill snap. Ruffed grouse produce sonations that are the functional equivalent of vocal song, and also lack vocal songs, as do *Smithornis* broadbills (Clark, pers. obs.). These taxa have reduced syringes (Prum 1993) and reduced vocal capacity (Rusch et al. 2000). Species such as vultures or mute swan (*Cygnus olor*) are relatively nonvocal and produce distinctive flight sounds, but it has not been established whether their distinctive sounds are sonations that have replaced a vocal function, or are simply adventitious. It also has not been established whether the examples just provided are rare exceptions or the general rule, and this could be tested phylogenetically. As evidence against this tradeoff hypothesis, a number of taxa that produce sonations are also highly vocal, such as Anna's hummingbird, or humans. Perhaps vocalizations and sonations trade off only in specific evolutionary contexts.

Another pattern observed in hummingbirds is similarity between phonations and sonations. Costa's and Anna's hummingbirds both have strong intraspecific similarities between their vocal songs and nonvocal dive sounds, but larger interspecific differences (Clark and Feo 2010). The most parsimonious phylogenetic reconstruction has in their ancestor a single similar song and dive sound, which have each diverged in the daughter taxa. What selective force would maintain two mechanisms of sound production that yield similar sounding sounds? Clark and Feo (2010) propose that a similar female preference selects for similar acoustic form via "sexual sensory bias," but a key prediction of this hypothesis, that females do prefer the similar components of sound in each display, has not yet been tested.

No other published examples of self-imitation between vocalizations and mechanical sounds are yet known, but additional instances could be easily overlooked. Baptista and Matsui (1979) concluded that the dive sounds of Anna's and Costa's hummingbirds were vocal because they sounded just like the obviously vocal song (Sect. 4.2.2). This argument is normally a reasonable one. If an animal produces two similar sounding sounds, the null hypothesis is that they are simply serial repetitions of the same signal. It takes specific, positive evidence to the contrary to conclude that two similar sounds in fact constitute different signals (Clark and Feo 2010).

4.3.4 Evolution

Communication sounds arise out of adventitious sounds that become salient to receivers (Darwin 1871; Ewing 1989). Once signaler–receiver coevolution begins, the signal may become modified in form, for instance, becoming louder or changing in pitch. It seems that many sonations are quiet, and perhaps there are fewer options for evolution to modify external animal morphology to add resonators or other mechanisms that amplify acoustic amplitude, than are available for vocal mechanisms. What is clear is that, just as Darwin (1871) hypothesized, subtle changes in morphology can have significant impacts on acoustic form (Clark et al. 2011). In the case of feathers and aeroelastic flutter, these sounds may evolve in either a linear or nonlinear fashion.

It is easy to imagine the linear case: over evolutionary time, a slight, gradual change in morphology of a resonator changes stiffness, and that produces a slight, gradual change in pitch. For instance, a slightly stiffer feather produces a slightly higher pitch. However, pitch may also evolve nonlinearly, jumping abruptly from one frequency to another. One example of how this can occur was described in Sect. 4.2.6. There is a second way, and it occurs because three dimensional resonant structures have multiple resonance frequencies that are not integer multiples. Changes in how the structure is excited can cause the system to cross a threshold, transferring excitation from one resonance frequency to another (Clark et al. 2012). The result is that, when such a threshold is crossed, the system may "jump" from one vibration frequency to a much different one. The difference between this mechanism and the one described in Sect. 4.2.6, is that this mechanism concerns frequency jumps *within* a single feather, whereas Sect. 4.2.6 described a switch *between* coupled adjacent feathers. It is trivial to make feathers fluttering in a wind tunnel (under artificial conditions) jump from one mode of vibration to another. A natural example is shown in Fig. 4.2c, of an Anna's hummingbird dive sound. The dive sound contains a prominent approximately 1.2 kHz sound that is produced by a tip mode of vibration, just before and after the typical approximately 4 kHz trailing vane mode (Clark et al. 2013a). Evolutionary changes in morphology may produce both linear and nonlinear changes in the types of sound produced. And to extend the analogy raised in Sect. 4.2, these constitute changes in the instrument itself.

The other major way sonations may vary among related taxa is behavioral, in "how the instrument is played." Manakins produce either single snaps, or "roll-snaps" that consist of a rapid-fire sequence of snaps. In either display, the physical acoustic mechanism is the same, differences in acoustic form arise from behavior. Woodpeckers (Stark et al. 1998) and various mammals (Randall 2001) drum using the same percussive mechanism, but behaviorally produce species-specific sequences. Hummingbirds produce a diverse array of sounds with their wings and tail, where evolutionary diversity in the form of the sound arises from changes both in morphology and in behavior; tail feather shape changes the pitch, amplitude, and harmonic structure of sound, while behaviorally the birds vary in the number and duration of pulses of sound that are produced by spreading the tail (Clark and Feo

2010; Clark et al. 2011). Whether sonations evolve more under the action of changes in the instrument (morphology), or changes in playing style (behavior), remains to be tested.

4.4 Limitations and Advantages

Sonations seem to evolve repeatedly and recently in comparison to the ancient origins of vocalizations deep in the vertebrate phylogeny. Though sonations may be produced by any part of the animal, and are therefore arguably morphologically more diverse than vocalizations, it seems they are acoustically less diverse than vocalizations. This is apparently due to the inflexible nature of the sound production mechanisms underlying many sonations. A key innovation of vocalizations, poorly replicated by sonations, is the ability for the acoustic form to be dynamically and actively modulated by the animal, both within the source (larynx/syrinx) and filter (vocal tract) (Taylor et al., Chap. 8). The inflexibility of sonations is also a potential advantage in some contexts, as it allows sonations to contain information about locomotion and performance that is not intrinsic to vocalizations, which have greater potential to be arbitrary in acoustic form.

4.4.1 Physical Limitations

A feature of birdsong is biphonation, two-voiced song, via production of two independent sounds on each side of the syrinx. By doubling the number of sounds that can be produced at any moment, biphonation greatly increases the diversity and complexity of bird vocalizations relative to other animals (Riede et al. 2006; Zeigler and Marler 2012). In this respect, sonations could be even more diverse, because the number of simultaneous sounds an animal can make is limited only by the number of appendages it has. For instance, a human acting as a "one-man band" can make several sounds at once, including clapping hands, stomping feet, and head motions. Allen's hummingbirds trisonate (produce three nonvocal sounds at once) during their dive, simultaneously producing one sound with the wings and another two with the tail (Clark et al. 2011). A number of other hummingbirds produce sounds with both wings and tail during display (Feo and Clark 2010), or coordinate vocal and nonvocal sounds. However, even though there are more possible sonations than phonations, claims such as that Allen's hummingbirds trisonate (Clark et al. 2011) are a bit superficial. The reason is that locomotion-induced sounds remain less acoustically diverse than vocalizations because they are not as subject to modulation as are vocalizations.

Sonations are less subject to modulation than vocalizations in two respects: Acoustically they are relatively fixed in form, and behaviorally there are fewer options available for them to be voluntary, produced independently of other behaviors such as locomotion. Intrinsic aspects of the sound production mechanisms (Sect. 4.2) limit the acoustic form of sonations. For instance, consider frequency. Many sonations, such as a hummingbird's wing trill, contain two frequencies—a higher frequency set by a mechanism such as flutter of individual wing feathers (usually measured in the frequency domain), and the lower pulse rate (usually measured in the time domain) set by the oscillation of the wing. One pulse of sound is produced by each flap of the wing. Both the higher frequency and the lower pulse rate are relatively fixed; the pulse rate is fixed by the muscular and other mechanical limitations of the limb, similar to how vocal trill frequencies are limited by the vocal tract, such as in Darwin's finches (Podos 2001; Podos et al. 2004). The higher frequency is limited by the mechanics of flutter, and in most cases flutter of a particular feather produces only a narrow frequency range, often varying by less than 20 %. As a result of these constraints, wing trills of any particular species do not and likely cannot express the same frequency range observed in the trilled vocalizations of passerine birds.

Some frequency modulation is possible. The highest known *potential* frequency modulation of a sonation is from *Chaetocercus* woodstars. In a wind tunnel, their tail feathers can vary in frequency by roughly 50 %, through changing the orientation of a tail feather relative to airflow (Clark et al. 2011, 2013b). The degree to which birds take advantage of this entire range of motion is unclear. While *Chaetocercus* tail feathers can change pitch through changes in feather orientation relative to airflow, most taxa modulate these sounds via changing flight speed during a dive such as in Costa's hummingbird (Clark and Feo 2010) or snipe (Reddig 1978). This means they are frequency modulated by the animal only with physical difficulty, and in only one behavioral context, diving. Greater sage grouse produces an FM swish during its display as the wings are brushed against stiff breast feathers (Koch et al. 2015). By comparison, vocalizations are acoustically more diverse because animals change pitch of vocalizations rapidly and easily via either vocal source or filter (Düring and Elemans, Chap. 5; Taylor et al., Chap. 8).

Beyond frequency, two additional acoustic parameters that could pose limitations on the form of sonations are intensity (amplitude) and directionality, though rigorous data are scarce. The only amplitude data are from ruffed grouse, with sound pressure levels (SPLs) at a reference distance of 1 m of 64–70 dB (Garcia et al. 2012b). Many of the best-known sonations (snipe winnowing, Anna's hummingbird tail sound, manakin wing snaps) can carry for 100 m or more under natural ambient conditions; lyre-tailed honeyguide (*Melichneutes robustus*) tail sounds and ruffed grouse wing sounds are audible several hundred meters away (Friedmann 1955). These sonations approach or exceed (Clark and Feo 2008) the loudness of similar vertebrate vocalizations—but these examples may also be exceptions, studied or well-described specifically because of how loud they are. Ordinary adventitious sounds are often quiet, and the problem of diagnosing sonations versus adventitious sounds (Sect. 4.2.1) renders problematic any assertion about the average amplitude of sonations, owing to intrinsically arbitrary decisions about which quiet sounds to include in such an analysis.

Acoustic directionality of sonations has been studied only in ruffed grouse (Archibald 1974; Garcia et al. 2012b). Garcia et al. (2012b) found that acoustic radiation from ruffed grouse is similar to a dipole, with nearly equal SPL levels directly in front of and directly behind the drumming bird, but significantly lower SPL to the sides. Because sonations are produced external to the animal, sonating animals may have fewer physical avenues available for beaming the sound in a single direction, as compared to vocalizations. Several of the mechanisms outlined in Sects. 4.2.3 and 4.2.4 seem likely to be inherently dipole like.

4.4.2 Behavioral Limitations

In addition to acoustic limitations, sonations may also be behaviorally limited. The animal must perform a discrete, obvious behavior to produce the sound. That this is a limitation is situational; this same feature is also an advantage (see Sect. 4.4.3) in different circumstances. Birds that broadcast undirected sonations in place of vocal song, for instance, tend to perch in one place and then either jump, or flap the wings vigorously, to produce sound. If visual crypsis is important, such as to avoid predators, a sonation will be at a relative disadvantage to a vocalization. Further, the sounds can be difficult for the animal to modulate strategically. Hingee and Magrath (2009) demonstrated that crested pigeons use the wing sounds of other individuals as an alarm signal, because, when alarmed, pigeons flap their wings at an audibly higher rate. Unlike an alarm vocalization, it is not possible for a crested pigeon to strategically signal alarm without taking flight; and they may also not have the capacity to strategically choose to flee silently, if this were advantageous.

Finally, a limit sometimes invoked is the "energetic cost" of vigorous displays that accompany some types of sonations (Clark 2012). The performance of a behavior that produces a sonation could result in energy costs that exceed the costs of a similar vocalization. But because energy costs scale with the time duration spent performing the behavior (Clark 2012), most if not all individual sonations cost only small to trivial amounts of energy relative to an animal's daily energy budget. For instance, Barske et al. (2014) placed heart rate monitors on golden-collared manakins (*Manacus vitellinus*) to document heart rate during display, which includes production of wingsnap and rollsnap sonations. Though heart rate during display was high, display only occupied approximately 5 min/day and accounted for only 1.2 % of the daily energy budget. Animals that produce sonations repeatedly throughout the day (such as ruffed grouse) may entail nontrivial energy costs, if the total amount of time spent performing the behavior is significant. As ruffed grouse produce sonations when otherwise sitting still, it might be possible to examine directly how much energy production of sonations cost. But for many species, such as manakins, it may be difficult to differentiate between calories expended producing a sonation, versus other vigorous components of display, because the two are performed at the same time. Regardless, there is no evidence or theoretical reason to believe that sonations consume any more than a trivial amount of energy, because most sonations are produced infrequently (Clark 2012).

4.4.3 Advantages

This 1:1 link between motions and sound, described in Sect. 4.4.1 as a limitation, is also an advantage in other ecological circumstances. As described in Sect. 4.3.1, the intrinsic link between kinematics and sound makes sonations an acoustic record of an animal's performance of a behavior. In some communication contexts such as courtship displays, it is thought that the quality of a performance is important to females (Byers et al. 2010). The 1:1 correspondence between motion and sound means that sonations are intrinsically suited to be index signals of performance quality because they are an acoustic record of the performance itself. For instance, a wing trill encodes wingbeat frequency (Sect. 4.3.1). Sound may be the better sensory modality for a female to evaluate locomotor variables such as frequency. To a human, the increase in wingbeat frequency of displaying hummingbirds is difficult to discern visually, as their normal wingbeat frequencies are above the human flicker–fusion frequency. Yet increases in frequency associated with displays are easy to detect by ear. Acoustic evaluation of displays may afford females the best ability to make subtle discriminations between potential mates, a hypothesis that has not yet been evaluated for any sonating species.

Another advantage of locomotion-induced sounds is experimental: sound of a behavior can be easier to record than video, so locomotion-induced sounds may be useful for studying locomotor performance and behavior. As described in Sect. 4.2.2, the sound production morphology is external to the animal, making experimental manipulations that affect the sound production mechanism especially feasible in some animals. Such experimental manipulations allow experimental tests of acoustic function, in ecologically relevant contexts, in ways largely unavailable to students of vocalizations.

4.5 Summary

Mechanisms of locomotion-induced sounds are diverse (Bostwick 2006). All locomotion produces sound via several possible physical acoustic mechanisms. Therefore locomotion-induced sound is an intrinsic component of virtually all animal behavior, though some locomotion produces so little sound as to be effectively silent. Locomotion-induced sounds tend to be loudest during rapid or vigorous behaviors, such as displays. These sounds can evolve from adventitious sounds that are an incidental byproduct of motion into communication signals called sonations. Because locomotion-induced sound is ubiquitous, sonations have evolved independently out of adventitious sounds repeatedly, particularly during displays (Fig. 4.1).

Mechanistically, production of sonations can be examined at the level of animal motions (kinematics) and the physical acoustic mechanism that generates sound. Air is a nearly uniform medium, so the acoustic form of flight sounds is dependent largely on an animal's morphology and kinematics. By contrast, many terrestrial locomotion-induced sounds (such as footsteps) come from interactions between the animal and a substrate, meaning that substrate variability produces an array of possible sounds that could be produced by a particular morphology and kinematics. Physical acoustic mechanisms of locomotion-induced sounds are diverse, taxon specific, and poorly described, and can be divided into aerodynamic and structural mechanisms.

There is a direct correspondence between locomotion-induced sounds and an animal's motions. This correspondence has two implications. The first is experimental: Sound recordings of a behavior are a perhaps underutilized source of data about an animal locomotion, and sound recordings are easier to obtain than video in some contexts. With sufficient prior information, a sound recording yields information about locomotion. However, in diagnosing the relationship between kinematics and sound, beware of spurious correlations, when an obvious motion is correlated with but does not cause the sound (Fig. 4.3). The second implication is behavioral: The correspondence between behavior and sound means that the locomotion-induced sound contains information of potential use to other animals, especially by encoding frequencies, such as that of the wingbeat. Therefore, sonations are intrinsically suited to contain information about performance ability, suggesting the hypothesis that animals use sonations to evaluate locomotor performance.

It can be difficult to diagnose which locomotion-induced sounds are sonations and which are adventitious. The simplest diagnosis of a sonation is provided by morphology evolved specifically for sound production, such as the rattle of a rattlesnake. The second simplest diagnosis comes from kinematics specialized for sound production (rattlesnakes rattle their tail). When the animal lacks obviously specialized morphology or behavior, then the criterion is whether the sound is produced voluntarily. Voluntariness, the degree to which an animal modulates locomotion-induced sound independent of other behaviors, can be difficult to assess. Locomotion-induced sounds may be produced only in certain behavioral contexts, but this is not proof of voluntariness. The hidden switch hypothesis states that the sound is voluntary and is controlled via a switch that is subtle and not easily detected ("hidden" to casual observation); the null is that the sound is not voluntary. Support for the hidden switch hypothesis entails positively identifying the switch (which may be kinematic or morphological). This will often be hard to do, and so many potential sonations will be difficult to diagnose as such.

4.5.1 Avenues for Future Research

Decades of research on how vertebrates vocalize have chipped away at questions of vocal function and vocal mechanism. In comparison, sonations seem disproportionately understudied. The physical acoustics of sonations remains wide open for future research; surely not all possible mechanisms have been identified in Sects. 4.2.4 and 4.2.5, and some of the mechanisms qualitatively sketched in this review are conjectural and warrant further empirical validation. At the same time, as some of the physical acoustic mechanisms seem likely to be specific to individual species or small clades, rather than broad or general, examination of physical acoustic mechanisms for their own sake may not be especially conceptually exciting.

Rather, physical acoustic mechanisms of sonations should be further studied because they have behavioral, ecological, or evolutionary implications. Understanding the physical mode of sound production reveals, for instance, that sound pitch can evolve both linearly and nonlinearly (Sects. 4.2.6 and 4.3.4). Understanding physical mode of sound production will reveal design constraints, which in turn provides a rigorous context for studies on the behavioral, ecological, or evolutionary relevance of acoustic communication. Sonations offer opportunities to examine certain types of questions with simpler experimental methods than are available for vocalizations. The physical mechanisms of sound production can be easy to manipulate experimentally on wild animals (e.g. bird feathers) and do not require surgery. Therefore, large sample sizes and subtle manipulations are available. Natural individual variation is present in sonations just as it is in vocalizations (Garcia et al. 2012a), and in some instances this variation can be extensive, such as when sound-producing feathers are damaged or have a species-atypical shape (Miskelly 2005; Clark 2011; Clark et al. 2013c). As the sound production mechanism is external, it may be easy to quantify how natural morphological differences in the production mechanism (e.g., caused by size, age, sex, wear, or other aspects of condition, such as parasite damage) contribute to variation in the signal, research that is far harder to do noninvasively on vocal morphology.

Another avenue of future research is the relationship between sound and display performance, because many sonations are produced during displays (in birds). There is a direct intrinsic link between sonation form and locomotor performance. Although this intrinsic link constrains acoustic form and makes sonations nonversatile, unlike vocalizations, it also means that sonations have the potential to be intrinsic index signals of locomotor capacity. This is particularly likely because acoustic sensory systems are intrinsically better tuned to measure display attributes such as frequencies, than are visual systems. Therefore, it appears that examining how females use sonations to assess male locomotor performance (Byers et al. 2010) is wide open to future study. That said, the "honesty" of courtship signals remains a debated topic (Prum 2010), and the idea that sonations are indicators of locomotor performance is an adaptive hypothesis to be tested, not assumed.

Finally, the phylogenetic diversity of sonations implies they have evolved hundreds of times independently. This phylogenetic diversity provides a large sample size of independent origins for analyses that explore how communication systems initially evolve, and then diversify. By contrast, vocalization, despite their preeminent place in acoustic biology of vertebrates, have evolved so few times independently that their early origins and context may never be understood fully. As ever-larger phylogenies become available, evolutionary hypotheses can be tested more robustly. The repeated origins of sonations out of adventitious sounds offer the opportunity to examine why and how acoustic communication evolves.

References

Alben, S., & Shelley, M. J. (2008). Flapping states of a flag in an invicid fluid: Bistability and the transition to chaos. *Physical Review Letters, 100*, 074301.

Aldrich, E. C. (1939). *Natural history of the Allen hummingbird (Selasphorus alleni)*. MA thesis, University of California, Berkeley, CA.

Archibald, H. L. (1974). Directional differences in the sound intensity of Ruffed Grouse drumming. *Auk, 91*, 517–521.

Au, W. W. L., & Suthers, R. A. (2014). Production of biosonar signals: Structure and form. *Springer Handbook of Auditory Research, 51*, 61–105.

Bae, Y., & Moon, Y. J. (2008). Aerodynamic sound generation of flapping wing. *Journal of the Acoustical Society of America, 124*, 72–81.

Baptista, L. F. (2001). The song and dance of hummingbirds. *California Wild, 54*, 14–20.

Baptista, L. F., & Matsui, M. (1979). The source of the dive-noise of the Anna's hummingbird. *Condor, 81*, 87–89.

Barlow, C. (1967). Ethological units of behavior. In D. Ingle (Ed.), *The central nervous system and fish behavior* (pp. 217–232). Chicago: University of Chicago Press.

Barrera, J. P., Chong, L., Judy, K. N., & Blumstein, D. T. (2011). Reliability of public information: Predators provide more information about risk than conspecifics. *Animal Behaviour, 81*, 779–787.

Barske, J., Fusani, L., Wikelski, M., Feng, N. Y., Santos, M., & Schlinger, B. A. (2014). Energetics of the acrobatic courtship in male golden-collared manakins (*Manacus vitellinus*). *Proceedings of the Royal Society B: Biological Sciences, 281*, 20132482. doi:10.1098/rspb.2013.2482.

BBC. (2011). Bizarre mammals filmed calling using their quills. Retrieved June 27, 2014, from http://news.bbc.co.uk/earth/hi/earth_news/newsid_9392000/9392070.stm.

Bennet-Clark, H. C. (1999). Resonators in insect sound production: How insects produce loud pure-tone songs. *Journal of Experimental Biology, 202*, 3347–3357.

Bhatt, M., & Bardolia, H. (2006). *Call of Indian birds*, Vol. 1 (CD). Nature Club Surat.

Blake, W. K. (1986). *Mechanics of flow-induced sound and vibration* (Vol. 1). Orlando, FL: Academic.

Bostwick, K. S. (2006). Mechanisms of feather sonation in Aves: Unanticipated levels of diversity. *Acta Zoologica Sinica, 52S*, 68–71.

Bostwick, K., Elias, D. O., Mason, A. C., & Montealegre-Z, F. (2009). Resonating feathers produce courtship song. *Proceedings of the Royal Society B: Biological Sciences, 227*, 835–841.

Bostwick, K. S., & Prum, R. O. (2003). High-speed video analysis of wing-snapping in two manakin clades (Pipridae: Aves). *Journal of Experimental Biology, 206*, 3693–3706.

Bostwick, K. S., & Prum, R. O. (2005). Courting bird sings with stridulating wing feathers. *Science, 309*, 736.

Bostwick, K., Riccio, M. L., & Humphries, J. M. (2012). Massive, solidified bone in the wing of a volant courting bird. *Biology Letters, 8*, 760–763.

Bostwick, K., & Zyskowski, K. (2001). Mechanical sounds and sexual dimorphism in the crested doradito. *Condor, 103*, 861–865.

Bro-Jørgensen, J., & Dabelsteen, T. (2008). Knee-clicks and visual traits indicate fighting ability in eland antelopes, multiple messages and back-up signals. *BMC Biology, 6*, 47.

Byers, J., Hebets, E. A., & Podos, J. (2010). Female mate choice based upon male motor performance. *Animal Behaviour, 79*, 771–778.

Clark, C. J. (2008). Fluttering wing feathers produce the flight sounds of male streamertail hummingbirds. *Biology Letters, 4*, 341–344.

Clark, C. J. (2009). Courtship dives of Anna's hummingbird offer insights into flight performance limits. *Proceedings of the Royal Society B: Biological Sciences, 276*, 3047–3052.

Clark, C. J. (2011). Wing, tail, and vocal contributions to the complex signals of a courting Calliope hummingbird. *Current Zoology, 57*, 187–196.

Clark, C. J. (2012). The role of power versus energy in courtship: What is the 'energetic cost' of a courtship display? *Animal Behaviour, 84*, 269–277.
Clark, C. J. (2014). Harmonic hopping, and both punctuated and gradual evolution of acoustic characters in *Selasphorus* hummingbird tail feathers. *Plos One, 9*, e93829.
Clark, C. J., Elias, D. O., Girard, M. B., & Prum, R. O. (2013a). Structural resonance and mode of flutter of hummingbird tail feathers. *Journal of Experimental Biology, 216*, 3404–3413.
Clark, C. J., Elias, D. O., & Prum, R. O. (2011). Aeroelastic flutter produces hummingbird feather songs. *Science, 333*, 1430–1433.
Clark, C. J., Elias, D. O., & Prum, R. O. (2013b). Hummingbird feather sounds are produced by aeroelastic flutter, not vortex-induced vibration. *Journal of Experimental Biology, 216*, 3395–3403.
Clark, C. J., & Feo, T. J. (2008). The Anna's hummingbird chirps with its tail: A new mechanism of sonation in birds. *Proceedings of the Royal Society of London B: Biological Sciences, 275*, 955–962.
Clark, C. J., & Feo, T. J. (2010). Why do *Calypte* hummingbirds "sing" with both their tail and their syrinx? An apparent example of sexual sensory bias. *American Naturalist, 175*(1), 27–37.
Clark, C. J., Feo, T. J., & Bryan, K. B. (2012). Courtship displays and sonations of a male broad-tailed × black-chinned hummingbird hybrid. *Condor, 114*, 329–340.
Clark, C. J., Feo, T. J., & van Dongen, W. (2013c). Sounds and courtship displays of the Peruvian sheartail, Chilean woodstar, oasis hummingbird, and a hybrid male Peruvian sheartail × Chilean woodstar. *Condor, 115*, 560–577.
Clark, C. J., & Mitchell, D. E. (2013). Allen's hummingbird (*Selasphorus sasin*). *Birds of North America Online*. http://bna.birds.cornell.edu/bna/species/501/articles/introduction
Darwin, C. (1871). *The descent of man, and selection in relation to sex*. Princeton, NJ: Princeton University Press.
Delacour, J., & Amadon, D. (1973). *Curassows and related birds* (2nd ed.). New York: The American Museum of Natural History.
Endo, H., Koyabu, D., Kimura, J., Rakotondraparany, F., Matsui, A., Yonezawa, T., et al. (2010). A quill vibrating mechanism for a sounding apparatus in the streaked tenrec (*Hemicentetes semispinosus*). *Zoological Science, 27*(5), 427–432.
Ewing, A. W. (1989). *Arthropod bioacoustics neurobiology and behaviour*. Ithaca, NY: Cornell University Press.
Feo, T. J., & Clark, C. J. (2010). The displays and sonations of the black-chinned hummingbird (Trochilidae: *Archilochus alexandri*). *Auk, 127*, 787–796.
Fletcher, N. H. (1992). *Acoustic systems in biology*. New York: Oxford University Press.
Fletcher, N. (2013). Shock waves and the sound of hand-clap—a simple model. *Acoustics Australia, 41*, 165–168.
Fletcher, N. H., & Rossing, T. D. (1998). *The physics of musical instruments*. New York: Springer.
Fournier, J. P., Dawson, J. W., Mikhail, A., & Yack, J. E. (2013). If a bird flies in the forest, does an insect hear it? *Biology Letters, 9*, 20130319. doi:10.1098/rsbl.2013.0319.
Friedmann, H. (1955). The honey-guides. *United States National Museum Bulletin, 208*, 1–292.
Garcia, M., Charrier, I., Rendall, D., & Iwaniuk, A. N. (2012a). Temporal and spectral analyses reveal individual variation in a non-vocal acoustic display: The drumming display of the ruffed grouse (*Bonasa umbellus*, L.). *Ethology, 118*, 292–301.
Garcia, M., Charrier, I., & Iwaniuk, A. N. (2012b). Directionality of the drumming display of the ruffed grouse. *Condor, 114*, 500–506.
Geyer, T., Sarradj, E., & Fritzsche, C. (2013). Silent owl flight: Comparative acoustic wind tunnel measurements on prepared wings. *Acta Acustica United with Acustica, 99*, 139–153.
Hingee, M., & Magrath, R. D. (2009). Flights of fear: A mechanical wing whistle sounds the alarm in a flocking bird. *Proceedings of the Royal Society B: Biological Sciences, 276*, 4173–4179.
Hunter, T. A. (2008). On the role of wing sounds in hummingbird communication. *Auk, 125*, 532–541.
Hunter, T. A., & Picman, J. (2005). Characteristics of the wing sounds of four hummingbird species that breed in Canada. *Condor, 107*, 570–582.

Kingston, T., & Rossiter, S. J. (2004). Harmonic-hopping in Wallacea's bats. *Nature, 429*, 654–657.

Koch, R. E., Krakauer, A. H., & Patricelli, G. L. (2015). Investigating female mate choice for mechanical sounds in the male greater sage-grouse. *Auk: Ornithological Advances, 132*, 349–358.

Konishi, M. (1973). How the owl tracks its prey. *American Scientist, 61*, 414–424.

Kroeger, R. A., Gruschka, H. D., Helvey, T. C., et al. (1972). Low speed aerodynamics for ultra-quiet flight (pp. 1–155). Air Force Flight Dynamics Lab, Wright-Patterson Air Force Base, OH.

Lane, D. F., Servat, G. P., Valqui, H. T., & Lambert, F. R. (2007). A distinctive new species of tyrant flycatcher (Passeriformes: tyrannidae: *Cnipodectes*) from southeastern Peru. *Auk, 124*, 762–772.

Lentink, D., Haselsteiner, A. F., & Ingersoll, R. (2015). *In vivo* recording of aerodynamic force with an aerodynamic force platform: From drones to birds. *Journal of the Royal Society Interface, 12*, 20141283.

Lighthill, M. J. (1952). On sounds generated aerodynamically. I. General theory. *Proceedings of the Royal Society of London A: Mathematical and Physical Sciences, 211*, 564–587.

Manson-Bahr, P. H., & Pye, J. D. (1985). Mechanical sounds. In B. Campbell & E. Lack (Eds.), *A dictionary of birds*. Vermillion, SD: Buteo Books.

Miller, S. J., & Inouye, D. W. (1983). Roles of the wing whistle in the territorial behavior of male broad-tailed hummingbirds (*Selasphorus platycercus*). *Animal Behaviour, 31*, 689–700.

Miskelly, C. M. (1990). Aerial displaying and flying ability of Chatham Island snipe *Coenocorypha pusilla* and New Zealand snipe *C. aucklandica. Emu, 90*, 28–32.

Miskelly, C. M. (2005). Evidence for 'hakawai' aerial displaying by Snares Island snipe (*Coenocorypha aucklandica huegeli*). *Notornis, 52*, 163–165.

Mitchell, D. E. (2000). Allen's hummingbird. In *The birds of North America,* No. 501. Vermillion, SD: Buteo Books.

Myhrvold, N. P., & Currie, P. J. (1997). Supersonic sauropods? Tail dynamics in the diplodocids. *Paleobiology, 23*, 393–409.

Nelson, S. K., & Hamer, T. E. (1995). Nesting biology and behavior of the marbled murrelet. In C. J. Ralph (Ed.), *Ecology and conservation of the marbled murrelet* (pp. 57–68). Berkeley, CA: Pacific Southwest Research Station, USDA.

Norberg, R. Å. (1991). The flappet lark *Mirafra rufocinnamomea* doubles its wingbeat rate to 24 hz in wing-clap flight display: A sexually selected feat. *Journal of Experimental Biology, 159*, 515–523.

Ortiz-Crespo, F. I. (1980). *Agonistic and foraging behavior of hummingbirds co-occurring in central coastal California.* PhD thesis, University of California, Berkeley.

Podos, J. (2001). Correlated evolution of morphology and vocal signal structure in Darwin's finches. *Nature, 409*, 185–188.

Podos, J., Southall, J. A., & Rossi-Santos, M. R. (2004). Vocal mechanics in Darwin's finches: Correlation of beak gape and song frequency. *Journal of Experimental Biology, 207*, 607–619.

Prum, R. O. (1993). Phylogeny, biogeography, and evolution of the broadbills (Eurylaimidae) and asities (Philepittidae) based on morphology. *Auk, 110*, 304–324.

Prum, R. O. (1998). Sexual selection and the evolution of mechanical sound production in manakins (Aves: Pipridae). *Animal Behaviour, 55*, 977–994.

Prum, R. O. (2010). The Lande-Kirkpatrick mechanism is the null model of evolution by intersexual selection: Implications for meaning, honesty, and design in intersexual signals. *Evolution, 64*, 3085–3100.

Pytte, C. L., & Ficken, M. S. (1994). Aerial display sounds of the black-chinned hummingbird. *Condor, 96*, 1088–1091.

Randall, J. A. (2001). Evolution and function of drumming as communication in mammals. *American Zoologist, 41*, 1143–1156.

Reddig, V. E. (1978). Der ausdrucksflug der Bekassine (*Capella gallinago gallinago*). *Journal für Ornithologie, 119*, 357–387.
Richardson, M. H. (1997). Is it a mode shape, or an operating deflection shape? *Sound and Vibration Magazine, 31*(1), 54–61.
Riede, T., Suthers, R. A., Fletcher, N. H., & Blevins, W. E. (2006). Songbirds tune their vocal tract to the fundamental frequency of their song. *Proceedings of the National Academy of Sciences of the U S A, 103*(14), 5543–5548.
Robillard, T., Montealegre-Z, F., desutter-Grandcolas, L., Grandcolas, P., & Robert, D. (2013). Mechanisms of high-frequency song generation in brachypterous crickets and the role of ghost frequencies. *Journal of Experimental Biology, 216*, 2001–2011.
Rodgers, T. L. (1940). The dive note of the Anna hummingbird. *Condor, 42*, 86.
Rusch, D. H., Destefano, M. C. R., & Lauten, D. (Eds.). (2000). *Ruffed grouse (Bonasa umbellus)* (Vol. 515). Ithaca, NY: Cornell Laboratory of Ornithology.
Sarradj, E., Fritzsche, C., & Geyer, T. (2011). Silent owl flight: Bird flyover noise measurements. *AIAA Journal, 49*, 769–779.
Schultz, J. S., Hertel, F., Bauch, M., & Schlinger, B. A. (2001). Adaptations for rapid and forceful contraction in wing muscles of the male golden-collared manakin: Sex and species comparisons. *Journal of Comparative Physiology A, 187*, 677–684.
Stark, R. D., Dodenhoff, D. J., & Johnson, E. V. (1998). A quantitative analysis of woodpecker drumming. *Condor, 100*(2), 350–356.
Sueur, J., Tuck, E. J., & Robert, D. (2005). Sound radiation around a flying fly. *Journal of the Acoustical Society of America, 118*, 530–538.
Trail, P. W., & Donahue, P. (1991). Notes on the behavior and evology of the red-cotingas (Cotingidae: *Phoenicircus*). *Wilson Bulletin, 103*, 539–551.
Tuck, L. (1972). *The snipes: A study of the genus Capella*. Ottawa, Canada: Canadian Wildlife Service.
Unsworth, A., Dowson, D., & Wright, V. (1971). 'Cracking joints': A bioengineering study of cavitation in the metacarpophalangeal joint. *Annals of the Rheumatic Diseases, 30*, 348–358.
Vogel, S. (1994). *Life in moving fluids*. Princeton, NJ: Princeton University Press.
Wilson, B., Batty, R. S., & Dill, L. M. (2003). Pacific and Atlantic herring produce burst pulse sounds. *Proceedings of the Royal Society B: Biological Sciences, 271*, S95–S97.
Wilson, T. A., Beavers, G. S., DeCoster, M. A., Holger, D. K., & Regenfuss, M. D. (1971). Experiments on the fluid mechanics of whistling. *Journal of the Acoustical Society of America, 50*, 366–372.
Zeigler, H. P., & Marler, P. (2012). *Neuroscience of birdsong*. Cambridge, England: Cambridge University Press.

Chapter 5
Embodied Motor Control of Avian Vocal Production

Daniel N. Düring and Coen P.H. Elemans

Abstract Recent developments in biologically inspired robotics and artificial intelligence emphasize the need for a systems view on motor control, termed embodied motor control. Embodied control systems consider the brain, body, and environment and incorporate mechanical and neural feedback. The control system thus forms a closed loop of which the biomechanics are an integral part. This chapter presents the motor control of avian vocal production in the framework of embodied control. Our conceptual framework identifies and discusses the forces produced in the three embodied motor subsystems [the respiratory system, the vocal organ (the syrinx), and the upper vocal tract] and various feedback mechanisms. It becomes evident that compared to other neuromechanical systems very essential quantitative information is lacking. However, the great advantage of the birdsong system is that it, in contrast to humans, provides much better experimental opportunities to quantify all aspects of the vocal motor control system. Neural activity and many physiological parameters can be monitored chronically in vivo during highly stereotyped song. Furthermore, complementary measurement techniques, for example, in vitro and ex vivo, and the development of computational models focusing on different levels of organization within the system, are essential to fill the gaps where experimental observations remain too challenging. The combination of exciting developments, and the large natural variation present in the system, make vocal motor control in birds an excellent model system where many discoveries are waiting to be made.

Keywords Bioacoustics • Biomechanics • Birdsong • Muscle • Songbird • Myoelastic-aerodynamic theory • Neural control • Neuromechanics • Respiration • Somatosensory integration • Song system • Sound • Syrinx • Vocal tract

---

D.N. Düring • C.P.H. Elemans (✉)
Department of Biology, University of Southern Denmark,
Campusvej 55, 5230 Odense M, Denmark
e-mail: danield@biology.sdu.dk; coen@biology.sdu.dk

R.A. Suthers et al. (eds.), *Vertebrate Sound Production and Acoustic Communication*, Springer Handbook of Auditory Research 53,
DOI 10.1007/978-3-319-27721-9_5

5.1 Introduction

Almost all of the 10,000+ bird species vocalize extensively to communicate. Brief vocalization or calls predominantly shape their complex social networks (Marler 2004; Vignal et al. 2004). Bird songs are more elaborate than calls and consist of stereotyped strings of syllables (Berwick et al. 2011), which, in some species, are accompanied by highly coordinated multimodal displays (Cooper and Goller 2004; see Clark, Chap. 4). The main function of song is considered to be attracting mates and negotiating agonistic and territorial disputes to avoid fighting (Marler and Slabbekoorn 2004). While in the temperate regions it is predominantly the males that sing, recent research showed that females may also sing, especially in the tropics, suggesting that the ancestral state is for both sexes to sing (Odom et al. 2014).

Next to being subject to intensive study in behavioral ecology (Slabbekoorn and Smith 2002), bird song gained much attention when Peter Marler and colleagues showed that song development in song birds requires imitation learning from a tutor, called vocal learning, much like human speech (Marler 1970). Since then, bird song has grown into an important neuroethological model system for understanding learned behavior and motor sequence generation (Konishi 1985; Fee and Scharff 2010) with many established parallels with human speech acquisition (Doupe and Kuhl 1999; Bolhuis et al. 2000). Three clades of birds are capable of vocal learning; the songbirds, hummingbirds, and parrots (Pfenning et al. 2014). Of these, the songbird clade has received most attention, and in particular one species: the zebra finch (*Taeniopygia guttata*). Juvenile songbirds learn to sing during a critical period comprising two phases, named sensory and sensorimotor learning. In the sensory phase the juvenile listens and memorizes tutor songs. In the subsequent sensorimotor phase, the juvenile starts producing sounds and aims to match these to the memorized song template over the course of weeks or months (Brainard and Doupe 2013). This phase often ends with the crystallization of a stereotyped song (Tchernichovski et al. 2001). The nonvocal learners do not learn their song by imitating a tutor and are thought to have an innate template of their vocalizations. Over the last decades, a complex network of song nuclei in the songbird brain has been unraveled responsible for song perception, learning, and production called the song system (for recent reviews see Fee and Scharff 2010; Brainard and Doupe 2013).

In birdsong research the song system (brain) and vocal periphery (body) have been investigated mostly in isolation from one another. Although this dualistic approach has been fruitful, recent developments in biologically inspired robotics and artificial intelligence emphasize the need for a systems view on motor control, termed embodied motor control (Pfeifer et al. 2007). Cognition cannot be analyzed and understood by looking at neural computational processes alone, but requires that the physical system and its interactions with the environment be taken into account (Pfeifer et al. 2014).

The relevance of embodied motor control can be illustrated by considering a generalized motor system (Fig. 5.1a). A motor pathway produces complex sequences of motor commands that activate muscles, which through the morphology and material characteristics of the body, interact with the environment. This environ-

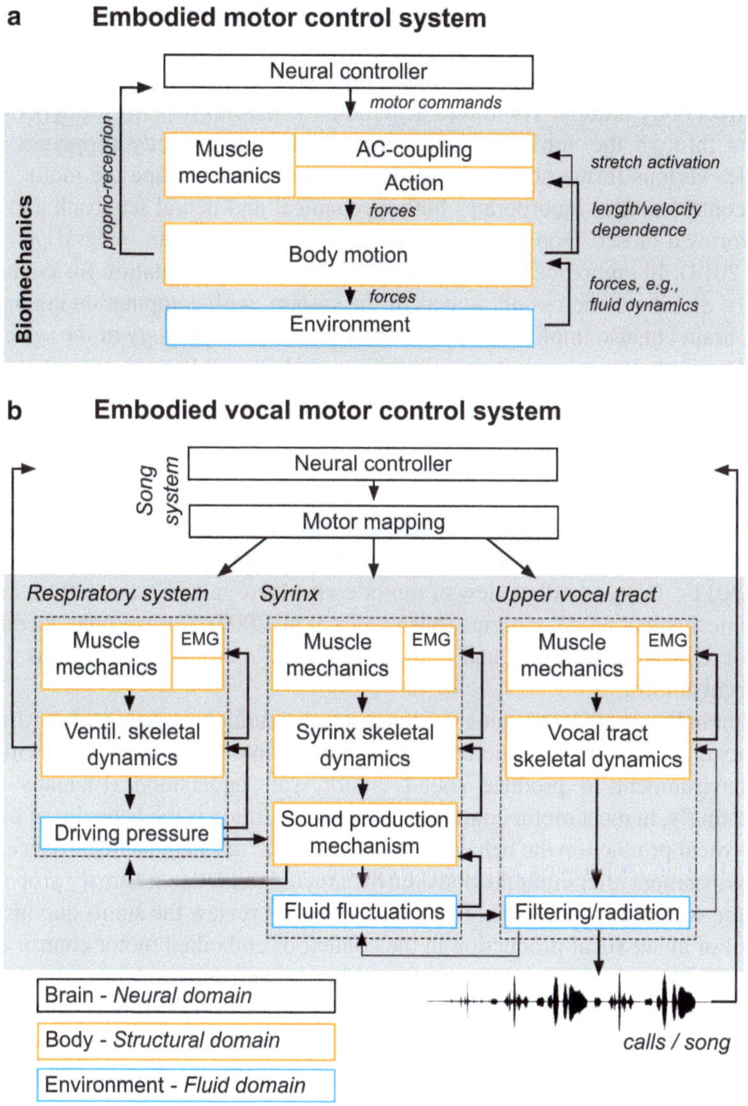

Fig. 5.1 The crucial role of biomechanics in shaping vocal behavior. (**a**) In a generalized motor control system a neural network produces motor commands, which by muscle action initiate motion of the body that interacts with the environment. The environment produces dynamic forces back on the body either in the structural (e.g., terrestrial locomotion) or fluid domain (e.g., swimming or sound production). Muscle force depends on the body motion according to the intrinsic nonlinear force–length and force–velocity properties of muscle. The behavior of the body is sensed by multiple sensory modalities (e.g., proprioreceptors, touch, vision) that provide feedback to the neural network, shaping future motor commands. Several mechanisms of feedback (or computation) are indicated with arrows and explained further in the text. (**b**) Embodied vocal motor control in birds. The neural output of the vocal network, the song system, is mapped on three motor subsystems: the respiratory system, vocal organ (syrinx), and upper vocal tract. Body posture could also influence vocal parameters, but this motor subsystem is left out for clarity. The forces produced within these subsystems and various feedback mechanisms are discussed throughout the text. The color coding refers to the neural (*black*), structural (*orange*), and fluid (*blue*) domains. AC-coupling, activation-contraction coupling; EMG, electromyogram

ment could be the ground's surface, as in terrestrial locomotion, but also fluid, resulting in more complicated structure–fluid interactions, as in swimming, flying, or sound production. The reaction forces produced by environment act on the body and affect body motion. The forces generated by muscles depend strongly on body motion through the nonlinear force–length and force–velocity properties of the muscle. Various forms of sensory feedback consequently shape the motor output. This control system incorporates both mechanical and neural feedback and therefore forms a closed loop of which the biomechanics form an integral part (Roth et al. 2014). In embodied motor systems the required computation for control can even be distributed across all aspects of the system, with computation taking place in the brain but also implemented inherently into the morphology of the system, so-called "morphological computation" (Pfeifer et al. 2014). For example, the complicated neural control of object grasping can be simplified by changing the morphology of the grasper by including soft materials that locally "compute" the control solutions. To conclude, neural circuits do not function in isolation, but interact strongly with the physical body and environment through sensory inputs and movement via muscles. The activity of neural circuits can therefore be understood only by also considering the biomechanics of muscles, bodies, and the exterior world (Tytell et al. 2011). This embodied view of motor control has gained increased attention in neuromechanical model systems (Nishikawa et al. 2007), for example, locomotion control (Cowan et al. 2014), and is applicable to all motor control systems, including vocal control.

Although our understanding of the central neural processing of birdsong is advancing rapidly, how the motor output of vocal motor circuits interacts with body and environment to produce sound is not well understood (Elemans 2014). Interestingly, in most motor control systems body motion is the behavior of interest, but in vocal production the behavior of interest is the mechanical disturbance left in the environment after complex body–environment interactions, namely propagating pressure waves, that is, sound. This chapter aims to review the status quo on motor control of avian vocal production in the context of embodied motor control theory. As such, this review does not aim to provide a complete overview of the literature (Suthers and Zollinger 2008; Riede and Goller 2010a), and instead outlines a conceptual framework including the translation of neural signals into sound and the various forms of feedback. This framework is outlined in Fig. 5.1b and explained and discussed throughout the text. To provide anatomical context, first the morphology of the sound production system is covered. Second, the contribution of these structures (body) to sound generation in air (environment) is discussed, because this affects the control parameters of the vocal organ. Third, muscle mechanics and body motion are reviewed, followed by neural control and sensory feedback. At the end of each section, feedback mechanisms are discussed. This chapter aims to promote the idea that neuroscience and biomechanics need to be integrated to close the loop of sensorimotor integration in vocal motor control.

5.2 The Avian Sound-Producing System

Whether making sound with a xylophone, violin, dog whistle, clarinet, or larynx, in general sound production in air requires four components; an energy source, sound source, filter, and radiator (Fletcher and Rossing 1998). The sound source converts some form of energy into propagating pressure waves in a medium, that is, sound. The filter modifies the amplitude spectrum of the sound by absorption or resonance, and finally the radiator couples the sound wave to the medium. In birds these four components are present as three morphologically separate, and highly adapted subsystems: the respiratory system, vocal organ, and upper vocal tract including larynx and beak (Fig. 5.2).

The following section describes the morphology of these subsystems. The body is considered to contain four types of tissue with regard to their distinct mechanical properties (Vincent 1990): (1) bone, (2) cartilage, (3) soft materials, and (4) muscle. Bone is stiff, brittle tissue that contains mineral deposits and can withstand high forces, but fractures when deformed sufficiently. Cartilage is a viscoelastic extracellular matrix (Young's modulus range [measure of stiffness of an elastic material] about 1–100 MPa) that can withstand limited deformation and store energy. Soft material is highly deformable viscoelastic extracellular matrix (Young's modulus range about 1–100 kPa) that contains, for example, collagen and elastin, and can store and dissipate energy. Muscle is a "smart" material of which the mechanical properties depend strongly on length, shortening velocity, activation state, composition, and so forth (Dickinson et al. 2000), and is treated separately in Sect. 5.4.1.

5.2.1 Morphology of the Respiratory System

The avian respiratory system differs substantially from those of mammals and other tetrapods. Whereas mammals have tidally ventilated, sac-like lungs that branch down to small, inflatable alveoli (Duncker 2004), avian lungs are ventilated unidirectionally by the bellowing action of air sacs (Fig. 5.2c) (Duncker 1972). Thus the functions of respiration, which is gas exchange between air and blood, and ventilation, which is the distribution of the air, are carried out by separate structures in the respiratory system. The dorsally situated parabronchial lungs are rigid structures in which cross-current gas exchange takes place (Maina 2006; Maina et al. 2010). They are shifted so far dorsally in the thorax that the ribs incise them deeply (Duncker 2004). Ventrally, the lungs are bounded by the tight horizontal septum. Avian air sacs are generally divided into two functional groups (Brackenbury et al. 1989): the cranial group, including the unpaired interclavicular, and paired cervical and cranial thoracic air sacs, and the caudal group, including the paired caudal thoracic and abdominal air sacs (Fig. 5.2c).

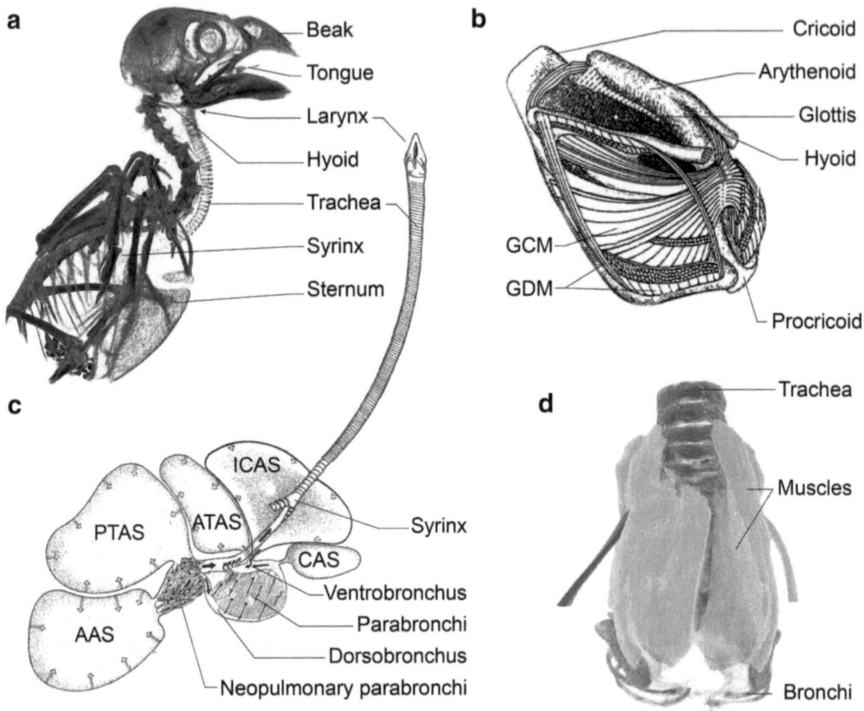

Fig. 5.2 Schematic overview of the avian sound production system. (**a**) Three-dimensional geometry of a zebra finch (*Taeniopygia guttata*) skeleton. [Adapted from Düring et al. (2013).] (**b**) The larynx of the rock pigeon (*Columba livia*). Indicated are the cartilages (cricoid, arytenoid, and procricoid), the glottal dilator muscle (GDM), and glottal constrictor muscle (GCM). [Adapted with permission from Zweers et al. (1981), Fig. 8.] (**c**) Schematic illustration of avian air sacs, lungs, and parabronchi during exhalation. Motion of the airsacs (*white arrows*) causes air flow (black arrows) over the parabronchi in the lung during exhalation. Note: Air sac terminology differs from that of Baumel et al. (1993), who prefers caudal/cranial over anterior/posterior and clavicular over interclavicular. Current terminology is used to avoid confusing abbreviations. [Adapted from Dürrwang (1974), Fig. 11.] (**d**) 3D reconstruction of a micro-CT scanned zebra finch syrinx. [Adapted from Düring et al. (2013), Fig. 8.] For more details see Fig. 5.4. *AAS* abdominal air sac, *ATAS* anterior thoracic air sac, *CAS* cervical air sac, *ICAS* interclavicular air sac, *PTAS* posterior thoracic air sac

5.2.2 Morphology of the Syrinx

In most nonavian vocalizing tetrapods, the larynx serves the dual function of separating the airway from the esophagus and sound production with vocal folds. In birds, a novel organ evolved dedicated to sound production, the syrinx. This organ is located at the fusion of the two primary bronchi into the trachea and inside the interclavicular air sac (ICAS; King 1989). The syrinx was named after the Greek nymph syrinx by Huxley (1871) to avoid confusion with the previous terminology of upper and lower larynx.

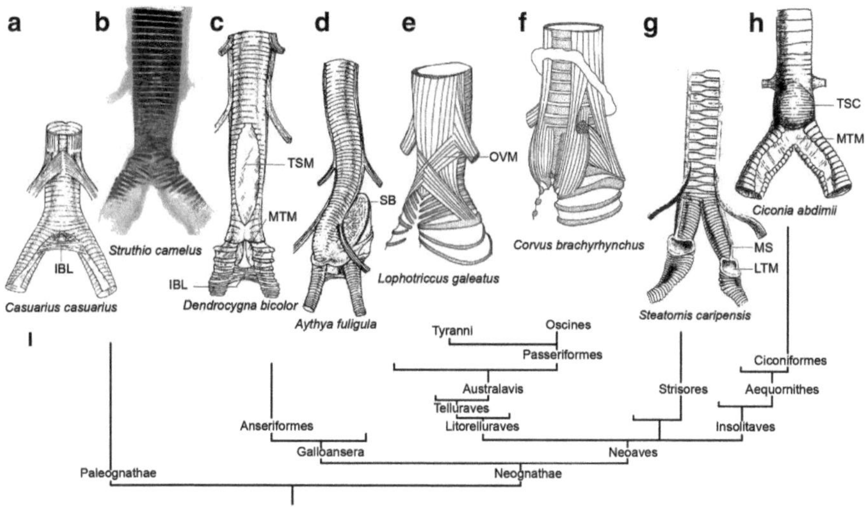

Fig. 5.3 Variation in syringeal morphology. (**a**) The syrinxes of the southern cassowary (*Casuarius casuarius*) and (**b**) common ostrich (*Struthio camelus*) are considered among the most regressive morphologies: intrinsic muscles, tympanum, and pessulus are not present and there are no distinct modifications of the first bronchial half-rings. King (1989) considers these syrinxes to be regressive rather than basal. (**c**) The syrinx of the fulvous whistling duck (*Dendrocygna bicolor*) contains about 14 tracheal rings that are dorsally incomplete and covered by a tracheosyringeal membrane (TSM). Medial tympaniform membranes (MTMs) are also present. (**d**) The syrinx of the tufted duck (*Aythya fuligula*) contains a syringeal bulla (SB), which typically is an asymmetric dilation of the tympanum. Whereas in this species it is extensively membranous, in other species it can be partially or completely ossified. (**e**) In the syrinx of the suboscine helmeted pygmy tyrant (*Lophotriccus galeatus*) the oblique ventral muscle (OVM) crosses the syringeal midline. (**f**) The syrinx of the American crow (*Corvus brachyrhynchus*) is an example of the typical oscine bipartite syrinx containing two sound sources, one in each bronchus. (**g**) In the syrinx of the oilbird (*Steatornis caripensis*) MTMs and lateral tympaniform membranes (LTMs) are positioned asymmetrical and deep in the bronchi. The m. syringealis runs from the last tracheal rings to a small tendon attaching to the LTM. (**h**) The syrinx of the white-bellied stork (*Ciconia abdimii*) contains a dilated region formed by tracheosyringeal cartilages (TSC). (**i**) Cladistic relationship of the aforementioned species using data from Jetz et al. (2014) and Jarvis et al. (2014). The tree was generated at www.birdtree.org using an online script by Jetz et al. (2014). [**a**, **c**, **d**, **g**, and **h** adapted with permission from King (1989); **b** adapted from Elemans et al. (2015); **e** and **f** adapted from Ames (1971).] *IBL* interbronchial ligament, *LTM* lateral tympaniform membrane, *MS* m. syringealis, *MTM* medial tympaniform membrane, *OVM* oblique ventral muscle, *SB* syringeal bulla, *TSC* tracheosyringeal cartilages, *TSM* tracheosyringeal membrane

The syrinx consists of an intricate system of joints, levers, springs, and muscles, and exhibits a tremendous morphological diversity among birds (Fig. 5.3). In fact, until the rise of molecular biology, syrinx diversity was an important tool to assist systematists and taxonomists in avian classification (Ames 1971). Syrinx morphology has been described extensively (King 1989; Düring et al. 2013), but few data are available on quantitative biomechanical aspects such as geometry and joint definitions.

The syringeal skeleton consists of modified tracheal rings and paired bronchial half-rings that are cartilaginous, but ossify partially or completely with age (Appel 1929; Hogg 1982). In many species several tracheal and bronchial rings are partly or completely fused to a tube-like structure called tympanum of which the ventral-dorsal part in the caudal end is often fused into a bridge-like structure called the pessulus (King 1993) (Fig. 5.4). A remarkable modification of the tympanum is the membranous or ossified bulla found in ducks (Frank et al. 2006) (Fig. 5.3d). In older morphological descriptions (Ames 1971), the first bronchial half-rings are commonly depicted as c-shaped and allow space for vibratory soft tissues (Fig. 5.3). However, in the oscines (i.e., songbirds), the first three bronchial bones B1–B3 are highly modified and have complicated shapes that contain the insertion sites for most syringeal musculature (Düring et al. 2013) (Fig. 5.4). On the ventral side, bronchial half-rings B2 and B3 end in a cartilaginous extension, of which the medial ventral cartilage (MVC) on B2 is most prominent.

Soft connective tissues are suspended between the cartilaginous/ossified syringeal skeleton, some of which form the vibratory tissues that generate sound. Historically, three types of syrinxes were recognized based on the location of these soft tissues (King 1989): the tracheal, bronchial, and tracheo-bronchial types. However, the identification of the sound-producing structures to make this classification has been inferred mostly from morphology (Wunderlich 1884) and needs to be readdressed using a functional approach in a comparative framework (Elemans et al. 2015). In some syringes the vibratory tissues are located around the tracheo-bonchial junction, with the exceptional cases of, for example, penguins and oilbirds, in which they are deep in the bronchi (Fig. 5.3g). The syrinx in many non-songbird species has one pair of lateral vibratory masses (LVMs) in the trachea and paired membranous medial tympaniform membranes (MTMs). The songbird syrinx has one pair of vibratory soft tissues in each bronchus, a lateral labium (LL) on the inside of B3 and a medial vibratory mass (MVM) that tapers from about 100 μm in the cranial part, the medial labium (ML), to about 10 μm in the membranous MTM (Riede and Goller 2010b).

The labia in songbirds are composed of connective tissue covered by epithelium. The extracellular matrix of connective tissue consists of fibrous proteins (elastin, collagen) and amorphous substances, such as glycosaminoglycans and proteoglycans (hyaluronan) (Riede and Goller 2010b). Those substances are organized in

Fig. 5.4 (continud) mainly dispersed randomly in layers 2 and 5, dorso-ventrally in layer 3, and cranio-caudally in layer 4. Layer numbers 2–4 are commonly used to identify similar organization in different species. Starlings do not have an elastin dominated layer that in other species is adjacent to layer 2 and numbered 1. [Adapted with permission from Riede and Goller (2014), Fig. 1. **a–e** adapted from Düring et al. (2013).] *B1–3* bronchial bones 1–3, *C* cartilage, *DDS* m. syringealis dorsalis profundus, *DTB* m. tracheobronchialis dorsalis, *DVTB* m. tracheobronchialis ventralis profundus, *E* epithelia, *IBL* interbronchial ligamentum, *LDC* lateral dorsal cartilage, *LDS* m. syringealis dorsalis lateralis, *LL* lateral labia, *LVC* lateral ventral cartilage, *MDC* medial dorsal cartilage, *MDS* m. syringealis dorsalis medialis, *ML* medial labia, *MVC* medial ventral cartilage, *P* pessulus, *ST* m. sternotrachealis, *STB* m. tracheobronchialis brevis, *SVTB* m. tracheobronchialis ventralis superficialis, *T* trachea, *TYM* tympanum, *VS* m. syringealis ventralis

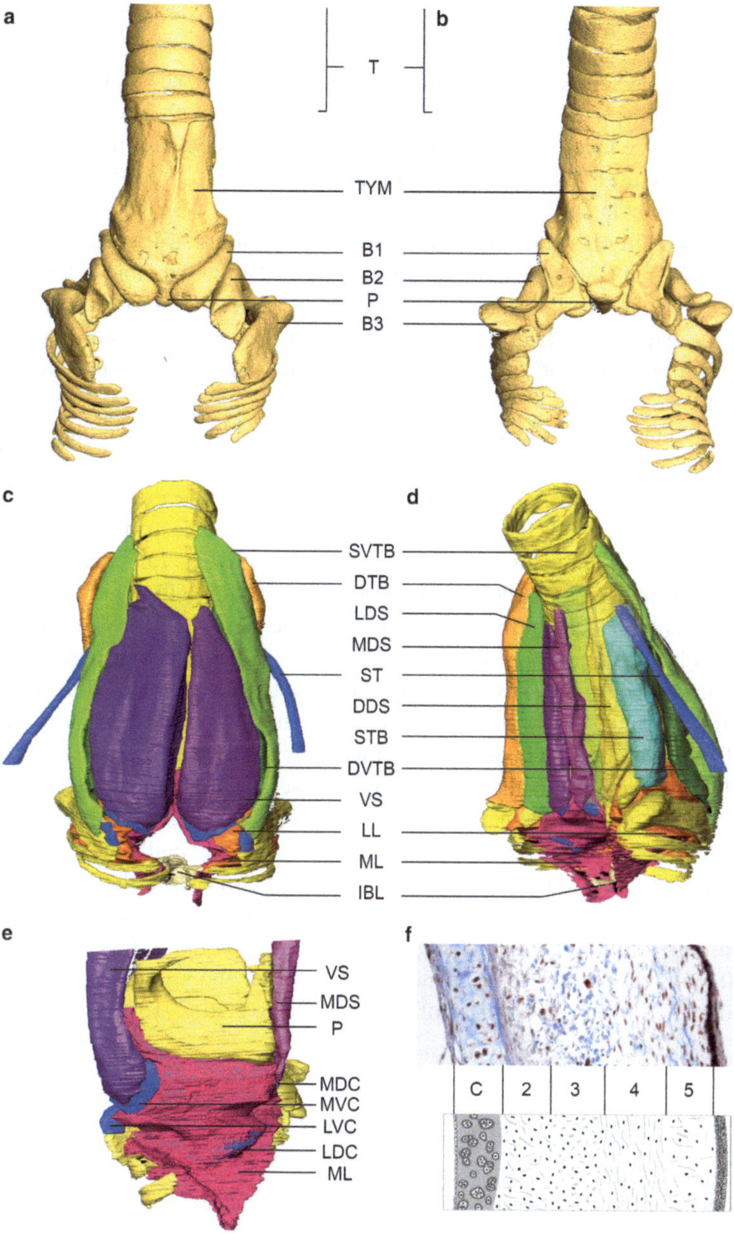

Fig. 5.4 Three-dimensional geometry and tissue components of the songbird syrinx. (**a**) Ventral and (**b**) dorsal aspect of the zebra finch syrinx skeleton consisting of bone or partially ossified cartilage. Imprints of the muscle insertion sites are clearly visible on the medio-cranial part of the tympanum and on the bronchial bones B1–B3. (**c**) Ventral and (**d**) dorsal aspects of the syringeal muscles. DTB and LDS are transparent in (**d**) to show DDS and STB. (**e**) Sagittal view on the soft tissue medial vibratory mass, which is suspended from the pessulus and forms the medial wall of the bronchus. Several elastic cartilages are embedded in the MVM. The left part of the syrinx and other muscles are removed for clarity. (**f**) Sagittal section (*top*) through the ML of a starling (*Sturnus vulgaris*) and corresponding schematic representation (*bottom*) illustrating the layered structure. Layers are composed of collagen (*blue*) and elastin fibers (*black*). Collagen fibers are

sagittal layers, where the relative quantities of these substances together with their spatial arrangement are used to identify the different layers. The thickness of the connective tissue layer is highly variable and tends to be greater in songbirds compared to other birds (King 1989; Riede and Goller 2014) and can be lateralized (Prince et al. 2011).

In addition, in the zebra finch two paired cartilaginous structures are suspended within the MVM (Fig. 5.4c); on the dorsal part of the medial labia is a small cartilaginous disc, the medial dorsal cartilage (MDC). On the border of the ML and MTM, the bar-shaped lateral dorsal cartilage (LDC) is embedded in the medial labia. Except for the LDC, all cartilages provide insertion sites for muscles (Düring et al. 2013).

The skeletal framework and the labia are controlled by syringeal muscles, which also vary greatly in quantity and size across bird taxa (Fig. 5.3). Syringeal muscles are considered extrinsic when one end inserts to the syrinx and one elsewhere, or intrinsic, when both ends insert onto the syrinx. Two paired extrinsic syringeal muscles are present in almost all birds, the musculus tracheolateralis (TL), which runs lateral to the trachea, and the musculus sternotrachealis (ST), which runs through the ICAS and connects the syrinx with parts of the sternum. The ST is native to all avian taxa but the tinamiformes (Garitano-Zavala 2009). Whereas basal birds have none or few intrinsic muscles, songbirds can have up to eight pairs of intrinsic muscles (Figs. 5.3 and 5.4c, d). In zebra finches, two muscles attach directly to cartilaginous elements embedded in the ML, the musculus syringealis ventralis (VS) and the musculus syringealis dorsalis (MDS) (Fig. 5.4c–e). The other six intrinsic muscles insert caudally onto bronchial half-rings B1–B3. Each ring has a parallel muscle pair that inserts closely together on the ring caudally and cranially inserts (1) directly on the tympanum without any tendon or (2) into connective tissue of the ICAS membrane on the cranial side, thus placed in series with the TL. An important control consequence is that length change of the first group of muscles could thus move the half-rings relative to the tympanum, while the second group could be additionally modulated by length changes of the TL (Düring et al. 2013).

5.2.3 *Morphology of the Upper Vocal Tract*

The upper vocal tract of birds is composed of the trachea, larynx, oropharyngeal-esophageal cavity (OEC), and the beak (Fig. 5.2a). The trachea is connected to the cranial part of the syrinx and consists of closely connected cartilaginous rings that, in contrast to mammalian tracheal rings, are fully closed. It exhibits large variation and specialization (Fig. 5.5). In about 60 species, the trachea is elongated 2–20 times compared to the expected length and coiled within the sternum or between the skin and the pectoral muscles (Fitch 1999). Some species exhibit a dilation of the trachea (Fig. 5.5b) or a tracheal sac or bulla (Fig. 5.5c) that opens from the trachea or an expansion in close proximity rostral to the syrinx (King 1993).

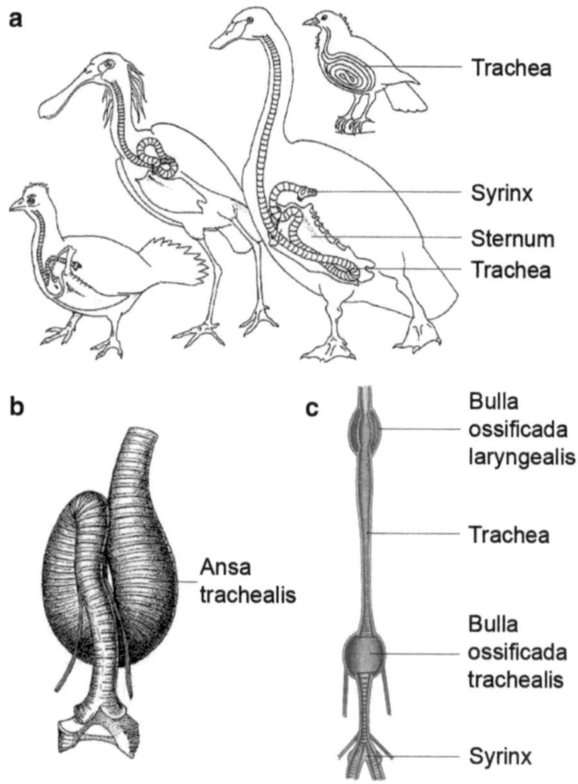

Fig. 5.5 Variation in vocal tract morphology. (**a**) Examples of tracheal elongation (from left to right): intraclavicular coiling in crested guinea fowl (*Guttera edouardii*), intrathoracic coiling in European spoonbill (*Platalea leucorodia*), intrasternal coiling in trumpeter swan (*Cygnus buccinators*), and subdermal coiling in trumpet manucode (*Manucodia keraudrenii*). [Adapted with permission from Cambridge University Press from Fitch (1999), Fig. 1.] (**b**) Tracheal widening in black curassow (*Crax alector*). (**c**) Upper vocal tract of velvet duck (*Melanitta fusca*) with two ossified bullae. Fused tracheal rings form an ossified tracheal bulla just cranially to the syrinx. A second ossified chamber is found caudally to the larynx [**b** and **c** adapted with permission from Rüppell (1933), Figs. 34 and 35]

The larynx forms the end of the trachea and is not known to contain vibratory tissue in birds. Its main function is protecting the trachea from food and other foreign particles (Häcker 1900). The avian larynx is composed only of cricoid, procricoid, and paired arytenoid cartilages (Fig. 5.2b) that ossify with age (Hogg 1982) and differ structurally from the mammalian laryngeal cartilages. In contrast to the mammalian larynx, no epiglottis or vocal folds are present. Two types of avian larynges are generally recognized, the passeriform larynx and the nonpasserifom larynx (Baumel et al. 1993). In passeriformes, the dorsal cricoid is separated from the rest of the cricoid cartilages (King 1993). The intrinsic muscles of the larynx are the dilator and constrictor of the glottis. The dilator is lateral and superficial, running essentially from the cricoid to the arytenoid cartilage. The constrictor runs typically from the procricoid cartilage in the caudal midline of the larynx to the arytenoid and cricoid cartilages (King 1993).

The larynx opens into the OEC and the beak. It is attached to the hyoid apparatus, which, in contrast to mammals, is not attached to the skull (Homberger and Meyers 1989). Therefore hyoid motion can modulate the volume of the oropharyngeal cavity. In addition, the cranial end of the esophagus can expand and form one continuous space with the oropharyngeal cavity, the oropharyngeal-esophageal cavity (Riede et al. 2006; Ohms et al. 2010). Beak shape and size differ tremendously among birds and are generally related to dietary specializations (Podos 2004).

5.3 The Environment: The Aero-Acoustics of Birdsong

When animals move in fluids, such as swimming in water or flying in air, strong interactions occur between body and fluid environment (Fig. 5.1a) that can leave signature flow structures in the fluid environment (Taylor et al. 2010). In avian vocal production, the three different morphological subsystems described in Sect. 5.2 all interact with the fluid environment with the goal to generate pressure patterns in the environment. This section reviews our knowledge of the fluid–structure interactions that occur within and between the subsystems.

5.3.1 Physical Mechanisms of Sound Production

The characteristics and motor control of specific vocal parameters are determined by the physical mechanism used to transfer some form of energy into sound. Small animals cannot radiate sound efficiently when the sound wavelength is larger than roughly twice their body size, a phenomenon called acoustic short-circuiting (Michelsen et al. 1987). This physical constraint predicts that small (10–15 cm) birds should preferably produce sounds above 1 kHz. Body motion at a frequency greater than 1 kHz would require musculature that can produce positive power over 1 kHz. However, such speeds are far above the maximum attainable observed in the fastest vertebrate synchronous muscle of about 250 Hz (Rome 2006; Elemans et al. 2008b). Some fish species can produce sound pulses up to 250 Hz by 1:1 muscle contraction (Elemans et al. 2014), but to radiate sounds efficiently, most small vertebrates therefore need to employ some form of mechanical frequency-multiplier that allows them to transform "slow" body motion into higher frequency sounds (Bennet-Clark 1999; Parmentier et al. 2006). As can be seen throughout this book, a plethora of sound production mechanisms has evolved in animals. Understanding this mechanism provides insight into the control space of the system and as such the possibilities and constraints that act on the motor control, from immediate (neural) computation, to ontogeny and evolution.

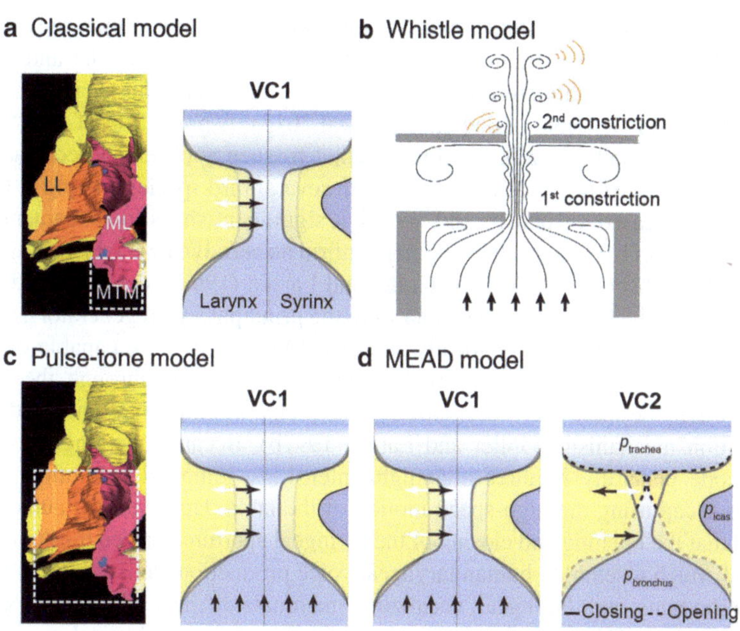

Fig. 5.6 Physical mechanisms of avian sound production. (**a**) In the classical model, the MTMs (*white dotted box*) were considered the structures that exhibited airflow-induced oscillation. (**b**) Whistle model, where air flow impinges on a constriction and forms a jet. This jet is stabilized by a second constriction and radiates sound (*orange lines*). [Adapted from Henrywood and Agarwal (2013)]. (**c**) In the pulse-tone model, the lateral and medial labia (white dotted box) were considered the structures that exhibited airflow-induced oscillation. (**d**) Myoelastic-aerodynamic (MEAD) model for self-sustained flow-induced oscillation due to two vibrational components (VC1 and VC2). Detailed explanation in the text of Sect. 5.3.1 [Adapted from Elemans et al. (2015)]

For more than 200 years researchers have been investigating the physical mechanisms and structures birds use to generate sounds. Based on the intensively studied syringeal anatomy different structures were proposed to be responsible for sound generation (Setterwall 1901; Ames 1971). The first experimental study of physical sound-producing mechanisms was conducted by Rüppell (1933), who could elicit sound from an excised syrinx preparation. A similar preparation was used to generate the first high-speed film of the MTMs from a frontal view during sound production in geese (Paulsen 1967). These and other studies led to the formulation of the so-called "classical model" in which oscillating thin MTMs were responsible for sound generation (Fig. 5.6). To overcome the experimental difficulties of studying the syrinx in situ, several sound-generating real-world physical models were developed to our knowledge all based on thin MTMs as sound generators (Dürrwang 1974; Elemans et al. 2009).

However, the often very tonal sound produced by birds could not be explained well by theoretical models based on edge-clamped membranes (Casey and Gaunt 1985; Fletcher 1988), and the aerodynamic whistle was proposed as an alternative mechanism (Gaunt et al. 1982). Aerodynamic whistles do not require the vibration of structures, but a constriction to generate a vortex-shedding air jet and aerodynamic feedback to stabilize the jet, for example, a second constriction as in the whole-tone whistle (Henrywood and Agarwal 2013). A prediction of this "whistle mechanism" hypothesis was that fundamental frequency of sound decreases with medium density. However, experiments with several bird species singing in experimentally decreased air density did not elicit significant shifts in fundamental frequency (Nowicki 1987) and thus did not confirm an aerodynamic whistle. Although the focus had been on the MTMs, Goller and Larsen (1997a) demonstrated that in songbirds not the MTMs, but the labia were the principal sound generators, as birds with either partial ablated or locally stiffened MTMs still were capable of sound production (Larsen and Goller 2002). Endoscopic imaging through the trachea showed that the labia were oscillating, which supported a vibration-based sound production mechanism (Goller and Larsen 1997b). Because imaging through an endoscope generally requires high light intensities, temporal resolution did not allow for capturing the high-speed motion of the labia. However, the authors suggested that the opening and closing of the syringeal aperture, termed the "pulse-tone model," much resembled human laryngeal voice production (Jensen et al. 2007).

For humans and laryngeal vocalizing mammals in general, the myoelastic-aerodynamic (MEAD) theory was formulated to explain the physical mechanisms underlying sound production (van den Berg 1958). This theory stated that self-sustaining laryngeal vocal fold oscillations were maintained through fluid–tissue interactions and (myo)elastic restoring forces generated within the tissues, without the need for periodic muscle contractions at the rate of tissue vibration or any other periodic input (see also Herbst, Chap. 6). MEAD is well-studied in humans: airflow from the lungs is mechanically converted by the vocal folds into pulse-like airflow feeding into the trachea. The flow rate variations cause pressure disturbances, constituting the acoustic excitation of the system. The original MEAD theory as formulated for humans by van den Berg (1958) and later extended to other mammals (Fitch 2006; Herbst et al. 2012) describes the prerequisites for human vocal fold oscillation to contrast the neurochronaxic hypothesis, which was the prevailing hypothesis at that time, and stated that vocal fold oscillation required periodic muscle contraction. The original formulation, however, did not include the physical mechanism underlying self-sustained oscillations (Titze 2000). This crucial component was added later to the MEAD theory (Titze 1988) to form the concurrent vital theoretical framework for studying mammalian vocal dynamics.

To retain self-sustaining oscillations, a system operating within the MEAD framework needs at least two degrees of freedom (DOF) that act within each oscillation cycle (Fig. 5.6). The first DOF consists of a lateromedial vibration of the inner edge of the vocal folds (DOF1), gating the airflow. The second DOF consists of an asymmetric forcing function over the opening and closing phases of vibration (DOF2). In mammals, DOF2 is maintained most commonly through phase differences along the caudo-cranial inner wall of the vocal folds, due to tissue rotation that results in alternating shapes (convergent vs. divergent) during opening and closing parts of the cycle, facilitating the transfer of aerodynamic energy into the tissue (Titze and Story 2002).

Several theoretical models assumed such labial shape configurations to occur in birds (Laje and Mindlin 2005; Elemans et al. 2008b), but experimental evidence was lacking (Riede and Goller 2010a). Endoscopic imaging in crows (*Corvus cornix*) in vivo demonstrated opening and closing of the syringeal air ducts (Jensen et al. 2007) (DOF1), but although filmed at 1000 frames/s still at insufficient temporal resolution to study the causal relation between dynamics of labial inner edges and sound generation events within an oscillatory cycle. Using stroboscopic illumination techniques (phantom high speed) on an excised preparation, Fee et al. (1998) showed syringeal structures of zebra finches exhibit wavelike motions in vitro, but this concerned the outer and not inner labial wall, and the dynamics were not related to acoustic data. In addition, these latter experiments were conducted at nonphysiological pressure conditions without an air sac system, potentially destabilizing syringeal oscillations leading to increased occurrence of chaotic oscillations (Elemans et al. 2010).

Recently an experimental paradigm was developed that allowed the quantification of syringeal labial motion under experimentally controlled conditions ex vivo (Elemans et al. 2015). Using transillumination, the motion of the inside of the syringeal structures could be visualized at high speeds (up to 35,000 frames/s) with simultaneous high-speed imaging of the syringeal aperture through the trachea allowing kinematic analysis of MVM and labia (Fig. 5.7). Large-amplitude oscillations were observed during sound production. Both DOF1 and DOF2 were present in a range of species with highly divergent syringeal morphologies, such as ratites, pigeons, parrots, and songbirds, suggesting that the MEAD framework is also applicable to birds (Elemans et al. 2015). Sound excitation occurred when the syrinx closed, opened or both (Elemans et al. 2015). Thus sound excitation events are caused by syringeal dynamics, and the frequency of oscillation sets the f_0 of sound. Given the large observed variation in syringeal morphology in the 10,000+ species of birds, it is possible that physical mechanisms for vocal production other than MEAD will be found. Nevertheless, the MEAD framework builds on a large body of literature on human sound production and as such provides a strong theoretical framework in which to study vocal control in birds.

5.3.2 *Ventilation Fluid Dynamics*

Respiratory fluid dynamics and in particular how birds maintain a near constant airflow over their lungs is still not fully understood and has been studied only in a few species (Bretz and Schmidt-Nielsen 1971; Mackelprang and Goller 2013). During inspiration, air enters the caudal air sacs, while a small part is directly gated over the parabronchi and air capillaries into the cranial air sacs. During expiration, stored air from the caudal air sacs is forced through the parabronchi and air capillaries into the major distal airways, where it is mixed and exhaled with the stored air from the cranial air sacs. Thus any inhaled volume of air needs two breathing cycles to pass the system completely (Bretz and Schmidt-Nielsen 1971; Maina 2006). Two

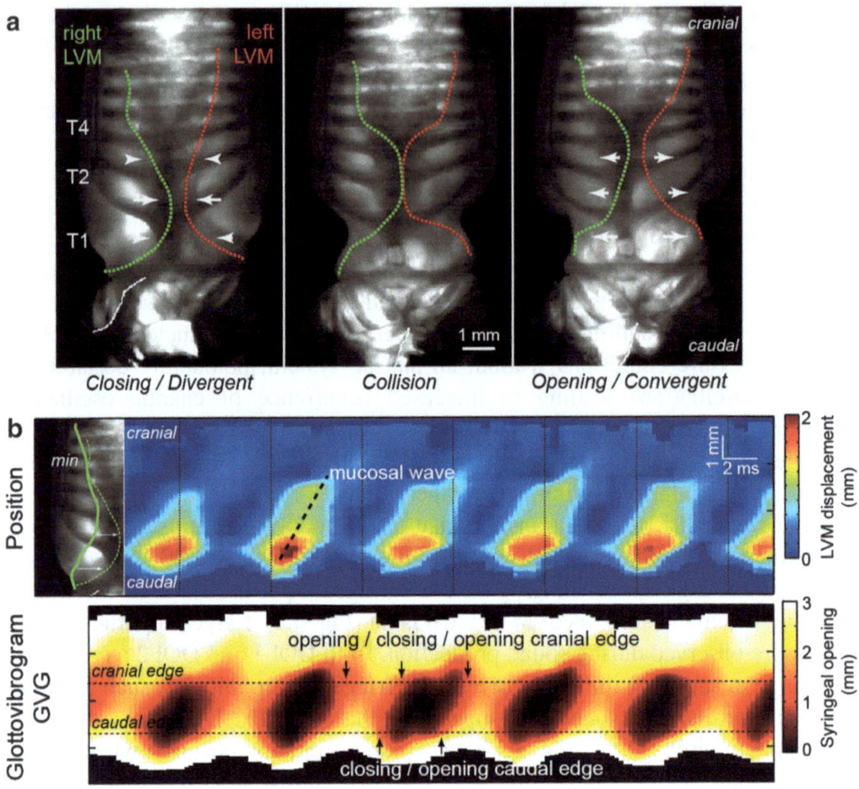

Fig. 5.7 The myoelastic-aerodynamic (MEAD) theory explains avian sound production. (**a**) Transilluminated syrinx of the rock pigeon (*Columba livia*) filmed at 4 kHz with traced outlines of right (*green*) and left (*red*) LVM during closing, collision, and opening phases of oscillation. White arrows indicate movement. (**b**) Spatiotemporal displacement analysis of the left LVM (*top*) and syringeal opening (also known as glottovibrogram, *bottom*) show that a mucosal wave is traveling from caudal to cranial (*dashed line*) [Adapted from Elemans et al. (2015)]

aerodynamic valves channel airflow and avoid air from taking the low-resistance pathways during inspiration and expiration. During inspiration, a narrowing in the ventrobronchus prevents air from flowing directly into the cranial air sacs. During expiration, a second valve in the dorsobronchus prevents direct outflow of air through the primary bronchus (Brown et al. 1995). These valves are not under muscular control, but recent modeling studies in ostriches suggest that, because of the local geometry of the stiff ducts, air jets are formed comparable to venturi valves (Maina et al. 2009).

During vocalizations, highly stereotyped pressure patterns occur in the air sacs (Schmidt et al. 2012), including very short minibreaths in between syllables (Suthers et al. 2012). Flow can be measured chronically with small thermistors and has even been applied successfully to measure bilateral flow in birds as small as the zebra finch (Goller and Cooper 2004). Flow amplitude modulation up to 80 Hz seems mostly driven by the respiratory system, and higher frequency modulations by syringeal dynamics (Goller and Riede 2013). During vocalization, pressure is modulated rapidly in the various air sacs and these dynamical scenarios can result in pressure differences up to 1 kPa between interclavicular and cranial thoracic air sacs (Beckers et al. 2003; Elemans et al. 2008b). These pressure differences seem not actively controlled by sphincter muscles connecting air sacs, but rather due to the flow patterns that arise due to the geometry and motion of the air sacs. However, sound is not produced without positive gauge pressure in the ICAS surrounding the syrinx (Hérissant 1753; Amador and Margoliash 2013). Because bronchial pressure is challenging to measure experimentally, the pressure in the anterior or cranial thoracic air sac is normally used as a proxy for bronchial pressure (Suthers et al. 2012).

5.3.3 Syringeal Soft Tissue Dynamics

At present, because of the difficulties in observing syringeal dynamics, little experimental data are available describing vibratory motion of the syringeal soft tissues, such as the amplitudes and frequency, excited by physiologically realistic pressure and flow excitation (Elemans et al. 2015). The available data have not been related to driving pressure during sound production. Most hypotheses regarding syringeal dynamics are therefore built on the human MEAD literature (Riede and Goller 2010a), remain conceptual, and need to be tested experimentally in birds.

Syringeal fluid–structure interactions already lay the foundation for several important acoustic parameters such as fundamental frequency (f_0), amplitude, and harmonic content of sound. First, what parameters determine the oscillation frequency of the vocal folds? A simple but successful approach to model the human vocal folds is using a spring-damper system or string (Titze 2000). The fundamental frequency at which such a system oscillates depends on its stiffness and mass. The stiffness of the vocal folds depends on their tension and dimensions (length and cross-sectional area). The mass would be defined as the mass taking part in vibration, or the effective mass. Although this is a rough approximation, this model predicts that any control parameters altering tension or mass must affect f_0. These parameters are closely related and are challenging to measure in vivo.

Tension in syringeal vibratory soft tissues is affected by (1) the viscoelastic properties of the tissue and (2) the forces that act on the tissue during oscillation (Goller and Riede 2013; Riede and Goller 2014). In several songbirds, the vibrating soft tissues consist of several layers of extracellular matrix that likely have different viscoelastic properties (Riede and Goller 2014). Owing to their small size the viscoelastic properties of the syringeal tissue (layers) are not easily determined using conventional one-dimensional stretching techniques (Alipour and Vigmostad 2012). However, using laser Doppler vibrometry, Fee (2002) measured the compliance of the MVM in zebra finches and canaries (*Serinus canaria*) in vitro using sound on top of static pressure to excite the MVM. The dominant resonant modes of the

MVM were mostly set by the heavier ML and corresponded well to the lower f_0 range found in these species (600 Hz and 1.7 kHz respectively). These measurement of elastic nonlinearity quantified the membrane spring constant due to static displacements, but not the dynamic nonlinearities associated with large-amplitude membrane oscillations as observed in vitro (Fee et al. 1998) and ex vivo (Elemans et al. 2015). For example, movement of the surrounding tissue that anchors the membrane could increase the effective-mass in oscillation (Fletcher 1988). In addition, owing to differential stiffness, adjacent tissue layers could also contribute differentially during small or large vibration amplitudes. A deep layer can serve as a base on which the more compliant outer layer can oscillate (Fee 2002; Riede and Goller 2010b), analogous to the "body-cover model" in human vocal fold oscillations (Titze 1988), changing the effective mass in oscillation.

The forces modulating the tension in syringeal soft tissues during oscillation are (1) aerodynamic forces and (2) forces exerted by position and motion of the syringeal skeleton. First, aerodynamic forces are fluid pressures from the bronchi, trachea, and surrounding the ICAS. For example, a two-mass model of sound production in doves suggests that pressure differentials occur between the ICAS and subsyringeal pressure leading transmural pressure on the LVMs (Elemans et al. 2008b). The magnitude of this transmural pressure is low compared to isometric muscle stress and has little effect on f_0. However, experimental manipulation of ICAS pressure and models in zebra finches show that pressure differentials affect tension and f_0 (Amador and Margoliash 2013). The pressure in the trachea is difficult to measure reliably, as the local flow around the measuring cannula is much higher than in an air sac. Therefore orientation and cannulae size strongly affect the measured pressure (Gaunt et al. 1982).

Second, the forces exerted by the syringeal skeleton modulate labial position, length, and thus tension. However, no experimental data are available as to how much force is required to stretch the LVM or how and if strain develops uniformly in the MVM under static let alone dynamic conditions during large-amplitude vibration.

While the sparse measurements have focused on the MVM in songbirds, the ipsilateral lateral labium is even harder to visualize. LL dynamics have not been quantified and thus its role in syringeal dynamics is not clear. Düring et al. (2013) hypothesizes that the ML sets the oscillation frequency and the LL is mechanically coupled to this oscillation. The position of the LL could modulate the phonation threshold pressure and as such the self-sustained oscillation could be stopped or started perhaps more precisely than by subsyringeal pressure fluctuations alone.

Because of highly nonlinear tissue properties and fluid–structure interactions, sound production is a highly nonlinear dynamical system (Fee et al. 1998; Laje and Mindlin 2005). Such systems have several stable modes of oscillation, comparable to human voice registers such as chest and falsetto (Titze 2000). During small-amplitude oscillations or falsetto only the top layer of the MVM could be involved in oscillation, analogous to soft human voice or the falsetto register. At higher amplitude vibration more mass could be entrained into the oscillation, resulting in a lower f_0. Nonlinear interaction such as biphonation can occur as a result of interaction between left and right sounds sources (Nowicki and Capranica 1986), but also on one side (Zollinger et al. 2008). However, evidence for nonlinear phenomena observed in nonlinear dynamical systems requires thorough analysis such as topology analysis (Fletcher 2000; Elemans et al. 2010).

5.3.4 Vocal Tract Filtering

The pressure and flow fluctuations generated in the syrinx flow into the upper vocal tract, which consists of resonant spaces that filter the pressure variations produced in the syrinx. If the minimal trachea resonance frequency is several times higher than the fundamental produced by the source, the length of the trachea can emphasize frequency bands or formants. More than 60 species of birds have remarkably elongated tracheas to achieve a closer formant spacing, which might be perceived as a larger body size (Fitch 1999). Songbirds can filter sound with a dynamic filter, as first suggested by Nowicki (1987) based on reduced-gas density experiments in several songbird species. The resonant cavity that acts as such a filter is the OEC as identified by Riede et al. (2006) using X-ray imaging of singing cardinals. Studies in several others species, such as zebra finch (Ohms et al. 2010; Riede et al. 2012), have shown that OEC volume is adjusted dynamically, so its resonance matches fundamental frequency to create tonal sounds (Riede et al. 2006), or supports certain harmonics (Riede et al. 2010), thus also acting as an amplifier.

In human speech, much information is contained in the relative spacing of frequency peaks, called formants, which are controlled mainly by suppressing certain harmonics through changes in lip, tongue, and jaw position. Birds do not possess any lips and the tongue is relatively immobile. However, in parrots (*Myiopsitta monachus*) the tongue can act as an articulator to modulate formant spacing (Beckers et al. 2004). Although beak gape was shown to correlate with f_0, this correlation varies substantially between individuals and species (Riede and Suthers 2008). However, there appears to be a beak-gape threshold, beyond which its acoustic effects remain constant (Fletcher and Tarnopolsky 1998; Nelson et al. 2005). Nevertheless the beak plays a crucial role in sound production, as it determines in which direction sound is radiated.

5.3.5 Mechanical Feedback from Environment to Body and Between Subsystems

In each subsystem aerodynamic forces of the environment act on the body and thus provide mechanical feedback within each subsystem of the embodied vocal control system (Fig. 5.1b). The magnitude of these forces has not been quantified and at this point one can only speculate on the exact strength of the fluid–structure feedback. First, the air sac pressures in the respiratory system range from −1 kPa during inspiration to 6 kPa during expiration (Brackenbury 1979). These pressures are generated by expansion and compression and therefore will affect the motion of the ventilation skeleton. Second, sound pressure fluctuations within the trachea have not been measured but are likely small compared to respiratory pressures. The resulting forces that would affect the motion of the upper vocal tract, trachea, and OEC will be negligible in many species. On the other hand, some species use tracheal pressure to inflate upper vocal tract structures such as the crop in pigeons and sage grouses (Riede et al. 2004), and the fluid forces clearly affect upper vocal tract structures mechanically. Third, strong fluid–structure interactions during sound production shape the large-amplitude oscillations of syringeal LVMs and MVMs (Fee et al. 1998; Elemans et al. 2015). These forces are sufficient to induce motion of syringeal skeletal elements during sound production in excised syrinx experiments, such as B3 in zebra finches (Fee 2002). To conclude, within each subsystem fluid forces act on the body, and these are expected to be most significant for the ventilation system and syringeal subsystem.

Forces acting in the fluid domain mechanically couple the three subsystems. First, the acoustic feedback of the upper vocal tract on syringeal dynamics can be considered weak, and in most cases the spectrum of the radiated sound is a linear superposition of the spectrum at the source and the filter properties of the vocal tract (Fant 1960). However, when vocal tract resonances are close to the syringeal vibration strong feedback interactions may occur from filter to source (Riede et al. 2006). Changes of vocal tract properties can enhance the driving pressures of the oscillating tissue and the glottal flow, thereby increasing the energy level at the source (Klatt and Klatt 1990; Titze et al. 2008). The avian larynx has no function in sound generation, but glottal opening can play a critical role in tuning radiated power output in doves (Fletcher et al. 2004) and perhaps also songbirds (Riede et al. 2006). In humans, increasing vocal tract resistance by singing through a straw that is held underwater has been shown to affect both vocal tract inertia and perhaps vocal fold dynamics (Enflo et al. 2013; Guzman et al. 2013).

Second, acoustic pressure fluctuations generated in the syrinx radiate both into the trachea, but can also be measured back in the thoracic air sacs (Goller and Cooper 2004). These pressure fluctuations are small compared to the respiratory pressures, and thus will have minor effects, but syringeal dynamics will directly affect driving pressure because it changes the resistance to flow through the vocal tract.

The strongest interaction between the three subsystems, however, occurs in the syrinx, where dynamic fluid–structure interactions cause tissue collisions and acoustic excitation. To be able to quantify the fluid–structure interactive forces at play, computational models will be indispensable. Current computational models of MEAD sound production built on highly successful models of human speech synthesis (Ishizaka and Flanagan 1972) that include an unilateral vocal cord consisting of one or two masses. These models are driven by lumped parameters such as tension and pressure and the flow in the acoustic output and can accurately describe radiated sound. To quantify the forces involved in complicated fluid–structure dynamics other computational approaches such as finite element models may be needed (Mittal et al. 2013).

5.4 The Body: Musculoskeletal Biomechanics

5.4.1 Muscle Mechanics

Muscles are often viewed as motors that produce movement by shortening to perform mechanical work (termed "actuation" in engineering and robotics). However, vertebrate skeletal muscles serve a variety of other functions during movement and stabilize motion at joints, store elastic energy in connective tissues, and absorb work as well as perform it (Dickinson et al. 2000). The most essential properties for embodied vocal control will be briefly outlined.

A vertebrate striated muscle is a hierarchically organized tissue consisting of multinucleated muscle fibers (cell) that contain multiple parallel myofibrils that consist of series of sarcomeres. The process of activation–contraction (AC) coupling describes how depolarization of the sarcolemma caused by an action potential initiates shortening and force development due to sliding filaments of actin and myosin. In brief, as a motor nerve action potential crosses the neuromuscular junction of a skeletal muscle, the action potential propagates over the sarcolemma and into the t-tubule system, where it releases calcium ions from the sarcoplasmic reticulum into the cell. These free calcium ions change the conformation of the troponin–tropomyosin complex, allowing myosin motor heads to form cross-bridges to bind to actin filaments. The hundreds to thousands of myosin motor heads on a myosin filament cycle through binding and unbinding steps as long as ATP and binding sites (and thus free calcium ions) are available. These steps cause myosin head rotation and produce force on and perhaps sliding of the actin filaments, leading to shortening of the sarcomere. The slowest process in this chain sets the speed of a muscle, such as the kinetics of calcium pumping or detachment rate of myosin from actin (Rome 2006). The currents due to membrane depolarization can be recorded with electromyography (EMG) electrodes as a proxy for activation of local muscle fibers (Loeb and Gans 1986).

The force development by muscle tissue depends strongly on imposed length changes through the intrinsic nonlinear force–length and force–velocity relationship of the muscle (Fig. 5.1a) (Dickinson et al. 2000). The force–length relationship is explained by the mechanism that sarcomere force development depends on the spatial overlap of actin and myosin proteins and thus the amount of cross-bridges that can be formed (Huxley and Hanson 1954; Huxley and Niedergerke 1954). This property can be scaled up to the length of the entire muscle and still hold well. With increasing load the shortening velocity of a muscle decreases up to a muscle-specific maximum shortening velocity at zero load, because the amount of cross-bridges formed and the force they generate per head are lower at higher speeds (Hill and

Howarth 1957). As a consequence of these intrinsic properties the action of a muscle can be to generate positive work when activated during shortening (concentric action), no work when activated without length change (isometric action), and negative work when activated during lengthening (eccentric action) (Knudson 2007). These intrinsic muscle properties have been measured successfully in vitro using the work-loop technique (Josephson 1985).

The mechanical context can thus lead to different functions of the muscle. For example, in inclined walking versus level walking, the length of the gastrocnemius muscle is slightly longer during the stance phase (Roberts et al. 1997). With identical activation patterns (EMG), the gastrocnemius muscle generates positive power during uphill running, but acts as a strut during level running, which allows the spring-like tendons to store and recover energy (Roberts et al. 1997). Thus although activation and force production can be nearly identical, the function of the muscle is different because of its mechanical context alone. Consequently, an identical neural command can produce variable mechanical outputs. This can be true in adjacent muscles but also in adjacent segments within a single muscle (Nishikawa et al. 2007). The intrinsic force–velocity and force–length properties can also play a role in stabilizing movement in response to perturbations and thus simplify control (Biewener and Daley 2007). The timing of activation, not even an experimental parameter in isometric or isotonic studies, emerges as an important variable by which the nervous system can regulate mechanical performance.

An additional intrinsic factor modulating force is that activated muscles can generate a delayed force increase after a stretch (Reger 1978). This phenomenon is called stretch-activation or residual force enhancement (Nocella et al. 2014). It has been suggested to depend on myosin isoform in vertebrate skeletal muscle (Galler et al. 1997), but a recent study suggests that sarcomeric stiffness increases as a result of calcium-dependent stiffening of the structural titin protein (Nocella et al. 2014).

Last, another important feature of muscle force is the differential recruitment of single motorneurons and their associated muscle fibers (i.e., the motor unit [MU]). The number of muscle fibers within an MU affects control resolution and ranges from three or four muscle fibers, as in extraocular muscle with fine control, to thousands of fibers, as in the thigh muscle m. quadriceps for coarser control resolution. A muscle model comprising MU that could be independently activated resulted in more accurate predictions of force than traditional Hill-type models (Wakeling et al. 2012), and different MU recruitment patterns are more or less suited to different movement tasks (Hodson-Tole and Wakeling 2009).

To conclude, the mechanical output of skeletal muscle depends on many factors, some of which are well understood, such as anatomy, kinematics, neural stimulation timing, contraction kinetics, force–velocity characteristics, and force–length relationships. But without a quantitative understanding of these parameters, a muscle's work production and thus its function cannot be properly assessed.

5.4.2 Body Motion

As argued in Sect. 5.4.1, muscle length changes as imposed by motion of the body strongly affect the action of a muscle. It is thus critical to quantify the body motion that affects muscle length to understand the mechanical effect of neural commands to that muscle. In locomotion studies the motion of limbs is measured relatively easily by reconstructing the path of external markers placed on the body. In stark contrast, in avian vocal production, any morphological markers for the three subsystems, with beak opening being the exception, are all either inside the body or covered by feathers. Therefore X-ray cineradiography and magnetic resonance imaging (MRI) are the best available options, but the fast modulations of small structures during vocalizations remain a challenge for the current temporal and spatial resolution of these techniques. The dynamics of body motion during avian vocalization has been only sparsely studied and thus forms a major lacuna in our knowledge of vocal motor control.

Skeletal kinematics has been quantified during breathing of larger birds such as emus, guinea fowl (Claessens 2009), and pigeons (Baumel et al. 1990), but not during vocalizations and not in songbirds. The air pressure in the air sacs depends on their shape changes, which is modulated by complicated skeletal kinematics caused by shortening of the intercostal and abdominal muscles (Tickle et al. 2007; Zimmer 1935). The generally accepted model for inspiration and airflow mechanics in an air sac flow-through system is the skeletal aspiration pump (Tickle et al. 2012). During inspiration, an initial caudoventral movement of the sternum followed by craniodistal displacement of the sternal ribs (Fig. 5.8) results in an expansion of the caudoventral air sacs. At the same time, expansion of the cranial air sacs is limited by the furcula, coracoid, and scapula, resulting in a negative pressure gradient from craniad to caudad, which is amplified through the greater lever of the caudal ribs. During expiration (Fig. 5.8), caudal air sacs are compressed through caudo-proximal displacement of the sternal ribs accompanied by a craniodorsal movement of the sternum. The simultaneous expansion of the cranial air sacs is amplified by furcular bending. Both inspiration and expiration are thus active processes that require work by respiratory muscles (King 1993; Claessens 2009). How birds maintain a near constant airflow during ventilation is still not fully understood (Mackelprang and Goller 2013).

At present, little quantitative information is available on the syringeal skeletal motion during vocalizations. Sparse observations include endoscopic images of syringeal motion induced by muscle or brain stimulation in situ in anesthetized songbirds, parrots, and pigeon (*Columba livia*) (Goller and Larsen 1997a; Larsen and Goller 2002) and schematic drawings of syringeal motion in crows (*Corvus brachyrhynchos*; Chamberlain et al. 1968) and oilbirds (*Steatornis caripensis*; Suthers and Hector 1985). The only record of quantified syringeal skeletal motion to our knowledge is manual B3 displacement and its effects on f_0 of sound in an

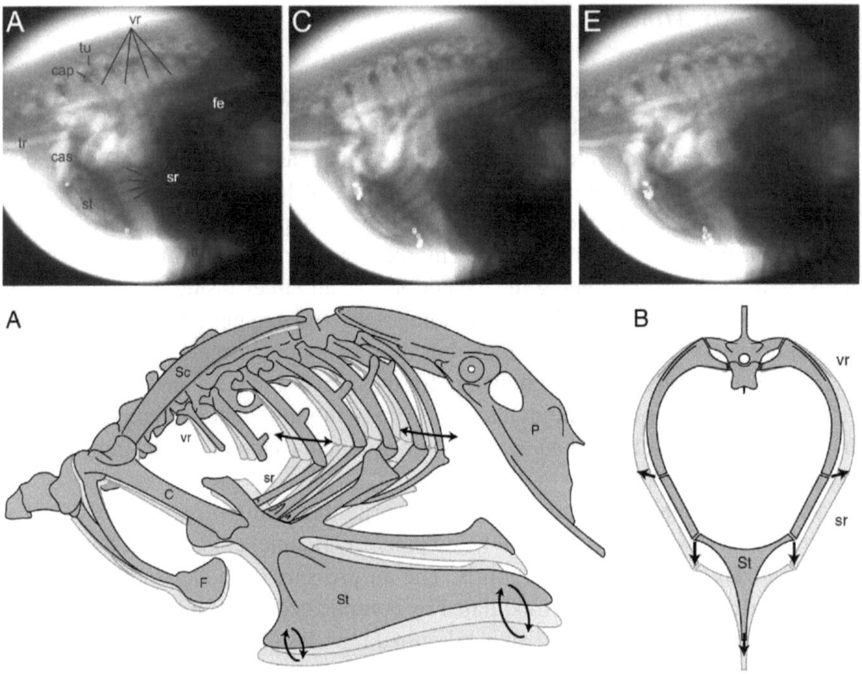

Fig. 5.8 Elliptical body motion during avian respiration. (**a**) Uniplanar X-ray cineradiography of heavily breathing juvenile emu showing the elliptical motion of the sternum (*arrows*) in three consecutive film frames. The excursions of the cranial and caudal margins of the sternum are marked with white dots. Inflation and deflation of the radiotranslucent interclavicular and thoracic air sacs can be observed in the consecutive frames. (**b**) Model of the avian skeletal aspiration pump [Adapted with permission from Claessens (2009)]

excised zebra finch syrinx preparation (Fee et al. 1998). Basic quantitative information regarding the syringeal skeleton such as material properties, joint definitions, joint stiffness, and degrees of freedom are also not available, and without these forces and torques cannot be estimated.

The best quantitative description of body motion during song is of the upper vocal tract length changes (Daley and Goller 2004) and OEC dimensions and beak gape in zebra finches (Fig. 5.9) (Ohms et al. 2010; Riede et al. 2012) and cardinals (*Cardinalis cardinalis*; Riede et al. 2006).

5.4.3 Vocal Muscle: Mechanical Performance and Architecture

Of the three vocal motor subsystems, some basic mechanical properties and architecture of syringeal muscles are known and have shown extreme specializations. The syringeal muscles of doves (Elemans et al. 2004), zebra finches, starlings

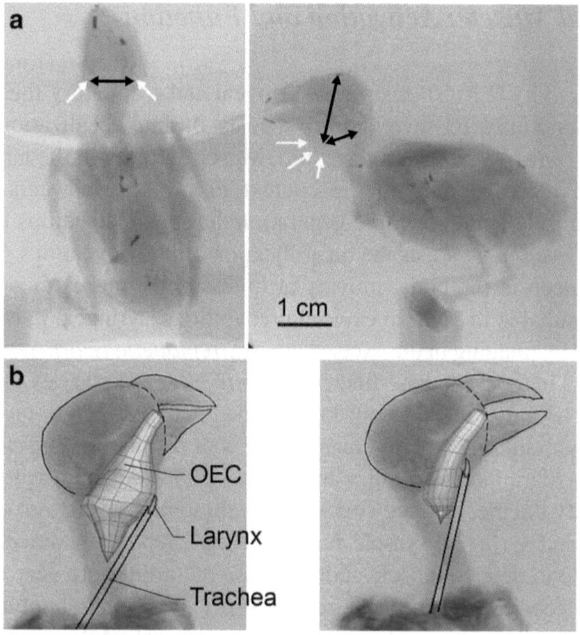

Fig. 5.9 Upper vocal tract motion during vocalization. The oropharyngeal-esophageal cavity (OEC) forms a dynamic filter that tracks the fundamental frequency of the sound produced at the syrinx. (**a**) Biplanar X-ray cineradiography of a zebra finch. The black arrows indicate the measurements taken to estimate OEC shape and volume. [Modified from Riede et al. (2012).] (**b**) Three-dimensional reconstructions of the OEC during singing in the northern cardinal. At the beginning of the syllable, the cranial end of the esophagus is inflated (*left*), but collapses at the end of the syllable (*right*), leading to an OEC reduction from 2.0 to 0.6 mL over the course of this syllable [Adapted from Riede et al. (2006)]

(*Sturnus vulgaris*; Elemans et al. 2008a), and probably most birds belong to the fastest vertebrate muscles, the superfast muscles, that can produce work greater than 100 Hz. Using the work-loop technique in vitro, syringeal muscles of starlings produce power up to 250 Hz. But also in situ with realistic loads, these muscles could modulate syringeal flow up to 250 Hz (Elemans et al. 2008a). Velocity–length properties including maximum shortening velocity were not measured. Superfast muscles have been associated mostly with sound-producing systems (Elemans et al. 2011), where modulations are fast, but the required forces are low, as superfast muscles trade speed for force (Rome 2006). Syringeal muscles in starling contain two fiber types (Uchida et al. 2010) and are organized mostly in sheets of parallel fibers (Düring et al. 2013). Generally, muscles with parallel architecture favor range of motion over force development (Knudson 2007). Tendons are mostly absent or extremely short, as many muscles insert directly on syringeal bones in zebra finches (Düring et al. 2013). The muscle mass can be lateralized and sexually dimorphic (Wade and Buhlman 2000).

5.4.4 Vocal Muscle: Activation and Function

The chronic activity (EMG) of several syringeal and respiratory muscles has been measured successfully in several bird species during vocalization (Goller and Cooper 2008; Suthers and Zollinger 2008). Muscle activity in the larynx and pharynx has been recorded only in the context of drinking (Zweers et al. 1994). Quantifying muscle length or force generation during vocalizations faces the same difficulties, or even more so, as the quantification of body dynamics. The small size of the syrinx deep in the body, moving at high speeds, makes most experimental techniques unsuitable in vivo. In excised preparations, syringeal muscle shortening was quantified by imaging in the excised dove (*Streptopelia risoria*) syrinx (Elemans et al. 2006) and in the Bengalese finch (*Lonchura striata*) syrinx ex vivo (Srivastava et al. 2015). The only record to our knowledge of muscle length measurements in combination with muscle activity during song is of respiratory muscles using sonomicrometry in zebra finches (Goller and Cooper 2004). Such measurements can be used to quantify the mechanical work done by these muscles in vitro.

To understand syringeal muscle function, EMG recordings have been the best proxy available to deduce muscle action. These recordings are very challenging to obtain during vocalization in any bird and especially small songbirds. Therefore most efforts have been invested in correlating syringeal EMG signal amplitude of energy with specific acoustic parameters. In some cases a rather clear function could be assigned to some muscles, such as frequency modulator (VS), adductor (DTB), and abductor (VTB) (Suthers and Goller 1997; Goller and Riede 2013). However, as argued earlier in this section, based on EMG alone one cannot deduce force or action. The correlations between EMG and acoustic parameters have also been mostly reported for specific syllables in specific species. A recent study showed that individual muscles (including VS) affects multiple sound parameters and that these relationships are also dependent on the syllable (Srivastava et al. 2015), and thus likely to be dependent on the mechanical context. Given the complexity describing the forward and feedback relations shown in Fig. 5.1b, this is perhaps not surprising, but nevertheless an important observation that supports the need for an integral approach to embodied vocal motor control. In addition to functions related to skeletal motion, syringeal muscle activation can also damp the action of other muscles (Wilson et al. 2001). Co-contraction of syringeal muscles could affect the stiffness of joints, which is known to be an important factor to modulate the accuracy of motion of the joint in locomotion (Gribble et al. 2003), or change the stability to perturbations or noise in the motor commands (Selen et al. 2006).

5.4.5 Mechanical Interaction Between Body and Muscle

To summarize, for all vertebrate skeletal muscles force development depends strongly on the intrinsic nonlinear force–length and force–velocity properties (Fig. 5.1a). In addition, muscle lengthening can cause force increase through the phenomenon of residual force enhancement. In each vocal motor subsystem length changes are imposed on the muscle by the motion of the skeleton (Fig. 5.1b).

Labial oscillations have been observed to cause motion of B3 (Fee 2002), and thus affect length of the muscles inserting on B3. These vibrations were above 200 Hz and thus past the performance limit of the muscles. In vivo coactivation of other syringeal muscles may dampen such motion and thus reduce the propagation of mechanical feedback.

5.5 The Brain: Neural Control

The brain nuclei associated with song production and learning, dubbed the song system, have, to a large extent, been identified and exist in several songbird species studied (Wild 1997). Current hypotheses regarding circuitry function within the song system are under active investigation and are not covered here (for recent reviews see Fee and Goldberg 2011; Brainard and Doupe 2013).

5.5.1 Motor Mapping: Anatomical and Functional Connectivity

The avian song system sends complicated sequences of motor commands to muscles in the three motor subsystems and various forms of sensory feedback shape the motor commands (Fig. 5.1b). The anatomical connectivity of the descending motor pathway spans from the telencephalon (nucleus HVC) to a brainstem vocal-respiratory network, and an anterior forebrain pathway traversing the telencephalon, striatum, and thalamus (Mooney et al. 2009).

This vocal-respiratory network controls the three motor subsystems essential to sound production: the respiratory system, vocal organ, and upper vocal tract (Wild 2004). In songbirds, the robust nucleus of the arcopallium (RA) projects onto the syringeal and tracheal musculature via motor neurons in the tracheosyringeal portion (nXIIts) of the hypoglossal nucleus (Vicario and Nottebohm 1988). RA also projects on the lingual musculature, which is important for hyoid motion (Zweers et al. 1994) and probably OEC modulation, via motor neurons in the lingual portion of the hypoglossal nerve (nXII-l). RA furthermore projects onto the two main respiratory nuclei in the ventrolateral medulla, the nucleus paraambigualis (PAm) and the nucleus retroambigualis (RAm), that drive expiration and inspiration through motor neurons in the spinal cord (Wild et al. 2009). Another main target of the RA is the dorsomedial nucleus of the intercollicular complex (DM), which is known to play an important role in innate vocalization (Vicario 1991). The close spatial relation and interconnections with the respiratory brainstem make the DM an ideal candidate to control respiration during vocalization (McLean et al. 2013). There is a myotopic organization in nXIIts (Vicario and Nottebohm 1988), some structure of which remains intact in RA (Vicario 1991). The motor neurons in NXIIts are physiologically similar between male and female zebra finches (Roberts et al. 2007). Motorneurons of the laryngeal muscles lie in nucleus ambiguous, near PAm, and their axons travel through the glossopharyngeal nerve (Wild 2004).

To understand how motor commands instruct musculature it is important to map the functional connectivity of the descending motor pathway. This has been studied mostly between HVC to RA, and such intracellular chronic recordings of HVC and RA neurons are technically demanding (Long et al. 2010; Amador et al. 2013). Intracellular recordings in motor neurons of the lower brainstem (nXIIts) have not been reported. Another approach would be to correlate recorded spikes in, for example, RA or HVC to simultaneously recorded EMG of muscles in the three motor subsystems, so-called spike-triggered EMG. Although many studies have shown that firing rates can predict variations in motor output, a recent paper demonstrated that in vocal motor circuits in Bengalese finches information is represented by spike timing, which suggests that variations in timing and not rate evoke differences in behavior (Tang et al. 2014).

The nervous system often does not control individual muscles, but activates flexible combinations of muscles, so-called muscle synergies (Ting and McKay 2007; Tresch and Jarc 2009). Muscle synergies define characteristic patterns of muscle activation to a group of muscle (fiber)s and reduce the dimensionality of control in arm reaching (D'Avella and Lacquaniti 2013) and finger movement control (Tresch and Jarc 2009). It is currently unknown if vocal motor commands from HVC are mapped into synergies (Elemans et al. 2014). A possible anatomical location for such mapping would be from RA to nXII. Because central pattern generator networks for patterning of locomotion and vocal production share a common developmental and evolutionary origin (Bass et al. 2008; Bass and Chagnaud 2012), birdsong may also be controlled with muscle synergies. A recent computational model of a virtual limb showed that the exact location of muscle placement resulted in significant changes in the control synergies and thus strongly affected the controller (Marques et al. 2014). These results emphasize the importance of an embodied approach to motor control: Each individual has to learn to interact with its own body.

5.5.2 Sensory Feedback

Auditory feedback is critical for song development and maintenance in songbirds (Brainard and Doupe 2013). This particular feature has made the birdsong system successful as an experimental animal model for human speech acquisition (Doupe and Kuhl 1999). However, additional modalities of sensory feedback may play a role in shaping vocal output (Wild 2004). The biomechanics of sensors is not included in this chapter.

The syrinx receives sensory innervation via nXIIts (Bottjer and Arnold 1982), but the type of sensor has not yet been identified; however, regular muscle spindles have not been observed in syringeal muscles (Wild et al. 2009; C. P. H. Elemans, pers. obs.). Anatomical evidence for sensory feedback from the upper vocal tract is sparse: It is unknown if there is somatosensory feedback coming from the larynx, but afferent nerves projecting from the beak-closing muscles through the trigeminal nerve are present (Gagliardo et al. 2006). Stretch activation of those nerves during beak opening could play a role in fine-tuning beak aperture (Gagliardo et al. 2006).

Somatosensory feedback is present to expiratory muscles of the respiratory system (Suthers et al. 2002). Anatomical evidence for afferent projections was found in the vagal nerve (nX) and in the nucleus tractus solitaries (nTS). Their origin was from unknown receptors in air sac walls (Wild 2004). The continuing projections of those afferents are, among others, PAm, nucleus uvaeformis (Uva), and the HVC. This suggest a close interaction between somatosensory feedback of the respiratory system and its control circuits, but also to motor control systems of vocal production (McLean et al. 2013; Wild et al. 2009). Experimental disruption of somatosensory feedback by vagal (Méndez et al. 2010) and hypoglossal (Bottjer and To 2012) lesions causes changes in the motor patterns of song and this feedback is thus likely to be functionally important to vocal learning.

5.6 Embodied Motor Control: Brain, Body, Environment

Understanding how behavior arises from the physiological complex of sensory, neural, and motor subsystems requires an understanding of how information flows through the closed loop network of motor control (Roth et al. 2014). From our conceptual framework for embodied vocal motor control (Fig. 5.1b), it becomes evident that mechanical feedback is present or can be expected in each motor subsystem. Furthermore, the three subsystems, but most dominantly the respiratory system and syrinx subsystem, are coupled mechanically by complex fluid–structure interactions during the production of sound. Compared to locomotion systems, where limb position is the control target, the vibratory motion of the sound producing tissues is orders of magnitude too fast for within-cycle neural feedback. Even modulation of the syringeal skeleton by superfast muscles is too fast (4–10 ms) for meaningful neural feedback, which could explain the absence of sensory feedback sensors such as muscle spindles. Local control thus consists predominantly of the intrinsic stabilizing properties of muscle (Nishikawa et al. 2007). The soft tissues at the fluid–structure interface "compute" local solutions, potentially simplifying neural control computations, a phenomenon termed morphological computation (Hoffmann and Pfeifer 2012).

Our conceptual framework for embodied vocal motor control (Fig. 5.1b) emphasizes anticipated difficulties in correlating muscle activity to acoustic properties, because forces in the body and environment need to be included. If muscle synergies are present (perhaps even individually unique ones) this adds yet another level of complexity to correlate premotor neural commands to sound. Owing to the experimental difficulties of measuring forces in this system in vivo, much data are lacking. A very fruitful approach comes from studies using computational models to model sound production driven by simple motor "gestures," which are lumped control parameters such as labial tension (Amador and Mindlin 2014). These low-dimensional models present yet the most integral efforts to model motor control and can be tuned to drive auditory activity similar to the bird's own song (Amador et al. 2013). These models purposely reduce dimensionality and for other aspects of the motor control, such as fluid–structure interaction, other modeling approaches may be fruitful to understand the interaction between forces and motion of the vocal body (Elemans 2014).

Vocal motor control in birds is a closed-loop control system, and not a feed-forward system, in which motor commands lead to a predictable output without error correction. In engineering, feedback control is used in systems in which the mechanics cannot be predicted accurately enough by an internal model of the controlled process. Thus the emphasis on auditory feedback in vocal motor control hints to the notion that because the internal mechanics are so complicated, intermediate steps are insufficiently reliable to compute the outcome; accurate control is useful only on the target parameter, that is, sound pressure. To understand how the brain controls vocal behavior, the brain, the body, and their interactions with the environment need to be taken into account.

5.7 Summary

This chapter aimed to provide a conceptual framework for embodied vocal motor control in birds. The song system produces motor commands to three motor subsystems: the respiratory system, the vocal organ—the syrinx—and the upper vocal tract. By muscle action these motor commands initiate motion of the musculoskeletal system that interacts with a fluid environment. This environment produces dynamic forces back on the body and the intrinsic properties of muscle tissue strongly depend on body motion. The behavior of the body is sensed by multiple sensory modalities providing feedback to the neural network, shaping future motor commands. The control system thus incorporates both mechanical and neural feedback and therefore forms a closed loop of which the biomechanics form an integral part. Our conceptual framework identifies and discusses the forces produced in the three embodied motor subsystems and various feedback mechanisms. It becomes evident that compared to other neuromechanical systems the most essential quantitative information is lacking, such as the biomechanical properties of tissues, definitions of skeletal motion, force production by muscle action, and relationships between driving parameters and fluid–structure interaction during sound production. However, the great advantage of the birdsong system is that it, in contrast to humans, provides much better experimental opportunities to quantify all aspects of the vocal motor control system. Neural activity and many physiological parameters can be monitored chronically in vivo during highly stereotyped song because of ongoing technological advances. Furthermore, complementary measurement techniques, for example, in vitro and ex vivo, and the development of computational models focusing on different levels of organization within the system, are essential to fill the gaps where experimental observation or perturbation in vivo remains too challenging. The combination of these exciting developments, and the large natural variation present in the system, make vocal motor control in birds an excellent model system in which exciting discoveries are waiting to be made.

References

Alipour, F., & Vigmostad, S. (2012). Measurement of vocal folds elastic properties for continuum modeling. *Journal of Voice, 26*(6), 816. doi:10.1016/j.jvoice.2012.04.010.

Amador, A., & Margoliash, D. (2013). A mechanism for frequency modulation in songbirds shared with humans. *Journal of Neuroscience, 33*, 11136–11144. doi:10.1523/JNEUROSCI.5906-12.2013.

Amador, A., & Mindlin, G. B. (2014). Low dimensional dynamics in birdsong production. *European Physical Journal B, 87*, 300. doi:10.1140/epjb/e2014-50566-5.

Amador, A., Perl, Y. S., Mindlin, G. B., & Margoliash, D. (2013). Elemental gesture dynamics are encoded by song premotor cortical neurons. *Nature, 495*, 59–64. doi:10.1038/nature11967.

Ames, P. L. (1971). The morphology of the syrinx in passerine birds. *Peabody Museum Natural History, Yale University, 37*, 1–194.

Appel, F. W. (1929). Sex dimorphism in the syrinx of the fowl. *Journal of Morphology, 47*, 497–517.

Bass, A. H., & Chagnaud, B. P. (2012). Shared developmental and evolutionary origins for neural basis of vocal–acoustic and pectoral–gestural signaling. *Proceedings of the National Academy of Sciences of the USA, 109*, 10677–10684. doi:10.1073/pnas.1201886109.

Bass, A. H., Gilland, E. H., & Baker, R. (2008). Evolutionary origins for social vocalization in a vertebrate hindbrain-spinal compartment. *Science, 321*, 417–421. doi:10.1126/science.1157632.

Baumel, J. J., King, A. S., & Breazil, J. E. (1993). *Handbook of avian anatomy: Nomina anatomica avium*. Cambridge, MA: Publications of the Nuttall Ornithological Club.

Baumel, J. J., Wilson, J. A., & Bergren, D. R. (1990). The ventilatory movements of the avian pelvis and tail: Function of the muscles of the tail region of the pigeon (*Columba livia*). *Journal of Experimental Biology, 151*, 263–277.

Beckers, G. J. L., Nelson, B. S., & Suthers, R. A. (2004). Vocal-tract filtering by lingual articulation in a parrot. *Current Biology, 14*, 1592–1597. doi:10.1016/j.cub.2004.08.057.

Beckers, G. J. L., Suthers, R. A., & Ten Cate, C. (2003). Mechanisms of frequency and amplitude modulation in ring dove song. *Journal of Experimental Biology, 206*, 1833–1843.

Bennet-Clark, H. C. (1999). Resonators in insect sound production: How insects produce loud pure-tone songs. *Journal of Experimental Biology, 202*, 3347–3357.

Berwick, R. C., Okanoya, K., Beckers, G. J. L., & Bolhuis, J. J. (2011). Songs to syntax: The linguistics of birdsong. *Trends in Cognitive Science, 15*, 113–121. doi:10.1016/j.tics.2011.01.002.

Biewener, A. A., & Daley, M. A. (2007). Unsteady locomotion: Integrating muscle function with whole body dynamics and neuromuscular control. *Journal of Experimental Biology, 210*, 2949–2960.

Bolhuis, J. J., Zijlstra, G. G., Den Boer-Visser, A. M., & Van Der Zee, E. A. (2000). Localized neuronal activation in the zebra finch brain is related to the strength of song learning. *Proceedings of the National Academy of Sciences of the USA, 97*, 2282–2285. doi:10.1073/pnas.030539097.

Bottjer, S. W., & Arnold, A. P. (1982). Afferent neurons in the hypoglossal nerve of the zebra finch (*Poephila guttata*): Localization with horseradish peroxidase. *Journal of Comparative Neurology, 210*, 190–197. doi:10.1002/cne.902100209.

Bottjer, S. W., & To, M. (2012). Afferents from vocal motor and respiratory effectors are recruited during vocal production in juvenile songbirds. *Journal of Neuroscience, 32*, 10895–10906. doi:10.1523/JNEUROSCI.0990-12.2012.

Brackenbury, J. H. (1979). Power capabilities of the avian sound-producing system. *Journal of Experimental Biology, 78*, 163–166.

Brackenbury, J. H., Darby, C., & Mahmoud, S. E.-S. (1989). Respiratory function in exercising fowl following occlusion of the thoracic air sacs. *Journal of Experimental Biology, 145*(1), 227–237.

Brainard, M. S., & Doupe, A. J. (2013). Translating birdsong: Songbirds as a model for basic and applied medical research. *Annual Reviews Neuroscience, 36*, 489–517. doi:10.1146/annurev-neuro-060909-152826.

Bretz, W. L., & Schmidt-Nielsen, K. (1971). Bird respiration: Flow patterns in the duck lung. *Journal of Experimental Biology, 54*, 103–118.

Brown, R., Kovacs, C., Butler, J., Wang, N. I. N. G., Lehr, J. O. H. N., & Banzett, R. (1995). The avian lung: Is there an aerodynamic expiratory valve? *Journal of Experimental Biology, 198*(11), 2349–2357.

Casey, R. M., & Gaunt, A. S. (1985). Theoretical models of the avian syrinx. *Journal of Theoretical Biology, 116*, 45–64.

Chamberlain, D. R., Gross, W. B., Cornwell, G. W., & Mosby, H. S. (1968). Syringeal anatomy in the common crow. *The Auk, 85*, 244–252.

Claessens, L. P. A. M. (2009). The skeletal kinematics of lung ventilation in three basal bird taxa (emu, tinamou, and guinea fowl). *Journal of Experimental Zoology, 311*, 586–599. doi:10.1002/jez.501.

Cooper, B. G., & Goller, F. (2004). Multimodal signals: Enhancement and constraint of song motor patterns by visual display. *Science, 303*, 544–546. doi:10.1126/science.1091099.

Cowan, N. J., Ankarali, M. M., Dyhr, J. P., Madhav, M. S., Roth, E., Sefati, S., Sponberg, S., Stamper, S. A., Fortune, E. S., & Daniel, T. L. (2014). Feedback control as a framework for understanding tradeoffs in biology. *Integrative and Comparative Biology, 54*, 223–237. doi:10.1093/icb/icu050.

D'Avella, A., & Lacquaniti, F. (2013). Control of reaching movements by muscle synergy combinations. *Frontiers in Computational Neuroscience, 7*, 42. doi:10.3389/fncom.2013.00042.

Daley, M., & Goller, F. (2004). Tracheal length changes during zebra finch song and their possible role in upper vocal tract filtering. *Journal of Neurobiology, 59*, 319–330. doi:10.1002/neu.10332.

Dickinson, M. H., Farley, C. T., Full, R. J., Koehl, M. A. R., Kram, R., & Lehmann, S. (2000). How animals move: An integrative view. *Science, 288*, 100–106. doi:10.1126/science.288.5463.100.

Doupe, A. J., & Kuhl, P. K. (1999). Birdsong and human speech: Common themes and mechanisms. *Annual Reviews Neuroscience, 22*, 567–631.

Duncker, H. R. (1972). Structure of avian lungs. *Respiration Physiology, 14*, 44–64.

Duncker, H. R. (2004). Vertebrate lungs: Structure, topography and mechanics. A comparative perspective of the progressive integration of respiratory system, locomotor apparatus and ontogenetic development. *Respiratory Physiology & Neurobiology, 144*(2–3), 111–124. doi:10.1016/j.resp.2004.07.020.

Düring, D. N., Ziegler, A., Thompson, C. K., Ziegler, A., Faber, C., Müller, J., Scharff, C., & Elemans, C. P. H. (2013). The songbird syrinx morphome: A three-dimensional, high-resolution, interactive morphological map of the zebra finch vocal organ. *BMC Biology, 11*, 1. doi:10.1186/1741-7007-11-1.

Dürrwang, R. (1974). *Funktionelle biologie, anatomie und physiologie der vogelstimme*. Basel, Switzerland: Aku-Fotodruck.

Elemans, C. P. H. (2014). The singer and the song: The neuromechanics of avian sound production. *Current Opinion in Neurobiology, 28*, 172–178. doi:10.1016/j.conb.2014.07.022.
Elemans, C. P. H, Düring, D. N., Herbst, C. T., Rasmussen, J. H., Zollinger, S. A., & Brumm, H. et al. (2015). Universal mechanism of sound production and control in birds and mammals. *Nature Communications, 6*, 1–13. doi: 10.1038/ncomms9978.
Elemans, C. P. H., Laje, R., Mindlin, G. B., & Goller, F. (2010). Smooth operator: Avoidance of subharmonic bifurcations through mechanical mechanisms simplifies song motor control in adult zebra finches. *Journal of Neuroscience, 30*, 13246–13253. doi:10.1523/JNEUROSCI.1130-10.2010.
Elemans, C. P. H., Mead, A. F., Jakobsen, L., & Ratcliffe, J. M. (2011). Superfast muscles set maximum call rate in echolocating bats. *Science, 333*, 1885–1888. doi:10.1126/science.1207309.
Elemans, C. P. H., Mead, A. F., Rome, L. C., & Goller, F. (2008a). Superfast vocal muscles control song production in songbirds. *PLoS ONE, 3*, e2581. doi:10.1371/journal.pone.0002581.
Elemans, C. P. H., Mensinger, A. F., & Rome, L. C. (2014). Vocal production complexity correlates with neural instructions in the oyster toadfish (*Opsanus tau*). *Journal of Experimental Biology, 217*, 1887–1893. doi:10.1242/jeb.097444.
Elemans, C. P. H., Muller, M., Larsen, O. N., & Van Leeuwen, J. L. (2009). Amplitude and frequency modulation control of sound production in a mechanical model of the avian syrinx. *Journal of Experimental Biology, 212*, 1212–1224. doi:10.1242/jeb.026872.
Elemans, C. P. H., Spierts, I. L. Y., Hendriks, M., Schipper, H., Müller, U. K., & Leeuwen, J. L. (2006). Syringeal muscles fit the trill in ring doves (*Streptopelia risoria* L.). *Journal of Experimental Biology, 209*(Pt 5), 965–977. doi:10.1242/jeb.02066.
Elemans, C. P. H., Spierts, I. L. Y., Müller, U. K., Leeuwen, J. L., & Goller, F. (2004). Bird song: Superfast muscles control dove's trill. *Nature, 431*, 146. doi:10.1038/431146a.
Elemans, C. P. H., Zaccarelli, R., & Herzel, H. (2008b). Biomechanics and control of vocalization in a non-songbird. *Journal of the Royal Society Interface, 5*, 691–703. doi:10.1098/rsif.2007.1237.
Enflo, L., Sundberg, J., Romedahl, C., & McAllister, A. (2013). Effects on vocal fold collision and phonation threshold pressure of resonance tube phonation with tube end in water. *Journal of Speech Language and Hearing, 56*(5), 1530–1538. doi:10.1044/1092-4388(2013/12-0040).
Fant, G. (1960). *Acoustic theory of speech production*. The Hague, the Netherlands: Mouton.
Fee, M. S. (2002). Measurement of the linear and nonlinear mechanical properties of the oscine syrinx: Implications for function. *Journal of Comparative Physiology A, 188*, 829–839. doi:10.1007/s00359-002-0349-z.
Fee, M. S., & Goldberg, J. H. (2011). A hypothesis for basal ganglia-dependent reinforcement learning in the songbird. *Neuroscience, 198*, 152–170. doi:10.1016/j.neuroscience.2011.09.069.
Fee, M. S., & Scharff, C. (2010). The songbird as a model for the generation and learning of complex sequential behaviors. *ILAR Journal, 51*, 362–377.
Fee, M. S., Shraiman, B., Pesaran, B., & Mitra, P. P. (1998). The role of nonlinear dynamics of the syrinx in the vocalizations of a songbird. *Nature, 395*, 67–71. doi:10.1038/25725.
Fitch, W. T. (1999). Acoustic exaggeration of size in birds via tracheal elongation: Comparative and theoretical analyses. *Journal of Zoology, 248*, 31–48.
Fitch, W. T. (2006). Production of vocalizations in mammals. *Visual Communication, 3*, 145.
Fletcher, N. H. (1988). Bird song: A quantitative acoustic model. *Journal of Theoretical Biology, 135*, 455–481.
Fletcher, N. H. (2000). A class of chaotic bird calls? *The Journal of the Acoustical Society of America, 108*, 821–826.
Fletcher, N. H., Riede, T., Beckers, G. J. L., & Suthers, R. A. (2004). Vocal tract filtering and the "coo" of doves. *The Journal of the Acoustical Society of America, 116*, 3750–3756. doi:10.1121/1.1811491.
Fletcher, N. H., & Rossing, T. D. (1998). *The physics of musical instruments* (2nd ed.). New York: Springer.

Fletcher, N., & Tarnopolsky, A. (1998). Acoustics of the avian vocal tract. *The Journal of the Acoustical Society of America, 103*, 2827.

Frank, T., Walter, I., Probst, A., & König, H. E. (2006). Histological aspects of the syrinx of the male mallard (*Anas platyrhynchos*). *Anatomy Histology Embryology, 35*, 396–401. doi:10.1111/j.1439-0264.2006.00701.x.

Gagliardo, A., Ioalè, P., Savini, M., & Wild, J. M. (2006). Having the nerve to home: Trigeminal magnetoreceptor versus olfactory mediation of homing in pigeons. *Journal of Experimental Biology, 209*, 2888–2892. doi:10.1242/jeb.02313.

Galler, S., Hilber, K., & Pette, D. (1997). Stretch activation and myosin heavy chain isoforms of rat, rabbit and human skeletal muscle fibres. *Journal of Muscle Research and Cell Motility, 18*, 441–448. doi:10.1023/A:1018646814843.

Garitano-Zavala, A. (2009). Evolutionary loss of the extrinsic syringeal muscle sternotrachealis in Darwin's Nothura (*Nothura darwinii*). *The Auk, 126*, 134–140. doi:10.1525/auk.2009.08026.

Gaunt, A. S., Gaunt, S. L. L., & Casey, R. M. (1982). Syringeal mechanics reassessed: Evidence from *Streptopelia. The Auk, 99*, 474–494.

Goller, F., & Cooper, B. G. (2004). Peripheral motor dynamics of song production in the zebra finch. *Annals of the New York Academy of Sciences, 1016*, 130–152. doi:10.1196/annals.1298.009.

Goller, F., & Cooper, B. G. (2008). Peripheral mechanisms of sensorimotor integration during singing. In H. P. Zeigler & P. Marler (Eds.), *Neuroscience of birdsong* (pp. 99–114). New York: Cambridge University Press.

Goller, F., & Larsen, O. N. (1997a). A new mechanism of sound generation in songbirds. *Proceedings of the National Academy of Sciences of the USA, 94*, 14787–14791. doi:10.1073/pnas.94.26.14787.

Goller, F., & Larsen, O. N. (1997b). In situ biomechanics of the syrinx and sound generation in pigeons. *Journal of Experimental Biology, 200*, 2165–2176.

Goller, F., & Riede, T. (2013). Integrative physiology of fundamental frequency control in birds. *Journal of Physiology, Paris, 107*, 230–242. doi:10.1016/j.jphysparis.2012.11.001.

Gribble, P. L., Mullin, L. I., Cothros, N., & Mattar, A. (2003). Role of cocontraction in arm movement accuracy. *Journal of Neurophysiology, 89*, 2936–2405. doi:10.1152/jn.01020.2002.

Guzman, M., Laukkanen, A. M., Krupa, P., Horáček, J., Švec, J. G., & Geneid, A. (2013). Vocal tract and glottal function during and after vocal exercising with resonance tube and straw. *Journal of Voice, 27*(4), 523. e19.

Häcker, V. (1900). *Der Gesang der Vögel: Anatomische und biologische Grundlagen*. Jena, Germany: Gustav Fischer.

Henrywood, R. H., & Agarwal, A. (2013). The aeroacoustics of a steam kettle. *Physical Fluids, 25*, 107101. doi:10.1063/1.4821782.

Herbst, C. T., Stoeger, A. S., Frey, R., Lohscheller, J., Titze, I. R., Gumpenberger, M., & Fitch, W. T. (2012). How low can you go? Physical production mechanism of elephant infrasonic vocalizations. *Science, 337*, 595–599. doi:10.1126/science.1219712.

Hérissant, F. D. (1753). Untersuchung über die Stimmwerkzeuge der Vierfüßer und der Vögel. *Mémoiares De l'Academie des Sciences, 1*(457).

Hill, A. V., & Howarth, J. V. (1957). Alternating relaxation heat in muscle twitches. *Journal of Physiology (London), 139*, 466–473.

Hodson-Tole, E. F., & Wakeling, J. M. (2009). Motor unit recruitment for dynamic tasks: Current understanding and future directions. *Journal of Comparative Physiology B: Biochemical, Systems, and Environmental Physiology, 179*, 57–66. doi:10.1007/s00360-008-0289-1.

Hoffmann, M., & Pfeifer, R. (2012). The implications of embodiment for behavior and cognition: Animal and robotic case studies. In W. Tschacher & C. Bergomi (Eds.), *The implications of embodiment: Cognition and communication* (pp. 31–58). Exeter, England: Imprint Academic.

Hogg, D. A. (1982). Ossification of the laryngeal, tracheal and syringeal cartilages in the domestic-fowl. *Journal of Anatomy, 134*, 57–71.

Homberger, D. G., & Meyers, R. A. (1989). Morphology of the lingual apparatus of the domestic chicken, *Gallus gallus*, with special attention to the structure of the fasciae. *American Journal of Anatomy, 186*, 217–57. doi:10.1002/aja.1001860302.

Huxley, T. H. (1871). *A manual of the anatomy of vertebrated animals*. New York: D. Appleton.
Huxley, H., & Hanson, J. (1954). Changes in the cross-striations of muscle during contraction and stretch and their structural interpretation. *Nature, 173*, 973–976.
Huxley, A. F., & Niedergerke, R. (1954). Structural changes in muscle during contraction; interference microscopy of living muscle fibres. *Nature, 173*, 971–973.
Ishizaka, K., & Flanagan, J. L. (1972). Synthesis of voiced sounds from a two-mass model of the vocal cords. *Bell System Technical Journal, 51*, 1233–1268. doi:10.1002/j.1538-7305.1972.tb02651.x.
Jarvis, E. D., Mirarab, S., Aberer, A. J., et al. (2014). Whole-genome analyses resolve early branches in the tree of life of modern birds. *Science, 346*, 1320–1331. doi:10.1126/science.1253451.
Jensen, K. K., Cooper, B. G., Larsen, O. N., & Goller, F. (2007). Songbirds use pulse tone register in two voices to generate low-frequency sound. *Proceedings of Biological Science, 274*, 2703–2710. doi:10.1098/rspb.2007.0781.
Jetz, W., Thomas, G. H., Joy, J. B., Redding, D. W., Hartmann, K., & Mooers, A. O. (2014). Global distribution and conservation of evolutionary distinctness in birds. *Current Biology, 24*, 919–930. doi:10.1016/j.cub.2014.03.011.
Josephson, R. K. (1985). The mechanical power output from striated muscle during cyclic contraction. *Journal of Experimental Biology, 118*, 185–208.
King, A. S. (1989). Functional anatomy of the syrinx. In A. S. King & J. McLelland (Eds.), *Form and function in birds* (pp. 105–192). New York: Academic Press.
King, A. S. (1993). Apparatus respiratorius. In J. J. Baumel (Ed.), *Handbook of avian anatomy* (2nd ed., pp. 257–300). Cambridge, MA: Nuttall Ornithological Club.
Klatt, D. H., & Klatt, L. C. (1990). Analysis, synthesis, and perception of voice quality variations among female and male talkers. *The Journal of the Acoustical Society of America, 87*, 820–857.
Knudson, D. (2007). *Fundamentals of biomechanics*. New York: Springer.
Konishi, M. (1985). Birdsong: From behavior to neuron. *Annual Review of Neuroscience, 8*, 125–170.
Laje, R., & Mindlin, G. (2005). *The physics of birdsong*. Berlin, Germany: Springer.
Larsen, O. N., & Goller, F. (2002). Direct observation of syringeal muscle function in songbirds and a parrot. *Journal of Experimental Biology, 205*, 25–35.
Loeb, G. E., & Gans, C. (1986). *Electromyography for experimentalists* (1st ed.). Chicago: The University of Chicago Press.
Long, M. A., Jin, D. Z., & Fee, M. S. (2010). Support for a synaptic chain model of neuronal sequence generation. *Nature, 468*, 394–399. doi:10.1038/nature09514.
Mackelprang, R., & Goller, F. (2013). Ventilation patterns of the songbird lung/air sac system during different behaviors. *Journal of Experimental Biology, 216*, 3611–3619. doi:10.1242/jeb.087197.
Maina, J. N. (2006). Development, structure, and function of a novel respiratory organ, the lung-air sac system of birds: To go where no other vertebrate has gone. *Biological Reviews, 81*, 545–579. doi:10.1111/j.1469-185X.2006.tb00218.x.
Maina, J. N., Singh, P., & Moss, E. A. (2009). Inspiratory aerodynamic valving occurs in the ostrich, *Struthio camelus* lung: A computational fluid dynamics study under resting unsteady state inhalation. *Respiratory Physiology & Neurobiology, 169*(3), 262–270.
Maina, J. N., West, J. B., Orgeig, S., Foot, N. J., Daniels, C. B., Kiama, S. G., Gehr, P., Mühlfeld, C., Blank, F., Müller, L., Lehmann, A., Brandenberger, C., & Rothen-Rutishauser, B. (2010). Recent advances into understanding some aspects of the structure and function of mammalian and avian lungs. *Physiology Biochemistry Zoology, 83*, 792–807. doi:10.1086/652244.
Marler, P. (1970). A comparative approach to vocal learning: Song development in white-crowned sparrows. *Journal of Comparative and Physiological Psychology, 50*, 83–97. doi:10.1037/h0029144.
Marler, P. (2004). Bird calls: Their potential for behavioral neurobiology. *Annals of the New York Academy of Sciences, 1016*, 31–44. doi:10.1196/annals.1298.034.

Marler, P. R., & Slabbekoorn, H. (2004). *Nature's music: The science of birdsong*. San Diego, CA: Elsevier Academic Press.

Marques, H. G., Bharadwaj, A., & Iida, F. (2014). From spontaneous motor activity to coordinated behaviour: A developmental model. *PLoS Computational Biology, 10*, e1003653. doi:10.1371/journal.pcbi.1003653.

McLean, J., Bricault, S., & Schmidt, M. F. (2013). Characterization of respiratory neurons in the rostral ventrolateral medulla, an area critical for vocal production in songbirds. *Journal of Neurophysiology, 109*, 948–957. doi:10.1152/jn.00595.2012.

Méndez, J. M., Dall'Asen, A. G., & Goller, F. (2010). Disrupting vagal feedback affects birdsong motor control. *Journal of Experimental Biology, 213*, 4193–4204. doi:10.1242/jeb.045369.

Michelsen, A., Towne, W. F., Kirchner, W. H., & Kryger, P. (1987). The acoustic near field of a dancing honeybee. *Journal of Comparative Physiology A, 161*, 633–643. doi:10.1007/BF00605005.

Mittal, R., Erath, B. D., & Plesniak, M. W. (2013). Fluid dynamics of human phonation and speech. *Annual Reviews of Fluid Mechanics, 45*, 437–467. doi:10.1146/annurev-fluid-011212-140636.

Mooney, T. A., Nachtigall, P. E., Taylor, K. A., Rasmussen, M. H., & Miller, L. A. (2009). Auditory temporal resolution of a wild white-beaked dolphin (*Lagenorhynchus albirostris*). *Journal of Comparative Physiology A: Neuroethology, Sensory, Neural, and Behavioral Physiology, 195*, 375–384. doi:10.1007/s00359-009-0415-x.

Nelson, B. S., Beckers, G. J. L., & Suthers, R. A. (2005). Vocal tract filtering and sound radiation in a songbird. *Journal of Experimental Biology, 208*, 297–308. doi:10.1242/jeb.01378.

Nishikawa, K., Biewener, A. A., Aerts, P., Ahn, A. N., Chiel, H. J., Daley, M. A., et al. (2007). Neuromechanics: An integrative approach for understanding motor control. *Integrative and Comparative Biology, 47*(1), 16–54.

Nocella, M., Cecchi, G., Bagni, M. A., & Colombini, B. (2014). Force enhancement after stretch in mammalian muscle fiber: No evidence of cross-bridge involvement. *American Journal of Physiology, 307*, C1123–C1129. doi:10.1152/ajpcell.00290.2014.

Nowicki, S. (1987). Vocal tract resonances in oscine bird sound production: Evidence from birdsongs in a helium atmosphere. *Nature, 325*, 53–55. doi:10.1038/325053a0.

Nowicki, S., & Capranica, R. R. (1986). Bilateral syringeal interaction in vocal production of an oscine bird sound. *Science, 231*, 1297–1299.

Odom, K. J., Hall, M. L., Riebel, K., Omland, K. E., & Langmore, N. E. (2014). Female song is widespread and ancestral in songbirds. *Nature Communications, 5*, 3379. doi:10.1038/ncomms4379.

Ohms, V. R., Snelderwaard, P. C., Ten Cate, C., & Beckers, G. J. (2010). Vocal tract articulation in zebra finches. *PLoS ONE, 5*, e11923. doi:10.1371/journal.pone.0011923.

Parmentier, E., Lagardère, J. P., Braquegnier, J. B., Vandewalle, P., & Fine, M. L. (2006). Sound production mechanism in carapid fish: First example with a slow sonic muscle. *Journal of Experimental Biology, 209*, 2952–2960. doi:10.1242/jeb.02350.

Paulsen, K. (1967). *Das Prinzip der Stimmbildung in der Wirbeltierreihe und beim Menschen*. Frankfurt am Main, Germany: Akademische Verlagsgesellschaft.

Pfeifer, R., Iida, F., & Lungarella, M. (2014). Cognition from the bottom up: On biological inspiration, body morphology, and soft materials. *Trends in Cognitive Science, 18*(8), 404–413. doi:10.1016/j.tics.2014.04.004.

Pfeifer, R., Lungarella, M., & Iida, F. (2007). Self-organization, embodiment, and biologically inspired robotics. *Science, 318*, 1088–1093. doi:10.1126/science.1145803.

Pfenning, A. R., Hara, E., Whitney, O., Rivas, M. V., Wang, R., Roulhac, P. L., et al. (2014). Convergent transcriptional specializations in the brains of humans and song-learning birds. *Science, 346*(6125), 1256846.

Podos, J. (2004). Vocal mechanics in Darwin's finches: Correlation of beak gape and song frequency. *Journal of Experimental Biology, 207*, 607–619. doi:10.1242/jeb.00770.

Prince, B., Riede, T., & Goller, F. (2011). Sexual dimorphism and bilateral asymmetry of syrinx and vocal tract in the European starling (*Sturnus vulgaris*). *Journal of Morphology, 272*, 1527–1536. doi:10.1002/jmor.11007.

Reger, J. F. (1978). A comparative study on the fine structure of tongue and cricothyroid muscle of the bat, *Myotis grisescens*, as revealed by thin section and freeze-fracture techniques. *Journal of Ultrastructure Research, 63*, 275.
Riede, T., Beckers, G. J. L., Blevins, W., & Suthers, R. A. (2004). Inflation of the esophagus and vocal tract filtering in ring doves. *Journal of Experimental Biology, 207*, 4025–4036. doi:10.1242/jeb.01256.
Riede, T., Fisher, J. H., & Goller, F. (2010). Sexual dimorphism of the zebra finch syrinx indicates adaptation for high fundamental frequencies in males. *PLoS ONE, 5*, e11368. doi:10.1371/journal.pone.0011368.
Riede, T., & Goller, F. (2010a). Peripheral mechanisms for vocal production in birds—Differences and similarities to human speech and singing. *Brain and Language, 115*, 69–80. doi:10.1016/j.bandl.2009.11.003.
Riede, T., & Goller, F. (2010b). Functional morphology of the sound-generating labia in the syrinx of two songbird species. *Journal of Anatomy, 216*, 23–36. doi:10.1111/j.1469-7580.2009.01161.x.
Riede, T., & Goller, F. (2014). Morphological basis for the evolution of acoustic diversity in oscine songbirds. *Proceedings of the Royal Society B: Biological Sciences, 281*, 20132306. doi:10.1098/rspb.2013.2306.
Riede, T., Schilling, N., & Goller, F. (2012). The acoustic effect of vocal tract adjustments in zebra finches. *Journal of Comparative Physiology A: Neuroethology, Sensory, Neural, and Behavioral Physiology, 199*, 57–69. doi:10.1007/s00359-012-0768-4.
Riede, T., & Suthers, R. A. (2008). Vocal tract motor patterns and resonance during constant frequency song: The white-throated sparrow. *Journal of Comparative Physiology A: Neuroethology, Sensory, Neural, and Behavioral Physiology, 195*, 183–192. doi:10.1007/s00359-008-0397-0.
Riede, T., Suthers, R. A., Fletcher, N. H., & Blevins, W. E. (2006). Songbirds tune their vocal tract to the fundamental frequency of their song. *Proceedings of the National Academy of Sciences of the USA, 103*, 5543–5548.
Roberts, T. J., Marsh, R. L., Weyand, P. G., & Taylor, C. R. (1997). Muscular force in running turkeys: The economy of minimizing work. *Science, 275*, 1113–1115.
Roberts, T. F., Wild, J. M., Kubke, M. F., & Mooney, R. (2007). Homogeneity of intrinsic properties of sexually dimorphic vocal motoneurons in male and female zebra finches. *Journal of Comparative Neurology, 502*, 157–169. doi:10.1002/cne.21310.
Rome, L. C. (2006). Design and function of superfast muscles: New insights into the physiology of skeletal muscle. *Annual Reviews of Physiology, 68*, 193–221. doi:10.1146/annurev.physiol.68.040104.105418.
Roth, E., Sponberg, S., & Cowan, N. J. (2014). A comparative approach to closed-loop computation. *Current Opinion in Neurobiology, 25*, 54–62. doi:10.1016/j.conb.2013.11.005.
Rüppell, W. (1933). Physiologie und Akustik der Vogelstimme. *Journal of Ornithology, 81*, 433–542.
Schmidt, M. F., McLean, J., & Goller, F. (2012). Breathing and vocal control: The respiratory system as both a driver and a target of telencephalic vocal motor circuits in songbirds. *Experimental Physiology, 97*, 455–461. doi:10.1113/expphysiol.2011.058669.
Selen, L. P. J., Beek, P. J., & Van Dieën, J. H. (2006). Impedance is modulated to meet accuracy demands during goal-directed arm movements. *Experimental Brain Research, 172*, 129–138. doi:10.1007/s00221-005-0320-7.
Setterwall, C. G. (1901). *Syrinx hos polymyoda passeres*. Lund, Sweden: Gleerupska Universitetsbokhandeln.
Slabbekoorn, H., & Smith, T. B. (2002). Bird song, ecology and speciation. *Philosophical Transactions of the Royal Society, B: Biological Sciences, 357*, 493–503. doi:10.1098/rstb.2001.1056.
Srivastava, K. H., Elemans, C. P. H., & Sober, S. J. (2015). Multifunctional and context-dependent control of vocal acoustics by individual muscles. *Journal of Neuroscience, 35*(42), 14183–14194.

Suthers, R. A., & Goller, F. (1997). Motor correlates of vocal diversity in songbirds. In V. Nolan Jr., E. D. Ketterson, & C. F. Thompson (Eds.), *Current ornithology* (pp. 235–288). New York: Springer.

Suthers, R. A., Goller, F., & Wild, M. (2002). Somatosensory feedback modulates the respiratory motor program of crystallized birdsong. *Proceedings of the National Academy of Sciences of the USA, 99*(8), 5680–5685.

Suthers, R. A., & Hector, D. H. (1985). The physiology of vocalization by the echolocating oilbird, *Steatornis caripensis. Journal of Comparative Physiology. A, Neuroethology, Sensory, Neural, and Behavioral Physiology, 156*, 243–266.

Suthers, R. A., Vallet, E., & Kreutzer, M. (2012). Bilateral coordination and the motor basis of female preference for sexual signals in canary song. *Journal of Experimental Biology, 215*, 2950–2959. doi:10.1242/jeb.071944.

Suthers, R. A., & Zollinger, S. A. (2008). From brain to song: The vocal organ and vocal tract. In H. P. Zeigler & P. Marler (Eds.), *Neuroscience of birdsong* (pp. 78–98). Cambridge, UK: Cambridge University Press.

Tang, C., Chehayeb, D., Srivastava, K., Nemenman, I., & Sober, S. J. (2014). Millisecond-scale motor encoding in a cortical vocal area. *PLoS Biology, 12*(12), e1002018.

Taylor, G., Triantafyllou, M. S., & Tropea, C. (2010). *Animal locomotion*. New York: Springer.

Tchernichovski, O., Mitra, P. P., Lints, T., & Nottebohm, F. (2001). Dynamics of the vocal imitation process: how a zebra finch learns its song. *Science, 291*, 2564–2569. doi:10.1126/science.1058522.

Tickle, P. G., Ennos, A. R., Lennox, L. E., Perry, S. F., & Codd, J. R. (2007). Functional significance of the uncinate processes in birds. *Journal of Experimental Biology, 210*, 3955–3961. doi:10.1242/jeb.008953.

Tickle, P. G., Norell, M. A., & Codd, J. R. (2012). Ventilatory mechanics from maniraptoran theropods to extant birds. *Journal of Evolutionary Biology, 25*, 740–747. doi:10.1111/j.1420-9101.2012.02465.x.

Ting, L. H., & McKay, J. L. (2007). Neuromechanics of muscle synergies for posture and movement. *Current Opinion in Neurobiology, 17*, 622–628. doi:10.1016/j.conb.2008.01.002.

Titze, I. R. (1988). The physics of small-amplitude oscillation of the vocal folds. *The Journal of the Acoustical Society of America, 83*, 1536–1552.

Titze, I. R. (2000). *Principles of voice production*. Upper Saddle River, NJ: Prentice Hall.

Titze, I. R., Finnegan, E. M., Laukkanen, A. M., Fuja, M., & Hoffman, H. (2008). Laryngeal muscle activity in giggle: A damped oscillation model. *Journal of Voice, 22*, 644–648. doi:10.1016/j.jvoice.2007.01.008.

Titze, I. R., & Story, B. H. (2002). Rules for controlling low-dimensional vocal fold models with muscle activation. *The Journal of the Acoustical Society of America, 112*, 1064–1076.

Tresch, M. C., & Jarc, A. (2009). The case for and against muscle synergies. *Current Opinion in Neurobiology, 19*, 601–607. doi:10.1016/j.conb.2009.09.002.

Tytell, E. D., Holmes, P., & Cohen, A. H. (2011). Spikes alone do not behavior make: Why neuroscience needs biomechanics. *Current Opinion in Neurobiology, 21*, 816–822. doi:10.1016/j.conb.2011.05.017.

Uchida, A. M., Meyers, R. A., Cooper, B. G., & Goller, F. (2010). Fibre architecture and song activation rates of syringeal muscles are not lateralized in the European starling. *Journal of Experimental Biology, 213*, 1069–1078. doi:10.1242/jeb.038885.

Van Den Berg, J. (1958). Myoelastic-aerodynamic theory of voice production. *Journal of Speech and Hearing Research, 1*, 227–244. doi:10.1044/jshr.0103.227.

Vicario, D. S. (1991). Organization of the zebra finch song control system: II. Functional organization of outputs from nucleus robustus archistriatalis. *Journal of Comparative Neurology, 309*, 486–494. doi:10.1002/cne.903090405.

Vicario, D. S., & Nottebohm, F. (1988). Organization of the zebra finch song control system: I. Representation of syringeal muscles in the hypoglossal nucleus. *Journal of Comparative Neurolog, 271*, 346–354. doi:10.1002/cne.902710305.

Vignal, C., Mathevon, N., & Mottin, S. (2004). Audience drives male songbird response to partner's voice. *Nature, 430*, 448–451. doi:10.1038/nature02645.

Vincent, J. (1990). *Structural biomaterials*. Princeton, NJ: Princeton University Press.

Wade, J., & Buhlman, L. (2000). Lateralization and effects of adult androgen in a sexually dimorphic neuromuscular system controlling song in zebra finches. *Journal of Comparative Neurology, 426*, 154–164.

Wakeling, J. M., Lee, S. S., Arnold, A. S., de Boef Miara, M., & Biewener, A. A. (2012). A muscle's force depends on the recruitment patterns of its fibers. *Annals of Biomedical Engineering, 40*, 1708–1720. doi:10.1007/s10439-012-0531-6.

Wild, J. M. (1997). Neural pathways for the control of birdsong production. *Journal of Neurobiology, 33*, 653–670.

Wild, J. M. (2004). Functional neuroanatomy of the sensorimotor control of singing. *Annals of the New York Academy of Sciences, 1016*, 438–62. doi:10.1196/annals.1298.016.

Wild, J. M., Kubke, M. F., & Mooney, R. (2009). Avian nucleus retroambigualis: Cell types and projections to other respiratory-vocal nuclei in the brain of the zebra finch (*Taeniopygia guttata*). *Journal of Comparative Neurology, 512*, 768–783. doi:10.1002/cne.21932.

Wilson, A. M., McGuigan, M. P., Su, A., & Van Den Bogert, A. J. (2001). Horses damp the spring in their step. *Nature, 414*, 895–899.

Wunderlich, L. (1884). *Beiträge zur vergleichenden Anatomie u. Entwickelungsgeschichte des unteren Kehlkopfes der Vögel*. Halle, Germany: Druck von E. Blochmann.

Zimmer, K. (1935). *Beiträge zur Mechanik der atmung bei den vögeln in stand und flug auf grund anatomisch-physiologischer und experimenteller Studien*. Stuttgart, Germany: Schweizerbart.

Zollinger, S. A., Riede, T., & Suthers, R. A. (2008). Two-voice complexity from a single side of the syrinx in northern mockingbird *Mimus polyglottos* vocalizations. *Journal of Experimental Biology, 211*, 1978–1991. doi:10.1242/jeb.014092.

Zweers, G., Bout, R., & Heidweiller, J. (1994). Motor organization of the avian head-neck system. In M. N. O. Davies & P. R. Green (Eds.), *Perception and motor control in birds* (pp. 201–221). Berlin, Germany: Springer.

Zweers, G. A., Van Pelt, H. C., & Beckers, A. (1981). Morphology and mechanics of the larynx of the pigeon (*Columba livia* L.): A drill-chuck system (Aves). *Zoomorphology, 99*, 37–69. doi:10.1007/BF00310353.

Chapter 6
Biophysics of Vocal Production in Mammals

Christian T. Herbst

Abstract Most mammals, including humans, produce sound in agreement with the myoelastic-aerodynamic theory (MEAD): by converting aerodynamic energy into acoustic energy via flow-induced self-sustaining oscillation of the vocal folds or other laryngeal tissue. The generated laryngeal sound is filtered by the vocal tract and radiated from the mouth and/or the nose.

In this chapter, some basic biophysical principles of the MEAD theory are explained, mostly based on research done in humans. Empirical evidence and concepts for nonhuman mammals are provided when available and applicable.

In particular, biomechanical properties of vibrating laryngeal tissue and respective vibratory modes are described, and the oscillatory components and forces necessary for flow-induced self-sustaining vibration are discussed. The notions of fundamental frequency and its control, periodicity, and irregularity are explored, followed by a basic description of nonlinear phenomena (NLP) such as bifurcations, subharmonics, or chaos. Subglottal pressure and glottal airflow are essential parameters of voice production, and their influence on the generated voice source spectrum is considered. Finally, linear and nonlinear effects of the vocal tract are reviewed, and the efficiency sound production is discussed.

Keywords Chaos • Glottal airflow • Fundamental frequency control • Glottal efficiency myoelastic aerodynamic theory (MEAD) • Nonlinear interaction • Nonlinear phenomena (NLP) • Periodicity • Self-sustaining oscillation • Source-filter theory • Subharmonics • Vibratory modes • Vocal fold vibration • Voice source • Voice source spectrum

C.T. Herbst (✉)
Voice Research Lab, Department of Biophysics, Faculty of Science, Palacký University, 17. listopadu 12, Olomouc 771 46, Czech Republic
e-mail: herbst@ccrma.stanford.edu

R.A. Suthers et al. (eds.), *Vertebrate Sound Production and Acoustic Communication*, Springer Handbook of Auditory Research 53,
DOI 10.1007/978-3-319-27721-9_6

6.1 Introduction

This chapter is concerned with the vocal production of mammals, and how it can be described in terms of physics. The voice production phenomenon is governed mainly by aerodynamics, mechanics, and acoustics. Interestingly, little would be known considering only research on nonhumans mammals, as direct observation of mammalian sound production in vivo is problematic owing to the limited accessibility to the sound generating organ and the typical lack of compliance of animals in vivo experimental situations. It is therefore not surprising that current insights into the physics of mammalian sound production are derived mainly from in vivo experiments involving humans, and from a few ex vivo and in vitro experiments with nonhuman mammals (mostly interpreted as models for human sound production, to be investigated in a medical context). Accordingly, the content of this chapter is largely compiled from literature concerned with human sound production, and, whenever possible, the respective concepts are related to nonhuman mammals.

6.2 Three Distinct Mammalian Sound Production Mechanisms

A disturbance (e.g., created by a mechanical oscillator), which is propagated as a longitudinal wave through an elastic medium, such as air and water, but also soil in the case of seismic P-waves (Attenborough 2007), causes a displacement of the particles in that material and pressure changes along the direction of motion of the wave (Serway 1990). A somewhat anthropocentric definition suggests that if these pressure changes are audible by humans (i.e., if they are in the frequency range of about 20–20,000 Hz), the phenomenon is called *sound* (Rossing 1990; Beranek 1996). Acoustic energy that is too high or too low in frequency to be perceived by humans is termed "ultrasonic" and "infrasonic," respectively.

The mammalian voice production organ has three subsystems: a power source, constituted by the pulmonary system; a sound generator, typically the larynx [but see e.g., Au and Suthers (2014) for a description of the nasal sound source in dolphins]; and a sound modifier, the (pharyngeal, oral, and/or nasal) vocal tract (see Fig. 6.1 for a schematic representation of the human vocal organ). From a physical perspective, sound production is the conversion of aerodynamic energy from the lungs into acoustic energy, requiring a sound generator that is able to create repeated acoustic excitations. Particularly for land-based mammals (less is known for marine mammals), three physically distinct mechanisms of sound production have been described.

The so-called myoelastic-aerodynamic (MEAD) mechanism was shown to apply to humans a half century ago (van den Berg 1958; Titze 2006). It can be considered to be the standard sound production mechanism in most mammals, spanning a range of body weights of more than five orders of magnitude (Herbst et al. 2012).

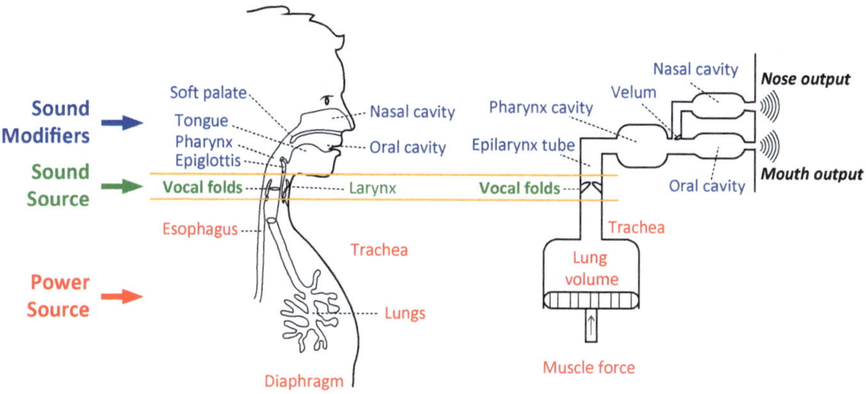

Fig. 6.1 Human vocal organs and a representation of their main acoustical features [Inspired by Rossing (1990), Fig. 15.1, modified by C. T. H.]

In MEAD, a steady airflow, as supplied by the lungs, is converted into a sequence of airflow pulses by the passively vibrating vocal folds (or other laryngeal tissue). The acoustic pressure waveform resulting from this sequence of flow pulses excites the vocal tract, which filters them acoustically, and the result is radiated from the mouth and sometimes from the nose/trunk (Fant 1960; Story 2002). This mechanism requires the interaction between the air stream and the laryngeal tissue, utilizing self-sustaining mechanical vibrations to generate the acoustical excitation (see later in this section for a detailed explanation). Examples of vocal fold vibration during human singing, documented with videostrobolaryngoscopy, are available online at http://dx.doi.org/10.1121/1.3057860 as supplementary material to Herbst et al. (2009).

In the active muscular contraction (AMC) mechanism, documented for purring cats, phonation is caused by a neurally driven periodic muscular modulation of respiratory flow (Sissom et al. 1991). It results from the intermittent activation and relaxation of intrinsic laryngeal muscles caused by very regular stereotyped patterns of electromyographic (EMG) bursts occurring about 20–30 times per second. Each of these muscle discharge bursts supposedly causes an interruption in the glottal airflow (the glottis being the airspace between the vocal folds), thus generating acoustic excitation events (Remmers and Gautier 1972). This phenomenon has so far only been documented via EMG recordings. Direct endoscopic observations of vocal fold motion and airflow measurements in vivo have not yet been made. In contrast to the MEAD mechanism, in the AMC mode the fundamental frequency is limited by muscle contraction speeds. Despite early claims (Husson 1950), there is no evidence of sound production based on AMC in humans (van den Berg 1958).

The whistle mechanism has been proposed to apply to ultrasonic sound production in rodents (as opposed to "sonic" rodent vocalizations in the audible range, which are believed to be produced by the MEAD mechanism). In this aeroacoustic phenomenon, the acoustic excitation is constituted by vibrating air alone; no mechanical oscillation is required. Experiments conducted by Roberts (1975) suggest that acoustic energy is produced by a mechanism that Powell (1995) terms a "Rayleigh bird call." The main empirical support for this hypothesis comes from two heliox experiments (Roberts 1975; Riede 2011) proposing that the dependency of the ultrasonic call fundamental frequencies on the speed of sound in the utilized gas mixture rules out sound production according to the MEAD theory. Unfortunately, the only study reporting direct empirical observation of the absence of vocal fold vibration during rodent ultrasound production was only published as an abstract in conference proceedings (Sanders et al. 2001), and no information about the utilized camera frame rate and image resolution was provided. Interestingly, experiments by Novick and Griffin (1961) showed that by and large no ultrasound production was possible in bats whose superior laryngeal nerve was bilaterally severed, and that action potentials in the cricothyroid muscle (CT) correlated with "emitted pulses of high frequency sound." CT activity stretches and thus increases tension in the bat's vocal folds/membranes. As muscle-induced tension increase is one of the main regulators of fundamental frequency according to the MEAD theory (see Sect. 6.3.6), this might constitute an argument against the "whistle" mechanism in ultrasound production, at least in bats.

Given the currently available evidence, the MEAD mechanism is assumed to apply to nearly all mammalian species. The physics of this sound production mechanism are quite well researched, mostly owing to research done in humans. Consequently, the remainder of this chapter concentrates on the MEAD theory.

6.3 MEAD: From Airflow to Sound

6.3.1 The Oscillator: Vibrating Laryngeal Tissue

Abundant evidence from direct in vivo observation in humans (Moore 1991), and from in situ (e.g., Berke et al. 1987; Chhetri et al. 2012) and ex vivo experiments with various mammalian species (e.g., Alipour and Jaiswal 2009; Welham et al. 2009), clearly show that the primary oscillator in MEAD sound production is constituted by vibrating vocal folds, in interaction with the glottal airflow. In some singing styles (e.g., Lindestad et al. 2001; Bailly et al. 2010) or in human voice pathology (Arnold and Pinto 1960), other laryngeal structures, such as the ventricular folds or the aryepiglottic folds (McGlashan et al. 2007), are engaged in oscillation in addition to or instead of vibrating vocal fold tissue. Similar phenomena have been observed in excised larynx experiments involving dogs (Finnegan and Alipour 2009), Siberian tigers (*Panthera tigris altaica*; Titze et al. 2010), and an African elephant (*Loxodonta africana*; Herbst et al. 2013b). A recent study describes "velar vocal folds" spanning the intrapharyngeal ostium in male koalas (*Phascolarctos cinereus*). This newly discovered nonlaryngeal oscillator enables the generation of low-frequency sounds (Charlton et al. 2013).

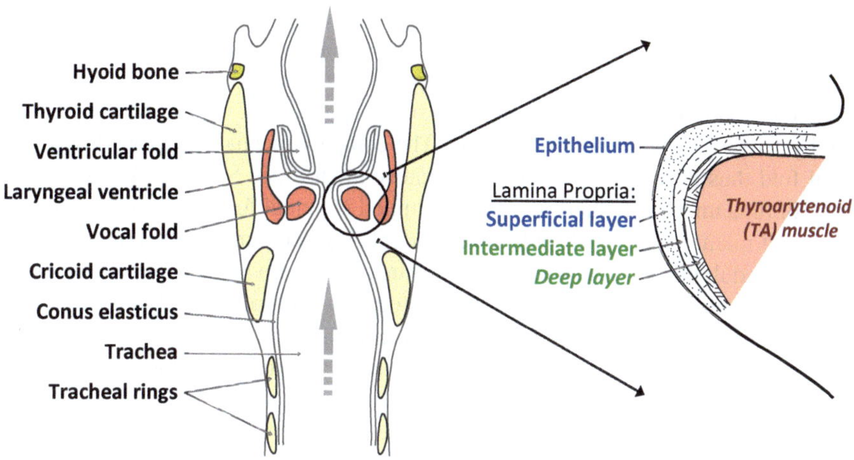

Fig. 6.2 Vocal fold structure in humans. (*Left*) Frontal section of the larynx. (*Right*) Schematic illustration of frontal section through the right vocal fold, showing tissue layers. Blue: mucosa; green: ligament; red: muscle. Vocal fold cover layers are labelled with regular characters; layers belonging to the vocal fold body are labeled with italics. [Inspired by Titze (2000), Fig. 1.13.] Note that Hirano (1981) proposed a slightly different classification system, where the "body" only consists of the vocalis (TA) muscle

Histological sections reveal that the vocal folds in adult humans are composed of five layers: the outer epithelium, three layers of the lamina propria (superficial, intermediate, and deep, the last two constituting the vocal ligament), and the innermost thyroarytenoid muscle (Zemlin 1998). When describing the vocal folds as a vibrating system, these structures are usually grouped into two (e.g., body and cover) or three layers (e.g., body, ligament, and cover); see Fig. 6.2 and Titze (2000). These layers have different biomechanical properties, which is an important fact when discussing the vibratory quality of the vocal folds.

The epithelium is squamous, representing a thin and stiff capsule that maintains the shape of the vocal fold (Zemlin 1998). The superficial and intermediate layers of the lamina propria are composed primarily of elastin fibres, allowing for ample elongation, similar to a rubber band (Titze 2000). In contrast, the deep layer of the lamina propria is composed mainly of collagen fibers, rendering this layer nearly inextensible. The vocal ligament (Fig. 2.17 in Titze 2006), mainly the deep layer of the lamina propria there, is thus the major tensile stress-bearing structure, allowing for high fundamental frequencies when the vocal fold is passively elongated (Titze and Hunter 2004), such as by the action of the cricothyroid muscle. The lamina propria has been found to be mechanically anisotropic, its tensile properties being directionally dependent (Kelleher et al. 2013). The thyroarytenoid (also known as vocalis) muscle embedded in the body of the vocal fold is capable of actively shortening the vocal fold, thus increasing tensile stress.

During vibration, the cover layer is typically only loosely connected to the body of the vocal fold (Story and Titze 1995). In all layers of the vocal fold, stress increases nonlinearly as a function of longitudinal elongation or strain (Alipour and Titze 1985). The corresponding stress–strain curves are time dependent, resulting in force-elongation hysteresis (Alipour and Titze 1991). The complex mechanical properties of the vocal fold thus allow for a variety of vibratory modes (see Sect. 6.3.3) and a wide range of fundamental frequencies, despite limited potential for change in length.

Not all mammalian vocal folds have a morphological composition comparable to that of the adult human. The vocal fold structure of newborn humans, for instance, has no ligament, and the typical layered structure found in adults most likely fully develops during puberty (Hirano et al. 1983). Various mammalian species exhibit a wide variety of vocal fold morphology, ranging from one layer (cats), two layers (dogs, pigs, rabbits, sheep) to three layers (horses, monkeys, guinea pigs, oxen, rats) (Kurita et al. 1983). More recent research suggests three vocal fold layers in cervids (Riede and Titze 2008; Frey and Riede 2013) and rhesus monkeys (*Macaca mulatta*; Riede 2010).

6.3.2 Self-Sustaining Oscillation

The larynx can be configured in two main ways: in the breathing configuration the vocal folds are separated (abducted), allowing for air to flow easily and noiselessly during inhalation and exhalation. In the vocalization configuration the vocal folds are adducted (in partial or full contact), thus typically sealing the laryngeal airway (Zemlin 1998). This laryngeal reconfiguration increases the glottal flow resistance opposing the exhalatory airflow. The resulting subglottal pressure buildup drives the vocal folds apart. A number of physical phenomena aid in closing the glottis again to establish a cyclical oscillatory pattern: elastic recoil forces of the vocal folds (van den Berg 1958), negative Bernoulli forces generated by the increased glottal flow (van den Berg et al. 1957), and a driving force asymmetry (Titze 1988b).

This driving force asymmetry is introduced by one or both of the following phenomena: (1) by a time-varying glottal shape: convergent (as seen in the frontal plane) in the opening phase, allowing for maximum energy transfer from the air stream into the tissue; and divergent in the closing phase, creating a drop of intraglottal pressure that facilitates vocal fold closure, possibly aided by intraglottal flow separation and vortices near the superior vocal fold edge (Oren et al. 2014); and (2) by an inertive supraglottal air column (i.e., a delayed vocal tract response caused by an inertive acoustic load), also aiding in the closure of the vocal folds (Titze 1988b).

In more theoretical terms, self-sustaining vocal fold vibration requires two oscillatory components (OCs) in the system; see Fig. 6.3. One OC is constituted by translatory motion, achieved by mediolateral oscillation of the vocal folds, typically attributed to the vocal fold body (Hirano 1974; Story and Titze 1995). The other OC is introduced (1) by the inertance of the supraglottal air column (i.e., the sluggishness of response of the air column, caused by the mass of its molecules), as suggested by an early one-mass model of vocal fold vibration (Flanagan and Landgraf 1968); and/or (2) by the time-varying glottal shape along the axis of air flow, as

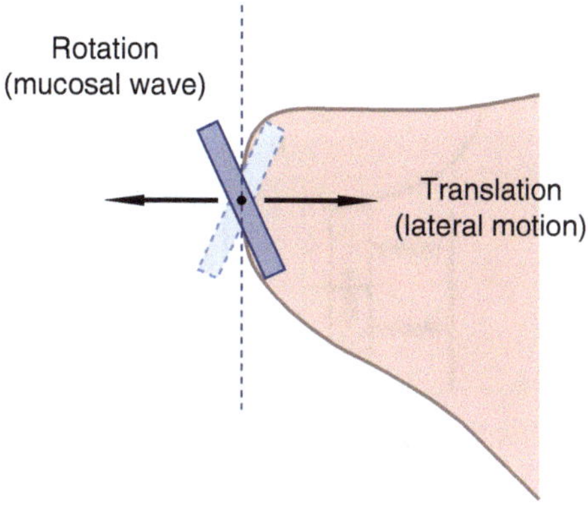

Fig. 6.3 Simplified schematic illustration of vocal fold motion. The rotation of the thin plate is intended to account for the mucosal wave while translation represents overall lateral motion [Inspired by Story (2002)]

explained by computational models with two or more masses (Ishizaka and Flanagan 1972). This time-varying glottal shape is typically attributed to the cover of the vocal fold (Hirano 1974; Story and Titze 1995), constituted by phase differences along the inferior–superior vocal fold surface (Baer 1981; Titze et al. 1993), often seen in the form of mucosal waves (Hirano 1981; Berke and Gerratt, 1993). Mucosal waves are airflow-driven traveling waves within the surface cover layer of the vocal fold tissue, initially moving along with the transglottal airflow from the inferior to the superior vocal fold edge and then propagating laterally across the upper vocal fold surface once every oscillatory cycle.

6.3.3 Vocal Fold Vibration Patterns

From a biomechanical point of view, vocal fold vibration can be understood as a superposition of independent characteristic vibratory patterns, called eigenmodes (Berry et al. 1994; Svec 2000). Any vibratory pattern can be described by its decomposition into multiple eigenmodes. Two particular low-order eigenmodes play a dominant role in stable phonation (see Fig. 6.4): the so-called *x-10* mode, in which the inferior and the superior portions of the vocal fold vibrate in phase, constituting the translational degree of freedom; and the *x-11* mode, in which the inferior and the superior portions of the vocal fold vibrate out of phase (Berry et al. 1994; Döllinger and Berry 2006), resulting in the force asymmetry in the second oscillatory component. Experiments with a periodically (see Sect. 6.3.4) vibrating human hemilarynx preparation suggest that about 90 % of the observed vibratory pattern can be accounted for by empirical eigenfunctions approximating these *x-10* and *x-11* modes (Döllinger et al. 2005), which are typically entrained (i.e., having the same frequency) in nearly periodic vibration. Desynchronization of a few of the low-order vibratory modes leads to irregular or chaotic vibration (Berry et al. 1994); see Sect. 6.3.5. Phase differences can occur not only along the inferior–superior dimension (as in the *x-11* mode), but also along the anterior–posterior dimension, resulting in a so-called zipper-like pattern of vocal fold contacting and decontacting (Childers et al. 1986).

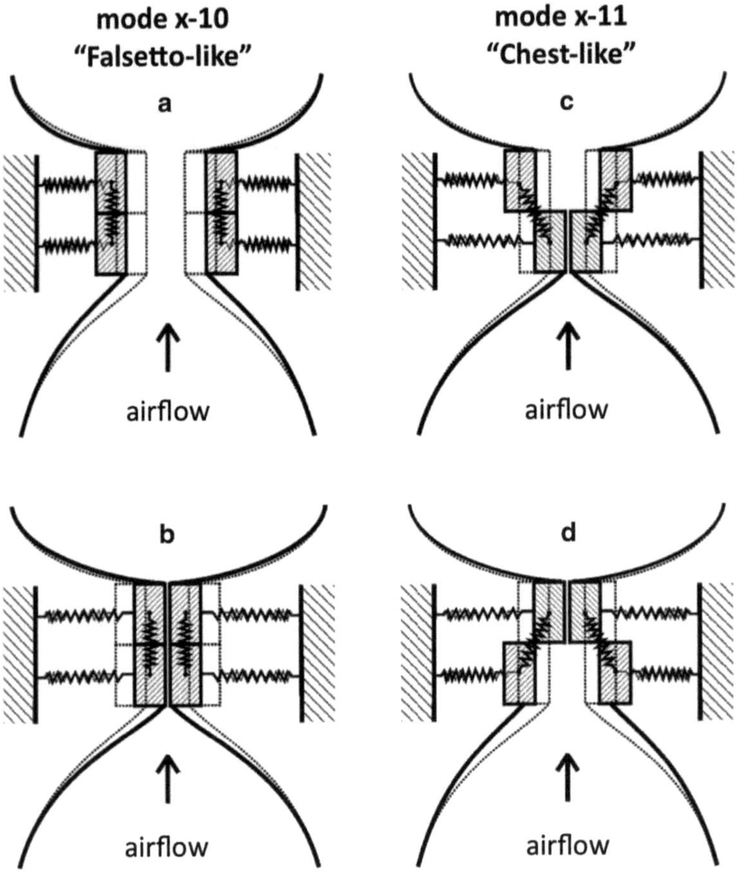

Fig. 6.4 Simplified schematic illustration of two characteristic eigenmodes of vocal fold vibration: *x-10* (**a**, **b**) and *x-11* (**c**, **d**). The vocal folds are shown at two opposite phases of the vibratory cycle [From Svec (2000), with permission]

Higher order modes of vibration can occur along both the inferior–superior and the anterior–posterior dimension. Their presence probably can induce subharmonic or irregular vocal fold vibratory patterns (Berry et al. 1994; see Sect. 6.3.5), or result in highly complex but still nearly periodic oscillatory patterns (Herbst et al. 2013b).

6.3.4 Periodicity

The vibration of laryngeal tissue during sound generation is self-sustaining and typically cyclic. The duration of each oscillatory cycle is the period (usually denoted as *T*). If the period of successive cycles is identical, satisfying the condition $T_n = T_{n-1}$ for all *n*, the oscillation is periodic. A signal generated by a periodically vibrating system has a fundamental frequency or f_0, which is constituted by the number of oscillations per second ($f_0 = 1/T$), measured in hertz (Hz) (Rossing 1990). f_0 can in most cases be perceived as a certain pitch, formally defined as "that attribute of auditory sensation in terms of which sounds may be ordered on a scale extending from low to high" (ANSI 1960). The concepts fundamental frequency and pitch should not be used interchangeably, as the first is a property of a vibrating physical system (and thus an empirical quantity), and the latter is a psychoacoustic (and thus psychological and subjective) perceptual quality (Howard and Angus 2009).

Most naturally occurring oscillatory phenomena do not exactly fulfill the strict criterion of periodicity. In the case of voice production, the period of consecutive cycles is at best approximately equal in duration, resulting in a nearly periodic (often also termed quasi-periodic) signal (Titze 1995). Deviations from the (nearly) periodic case (Titze 1995) manifest themselves as *perturbations* (i.e., temporary changes from an expected behavior); *fluctuations* (more severe deviations from a pattern); *jitter* (a short-term, cycle-to-cycle perturbation in the fundamental frequency); *tremor* (low-frequency fluctuation); and, in the special case of the singing voice, *vibrato*, which is a regular modulation of the fundamental frequency at rates of 4–7 Hz (Hirano et al. 1995).

Deviation from periodicity is an important trait in mammalian vocal communication (Fitch et al. 2002), but can also constitute evidence for vocal health impairment in humans. The basic protocol for functional assessment of voice pathology of the European Laryngological Society (ELS; Dejonckere et al. 2001) considers a cycle-to-cycle variation (jitter) of more than 1 % of the respective period in sustained phonation as an indicator for vocal pathology (Friedrich and Dejonckere 2005). The ELS protocol further advises that jitter measures that exceed 5 % are not meaningful. This suggestion is based on the observation that conventional period estimation techniques (performed in either the time and frequency domain) turn out to be unreliable as the analyzed signal deviates from periodicity and becomes increasingly irregular. In such cases, assessment and analysis methods from the domain of nonlinear dynamics are more appropriate, as discussed in the next section.

6.3.5 Nonlinear Phenomena

The topic of nonlinear dynamics was introduced relatively recently to acoustics (Lauterborn and Cramer 1981; Lauterborn and Parlitz 1988) and to voice science (Mende and Herzel 1990; Herzel 1993; Titze et al. 1993), but has subsequently received much attention. Research has shown that the mammalian larynx is able to generate highly complex and apparently unpredictable vocalizations without requiring equivalently complex neural control mechanisms (Fitch et al. 2002) in the form of bifurcations, subharmonics, and/or deterministic chaos. These so-called *nonlinear phenomena* (NLP) represent an important class of vocalizations in the call repertoire of various mammalian species. Wilden et al. (1998) suggested three communicative potentials of NLP: individuality, motivation, and status, Riede et al. (2007)

proposed that the relative prevalence of NLP during vocalization is a signal of physical condition, while Fitch et al. (2002) highlighted the role of chaos as a source of loud broadband signals. For an accessible general introduction to nonlinear dynamics and chaos, readers are referred to works by Gleick (1987), Glass and Mackey (1988), and Strogatz (2000). The first two of these are more elementary, while the last is most mathematically sophisticated. Here, only a limited discussion of the topic, as relevant to voice production, is presented.

At the heart of nonlinear dynamics is the notion of coupled nonlinear oscillators. MEAD sound production is the result of a complex interaction among several subsystems, each having their own physical properties: the paired vocal folds (Svec et al. 2000) and other laryngeal vibrators such as the ventricular folds; and both the supra- and subglottal vocal tract (Titze 2004b, 2008). Each of these components has a general tendency to vibrate/exhibit resonance in its own set of frequencies (i.e., eigenmodes). Because these subsystems are coupled, they are in constant negotiation with each other and constitute a complex nonlinear system. Depending on the properties of the individual subsystems, the state of vocal fold vibration is either stable or tends to exhibit abrupt changes even when changes in physiological laryngeal parameters are smooth (e.g., Svec et al. 1999). These abrupt transitions in the oscillatory state of the sound production system are called *bifurcations* and constitute an important class of phenomena among the NLP discussed in bioacoustics (Herzel 1993; see Strogatz 2000 for a mathematical introduction to bifurcations). From a mechanical point of view, a bifurcation can be explained as a spontaneous shift in dominance of the different eigenmodes of the involved oscillatory subsystems (e.g., Tokuda et al. 2007; Zhang 2009).

In a nonlinear system, the output is not directly proportional to the input. A typical example for a nonlinear system is illustrated in Fig. 6.5, depicting the vocal fold vibratory behavior of an excised fallow deer (*Dama dama*) larynx on the laboratory bench during a linear increase of subglottal air pressure (the input). As can be seen in this figure, the variation of driving pressure not only affects the generated sound level (Fig. 6.5a), but also results in the emergence of distinct vibratory regimes, as indicated by the terms above the spectrogram in Fig. 6.5b. In particular, three stereotypical signal types are observed—periodic, subharmonic, and irregular/chaotic—which are shown in Fig. 6.5c.

In the *periodic* case, one vibratory cycle per period occurs. In a *subharmonic* signal, consecutive cycles have different amplitudes and/or waveshapes, thus forming periodically repeating groups of subcycles. In the subharmonic signal in Fig. 6.5c, two cycles per period are found, resulting in the introduction of additional harmonics into the harmonic series as seen in the spectrogram (see Fig. 6.5b around $t=1.8$ s) and thus a halving of the fundamental frequency. This phenomenon, producing a so-called period-2 sequence, is sometimes called "period doubling." Other subharmonic cases are possible, such as period-3 or period-4 sequences, etc. In human voice terminology, such subharmonic phonation is sometimes referred to as diplophonia, triplophonia, and quadruplophonia (Titze 1995).

In the irregular signal in Fig. 6.5c (right), no clear periodicity is visible. When analyzed in the frequency domain, an irregular signal does not have the typical

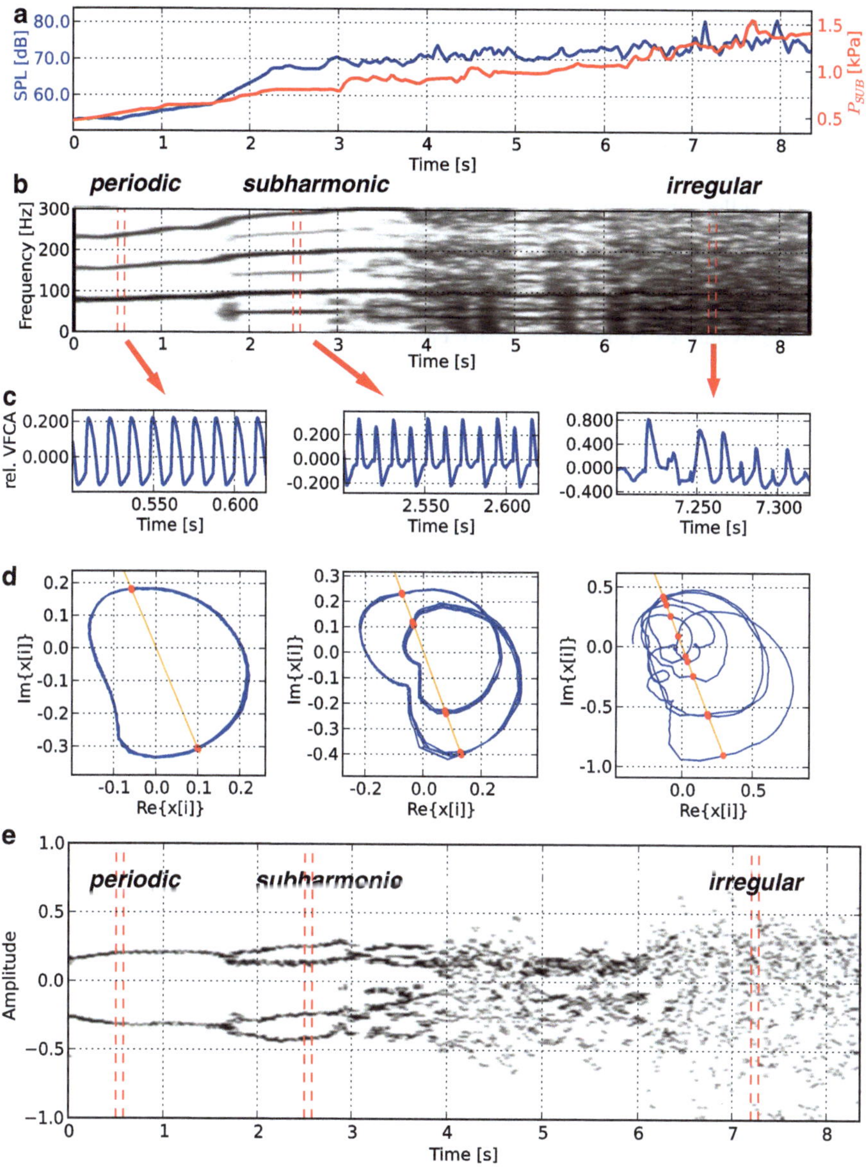

Fig. 6.5 Three stereotypical signal types in an excised fallow deer larynx phonated at linearly varying subglottal pressure conditions. (**a**) Subglottal pressure (red) and generated sound pressure level, measured at 30 cm distance (*blue*). (**b**) Spectrogram of the electroglottographic (EGG) signal. The EGG signal is a correlate of the relative vocal fold contact area (VFCA), thus being a physiological representation of vocal fold vibration (Baken 1992). (**c**) Stereotypical EGG signals: (nearly) periodic (extracted at t = 0.5 s), subharmonic ($t = 2.5$ s) and irregular ($t = 7.2$ s). (**d**) Two-dimensional phase space embedding of the signals from **c**, based on the Hilbert-transformed EGG signal (for details see Roark 2006; Herbst et al. 2013a). The orange lines represent Poincaré sections through the phase space at an angle of 0.6π radians, and the red dots are intersections between the phase space trajectories and those Poincaré sections. (**e**) Phasegram of EGG signal (see text)

harmonic structure found in the periodic or subharmonic case, but it is represented by a broadband energy distribution. However, in irregular signals generated by the vocal apparatus, some residual traces of harmonics are often found, as is seen in the spectrogram in Fig. 6.5b around 100 Hz.

In addition to assessing the signal type in the time domain (Fig. 6.5c) and in the frequency domain (Fig. 6.5b), the nature of a given time series can also be revealed in the so-called phase space by creating a "phase portrait" (i.e., a geometric representation of the trajectories of a dynamical system), thus describing the system's evolution in time (Bergé et al. 1984). If the governing differential equations of a dynamical system are not known (which is usually the case when analyzing biosignals), the dynamics of the phase space can be analyzed by attractor reconstruction (Packard et al. 1980; Roux et al. 1983), where the attractor is defined as a set on the phase plane to which all neighboring trajectories converge (Strogatz 2000). In the simplest form of attractor reconstruction, a two-dimensional vector

$$x(t) = (B(t), B(t+\tau))\ldots\tau > 0, \tag{6.1}$$

is extracted, based on the analyzed signal. The time series $B(t)$ then provides a trajectory $x(t)$ in two-dimensional phase space (Strogatz 2000, p. 438); in other words, the signal is projected against a version of itself, delayed by τ. Equation (6.1) produces a two-dimensional vector; embedding in more dimensions is also possible, but cannot be easily visualized with vectors having more than three dimensions.

Two-dimensional phase space embedding is illustrated in Fig. 6.5d, where three stereotypical trajectories are seen: a *limit cycle* in the case of a periodic signal, in which each period consists of exactly one oscillatory cycle in the analyzed signal (the limit cycle in the left panel of Fig. 6.5d is made up of nine revolutions that are plotted atop each other, overlapping nearly perfectly owing to the periodic nature of the underlying signal); a subharmonic limit cycle for the period-2 sequence, in which each system period encompasses two oscillatory subcycles; and an irregular phase space trajectory, termed a *strange attractor* in the case of chaos (see later in this section).

Signal type assessment using phase portraits is a useful alternative to spectrogram analysis. A recently developed visualization method has developed this approach further by creating cuts (so-called Poincaré sections; see Bergé et al. 1984) through consecutive two-dimensional phase portraits from longer signals. The intersection points between successive phase space trajectories and Poincaré sections are depicted along the time axis, forming the so-called *phasegram* (Herbst et al. 2013a; see www.phasegram.org for more information and open source software). The phasegram provides an intuitive representation of time-varying system dynamics. In a phasegram, the number and stability of lines perpendicular to the (vertical) *y*-axis indicate the system state at a particular point in time (see Fig. 6.5e): one line: no oscillation (stasis); two locally stable lines: periodic oscillation; more than two locally stable lines: subharmonic patterns; no continuous lines, rugged appearance: irregular system behaviour, potential indicator for chaos. Thus, the phasegram is a useful tool for voice signal classification (Herbst 2014).

Chaos is aperiodic long-term behavior here and in a deterministic system that exhibits sensitive dependence on initial conditions (Strogatz 2000, p. 323). Any chaotic signal or time series data is irregular, but not all irregular signals are chaotic. Chaos is created by simple, low-dimensional systems that can, under certain circumstances, generate complex irregular behavior. This can be observed in many natural phenomena: weather, cardiac rhythm, models for population growth, economic data, some chemical reactions, or the voice of humans and animals. The long-term predictability of such systems is difficult if boundary conditions change or cannot be precisely determined in the first place (just consider the weather forecast!).

Frequency analysis cannot distinguish between a chaotic and a purely random process. This is demonstrated in Fig. 6.6. An irregular sequence was generated from the logistic map equation, the simplest system that is capable of generating chaotic behavior (Fig. 6.6a, left). As is expected, the spectrum of that irregular sequence is a broadband distribution of energy and no harmonic series is present (Fig. 6.6b, left panel). However, phase space embedding reveals an attractor resembling a parabola (Fig. 6.6c, left), as expected, given the mathematical definition of the logistic map (Strogatz 2000). Contrasting this example, a stochastic time series has been created with identical frequency characteristics (but with randomized phases; see Fig. 6.6, left column). When embedding this signal in phase space, no clear attractor but a random cloud of data points emerges. However, the power spectra are identical. This example illustrates that conventional analysis methods, operating in the time or the frequency domain, may not be sufficient to detect low-dimensional chaos. For this task, quantitative methods from nonlinear systems analysis, such as the correlation dimension (Grassberger and Procaccia 1983), Lyapunov exponents (Eckmann et al. 1986), or Tokuda et al. (2002) low-dimensional nonlinearity measure should be utilized. Such analysis methods have been successfully applied in the analysis of physiological voice signals, such as the acoustical waveform (e.g., Herzel et al. 1998; Jiang et al. 2006), electroglottography (Baken 1990; Behrman and Baken 1997), or data derived from high-speed video recordings of vocal fold vibration (e.g., Mergell et al. 2000; Zhang et al. 2007).

6.3.6 Control of Fundamental Frequency

As a crude first approximation, a vibrating vocal fold can be modeled as a simple string (Titze 2000) using the equation

$$f_0 = \frac{1}{2L}\sqrt{\frac{\sigma}{\rho}}[Hz] \tag{6.2}$$

where f_0 is the fundamental frequency, L is the length, σ is the (tensile) stress in the string (defined as tension force distributed over an area), and ρ is the tissue density. As the tissue density cannot be changed in the vocal fold, the vibratory frequency

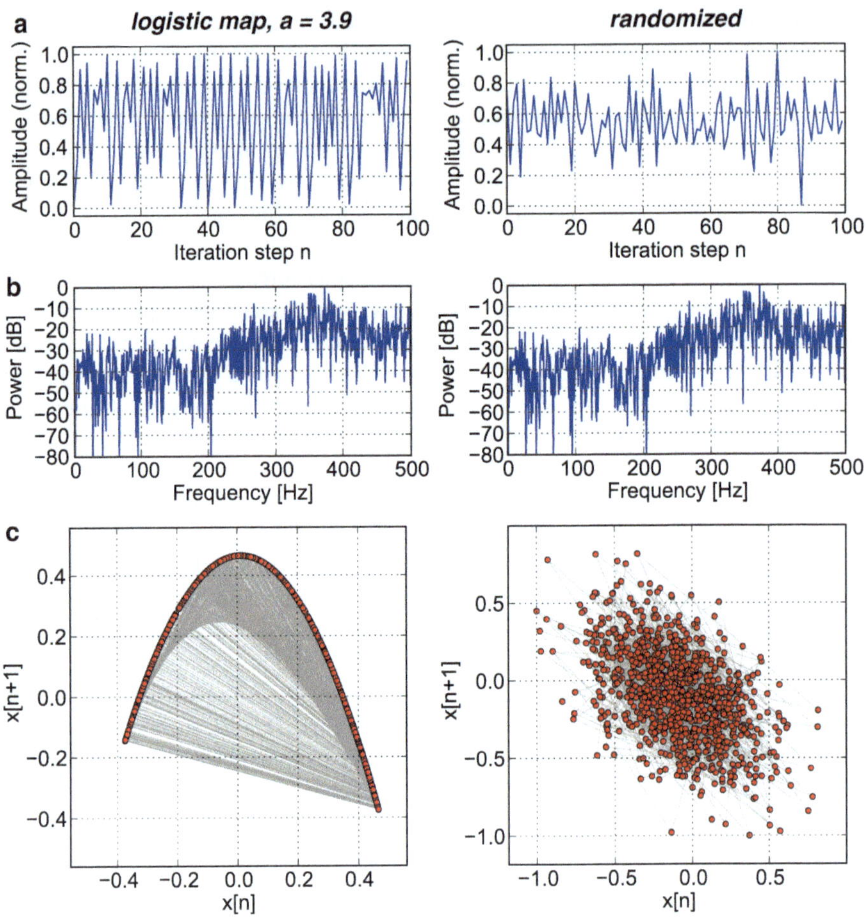

Fig. 6.6 Spectral and phase space analysis of a deterministic (i.e., chaotic) and a stochastic (i.e., random) signal. (**a**) The left panel shows a deterministic time series, derived from the logistic map equation $x[i+1] = ax[i](1-x[i])$, $a = 3.9$, sampled at $F_S = 1000$ Hz. The right panel shows stochastic (random) time series, derived from deterministic time series, but phases randomized in the frequency domain via a forward and an inverse Fourier transform. (**b**) Power spectrum. (**c**) Two-dimensional phase portraits of both sequences created by attractor reconstruction

according to this model is influenced by the vocal fold length (longer vocal folds vibrate at lower frequencies) and the longitudinal stress in the vocal fold (higher stress results in a higher oscillatory frequency). The intuitive assumption that the vocal fold mass is inversely related to fundamental frequency has been challenged recently (Titze 2011). If stress is applied to the vocal fold, this will also increase its length (resulting in strain, that is, a measure of normalized elongation), but the contribution of the stress typically overrides that of the length increase, resulting in an increasing frequency. The underlying relation between stress and strain in vocal fold tissue is non-linear and different for the individual vocal fold layers, and vocal fold anisotropy introduces further nonlinearities. Consequently, complex extensions of the string model have been proposed when theoretically predicting f_0 (Titze 2000; Eq. 22 in Titze and Hunter 2004).

The f_0 control in the mammalian larynx is influenced by multiple factors. The primary mechanism involves stretching the vocal folds along their anterior–posterior axis (and thus increasing tensile stress), typically via the cricothyroid (CT) muscle (Baer et al. 1976; Vilkman 1987). However, additional, more speculative mechanisms have been proposed, for instance, via the suggested contraction of the strap muscles in North American wapiti or "elk" (Frey and Riede 2013). Redundancy of f_0 control can be introduced by fine-tuning the stress in different vocal fold layers (Titze 2000, Chap. 8; Zhang et al. 2009), a concept that is utilized in singing for controlling vocal registers (van den Berg 1963; Herbst et al. 2011). Differential control of the mechanical properties in the different vocal fold layers also introduces variations into the spectral characteristics of the voice source (see later in this section), a concept which has to the author's knowledge been explored in detail only in humans, and not in other mammals. Indeed, little is known about physiological laryngeal fine control in nonhuman mammals.

Empirical research in humans shows that f_0 is also influenced by subglottal pressure (Ladefoged and McKinney 1963; Gramming et al. 1988). The increase in fundamental frequency with subglottal pressure can be understood in terms of nonlinear vocal fold tension, dependent on the amplitude of vocal fold vibration. A corresponding model predicts an increase of 5–60 Hz per kPa of subglottal pressure (Titze 1989). Evidence for the dependency of fundamental frequency on subglottal pressure is also available for other mammals: Häusler (2000) found that pulmonary activity affects both sound level and f_0 in squirrel monkeys. In rats, excised larynx data from Johnson et al. (2010) show that f_0 is partially dependent on air flow rates (and by inference also on subglottal pressure), and Riede found a weak dependency of f_0 on subglottal pressure in vivo (2011).

Finally, the fundamental frequency could also be influenced by non-linear interactions between the source and the vocal tract, a concept that is discussed in Sect. 6.3.10.

6.3.7 Subglottal Pressure and Glottal Airflow

In mammalian voice production, as is typical among terrestrial vertebrates, the respiratory apparatus is utilized as the power source. Most mammalian vocalizations are egressive, but rare examples of ingressive phonation are found for both humans and other mammals (Eklund 2008). For egressive vocalizations, muscle contraction (e.g., primarily the abdominals and the internal intercostals in humans) and/or passive recoil forces in the rib cage generate a positive pressure gradient between lung pressure and atmospheric pressure during exhalation (Hixon 1987). This pressure gradient, usually termed *subglottal pressure*, generates a DC (i.e., time-invariant) airflow in the trachea, which is converted to a fluctuating *glottal airflow* by the passively vibrating vocal folds.

In humans, subglottal pressure is typically in the range of about 0.5–1.5 kPa for soft and loud conversational speech, respectively (Baken and Orlikoff 2000). Minimal subglottal pressure values in human speech were found around 0.2–0.3 kPa (Jiang et al. 1999; Titze 2009), constituting the so-called phonation threshold pressure (PTP), an important determinant for healthy speech production (Titze 1992). In very loud human voice production, subglottal pressure was reported to reach extreme values of 10 kPa (Snelleman et al. 2007). Different normative values for time-averaged airflow rates in healthy humans are suggested in the literature (Hirano 1981; Baken and Orlikoff 2000), being largely in the range of 100–150 mL/s.

Subglottal pressure and glottal airflow data for in vivo mammalian sound production is scarce. Suthers and Fattu found comparatively high subglottal pressures of up to 7 kPa in the North American Vespertilionid bat (*Eptesicus fuscus*) (1973) and provide insights into subglottal pressure variation caused by neural and muscular control (1982). Riede (2011) reports subglottal pressures of 0.8–1.9 kPa in rat ultrasound production. Owing to the lack of available empirical data in many species, current modeling and ex vivo research is limited to hypothetical subglottal pressure values (compare, e.g., Titze and Riede 2010; Herbst 2014), potentially leading to incongruence that can be resolved only as empirical data become available.

Time-averaged glottal airflow and subglottal pressure are related to each other via the *glottal flow resistance*, by applying Ohm's law to fluids (van den Berg et al. 1957):

$$R = \frac{P_{\mathrm{SUB}}}{\Delta V} \tag{6.3}$$

The glottal flow resistance R equals the time-averaged subglottal pressure P_{SUB} (indicated in Pa, kPa, or $\mathrm{cmH_2O}$) divided by the time-averaged airflow volume velocity ΔV (indicated in $\mathrm{m^3/s}$, L/min or L/s) (Alipour et al. 1997). The glottal flow resistance is influenced by both anatomical constraints (i.e., the geometry of the laryngeal airway) and the detailed aspects of vocal fold adduction, which are an important aspect of the prephonatory configuration of the larynx linked to voice quality. In "breathy" human phonation, for instance, a so-called "posterior glottal chink" through which air can flow is present throughout the entire glottal cycle. Consequently, glottal flow resistance is low, resulting in large glottal airflow volumes. In "pressed" phonation, on the other hand, complete glottal closure caused by firm adduction of the vocal folds results in a high glottal flow resistance and relatively lower airflow volume (Sundberg 1981; Scherer 2014).

The effect of different sound production types in humans on the glottal airflow waveform is illustrated in Fig. 6.7, highlighting three acoustically relevant features of the glottal flow waveform, which are discussed in detail later in this section: (1) the amplitude of the waveform (also called "pulse amplitude"; see black arrows in Fig. 6.7); (2) the steepness of the waveform during the closing phase (dashed gray lines), representing the abruptness of glottal airflow deceleration; and (3) a DC or baseline flow component (see the gray arrow in the "breathy" case in Fig. 6.7).

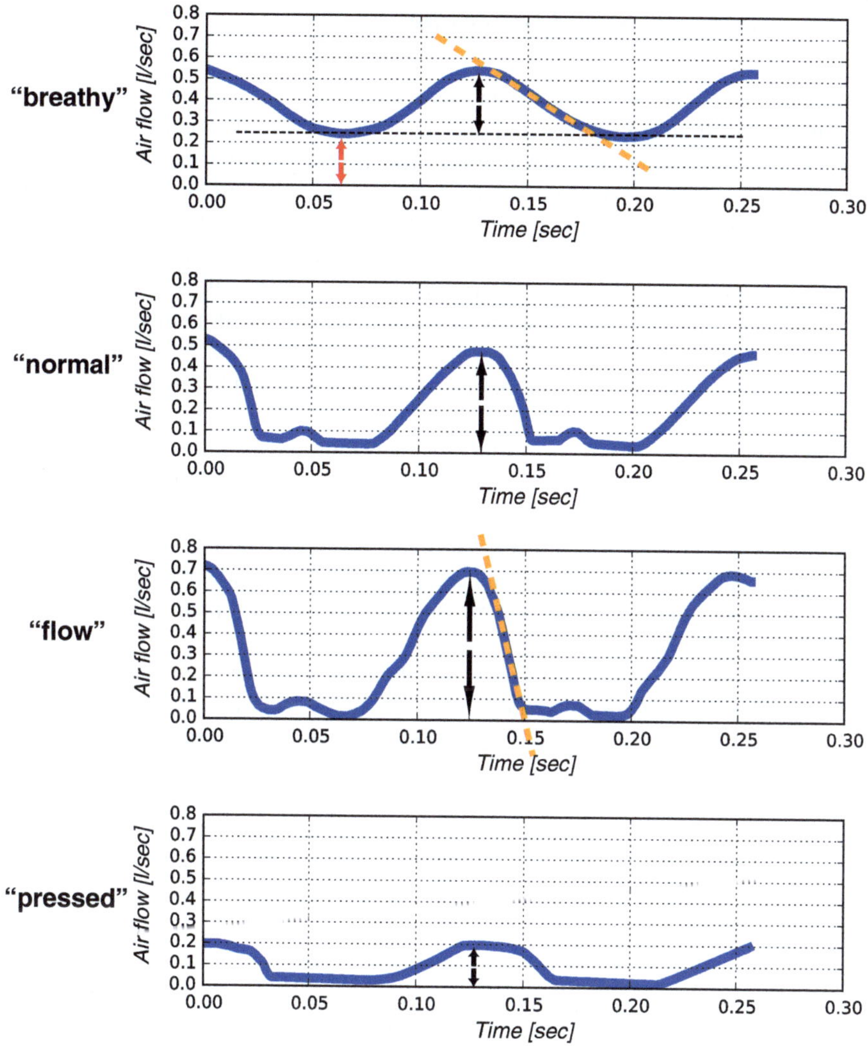

Fig. 6.7 Airflow rates for one cycle of phonation in four different phonatory qualities: "breathy," "normal," "flow," and "pressed" (data were obtained by direct measurement and subsequent inverse filtering of the air flow waveform). The pulse amplitude (*vertical arrows*) changes across the different phonation types. Note that in the "breathy" case the airflow never falls below 0.2 L/s (*red arrow*), indicating incomplete glottal closure. The rate of change in the closing phase (i.e., the steepness of the waveform, illustrated here by dashed gray lines for "breathy" and "flow" phonation) varies considerably, which is relevant for the spectrum of the generated sound (see text) [Data from Sundberg (1981), with permission]

6.3.8 Laryngeal Sound Generation

The actual physical sound generation process is complex: Both airflow related and mechanical (vibratory) sources add energy to the generated sound (Alipour et al. 2011). Vortices arising after separation of the airflow jet from the glottis (Khosla et al. 2007) may also contribute to sound generation, owing to interaction with the supraglottal vocal tract (Barney et al. 1999). Because the fluid-induced aspect of sound production is considerably more important for generating acoustic energy compared to the vibration-induced part (Alipour et al. 2011), the remainder of this section concentrates on this aspect of sound production, drawing on concepts established in the last few decades.

During phonation, interactions between glottal airflow and laryngeal tissue result in synchronized tissue velocity and glottal air pressure variations. The interruption of the flow at the instant of vocal fold closure creates damped acoustic oscillations in the supraglottal as well as in the subglottal vocal tract (Fant 1979). The waveform shape of these airflow pulses largely determines the amplitude and the spectral composition of the created sound at the laryngeal level. The relation between glottal airflow and acoustic air pressure can be approximated by mathematical differentiation: the voice source acoustic air pressure can be roughly predicted as the first derivative of the glottal flow.

This is illustrated in Fig. 6.8 with a model representing stereotypical glottal flow patterns from human speech (Klatt and Klatt 1990). This model suggests that in human speech the closing phase is usually shorter than the opening phase, resulting in a skewed glottal waveform (Rothenberg 1973). Owing to this asymmetry in the glottal flow waveform, its first derivative has a strong negative peak at the instant of glottal closure, constituting the major sound generation event in humans (Miller and Schutte 1984; Schutte and Miller 1988). Whether this can be generalized to all mammals is not clear. Recent research suggests for instance that the major sound generation event in elephant infrasound rumbles occurs at the instant of glottal *opening* in an excised larynx setup (Herbst et al. 2013b).

The amplitude of the negative peak in the derivative of the glottal airflow waveform, termed maximum flow declination rate (MFDR), is a key quantity in laryngeal sound production. As it strongly correlates with both subglottal pressure (Holmberg et al. 1988) and the acoustic level of the generated sound (Titze and Sundberg 1992; Alku et al. 1999), the MFDR can be considered to be the crucial link between input and output in the laryngeal sound production system. The MFDR can be increased by (1) increasing the intracycle amplitude of the flow waveform (also termed the pulse amplitude); (2) moving the moment of vocal fold contact earlier in time, closer to the centre of the pulse; and (3) skewing the pulses (Sundberg et al. 1993). These changes are mainly brought about by increasing subglottal pressure (1) (Holmberg et al. 1988) and adjustments in the quality of vocal fold adduction (2, 3) (Herbst et al. 2015).

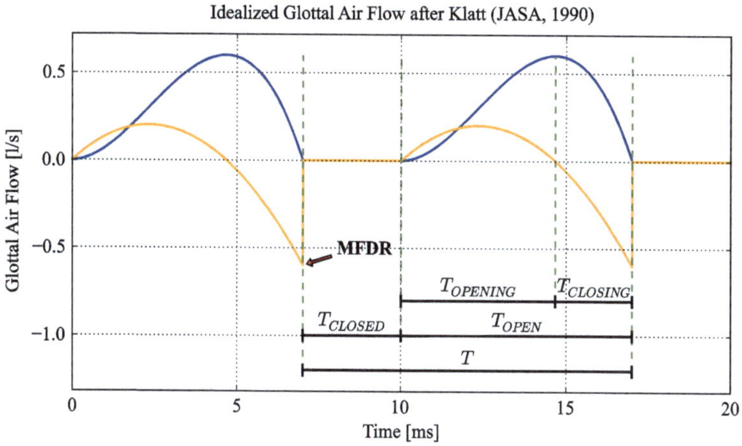

Fig. 6.8 Idealized glottal flow waveform, synthesized at 100 Hz with the model created by Klatt and Klatt (1990). Time-varying glottal air flow (*blue*); first derivative of glottal airflow, normalized to air flow amplitude (orange); MFDR: maximum flow declination rate; period (T) = 10 ms; duration of closed phase (T_{CLOSED}) = 3 ms; duration of open phase (T_{OPEN}) = 7 ms; duration of opening phase (T_{OPENING}) = 4.5 ms; duration of closing phase (T_{CLOSING}) = 2.5 ms; open quotient (OQ): $T_{\mathrm{OPEN}}/T = 0.7$; closed quotient ($CQ$): $1 - OQ = 0.3$; skewing quotient: $T_{\mathrm{OPENING}}/T_{\mathrm{CLOSING}} = 1.8$

6.3.9 *Spectral Composition of the Voice Source*

A crucial aspect of the voice source that has received much scientific attention in humans, but little in other mammals, is the spectral composition of the voice source (not to be confused with the spectral composition of the radiated sound, which is filtered by the vocal tract transfer function; see Sect. 6.3.10). The voice source spectrum greatly influences the perceptual *timbre* of the generated sound. [Timbre is formally defined as that attribute of auditory sensation in terms of which a listener can judge two sounds similarly presented and having the same loudness and pitch as dissimilar (ANSI 1960)].

The periodicity of vocal fold vibration is reflected in the voice source spectrum. Whereas irregular vocal fold vibration results in a broadband distribution of acoustic energy, a *harmonic series* emerges for (nearly) periodic vibratory patterns. In a harmonic series, the frequencies of all energy components (called *harmonics*, *partials*, and sometimes *overtones*) are integer multiples of the fundamental frequency, and typically no or very little acoustic energy is found between the individual harmonics. In the case of incomplete glottal closure (resulting in a DC offset in the glottal flow waveform), additional broadband energy is introduced into the voice source spectrum by the airflow turbulence caused by the glottal chink.

The number and strength of perceptually relevant harmonics contained in the harmonic series emerging from (nearly) periodic vocal fold vibration are crucially

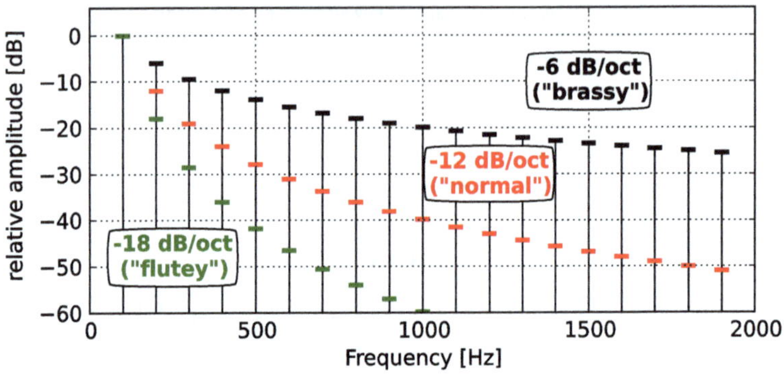

Fig. 6.9 Idealized source spectrum for three different spectral slopes at a fundamental frequency of 100 Hz [Modified after Titze (2000)]

dependent on the abruptness of glottal air flow modulation (in the human case mostly the termination of air flow during each glottal cycle). This affects the proportion of high-frequency energy in the resulting harmonic series, which can be approximately quantified by the *spectral slope*, expressed in negative dB per octave (an octave is a doubling of fundamental frequency). A typical spectral slope for normal human phonation is considered to be −12 dB/octave, with theoretical extremes at (1) −6 dB/octave, resulting in a "brassy" quality where many higher harmonics contribute significantly to the perceived sound; and (2) −18 dB/octave, resulting in a "flutey" or nearly sinusoidal quality where the sound is dominated by the first harmonic (Titze 2000); see Fig. 6.9. A more detailed discussion of sound source characteristics is found in Herbst et al. (2015).

Physiologically, three alternatives for influencing the shape of glottal flow waveform and thus the spectral composition of the voice source exist: (1) by adjustments of the intrinsic laryngeal musculature (Herbst et al. 2011); (2) as a by-product of intensity regulation via the respiratory system—an increase of subglottal pressure is likely to result in an increased closed quotient (a measure of the relative duration of glottal closure during an oscillatory cycle) and an increase of the maximum flow declination rate (Sundberg et al. 2005), thus introducing stronger high-frequency partials into the sound source spectrum; or (3) via nonlinear source-filter coupling (see Sect. 6.3.10).

6.3.10 *Source-Filter Theory and Nonlinear Extensions*

The basic operational principle of the vocal tract as an acoustic filter is explained in Taylor, Charlton, and Reby, Chap. 8, and in another article by this author (Herbst and Svec, 2016). An early and highly influential model for such filtering is the "source-filter theory" (Fant 1960), which proposes a simple linear superposition of source and filter. According to this theory, the acoustic output of the voice source is linearly affected by the vocal tract *transfer function*. This process results in frequency-dependent amplitude scaling of the voice source frequency components, generated by the vocal tract resonances[1] and radiation characteristics. First applications of the source-filter concept to human speech date back many decades (Chiba and Kajiyama 1941; Arai 2001), while its utilization for explaining voice production in nonhuman mammals only occurred more recently (Pye 1967; Carterette et al. 1979; Hartley and Suthers 1988; Fitch and Hauser 1995; Fitch 1997).

In traditional source-filter theory the vibratory behavior of the voice source is suggested to be unaffected by the filter. The (linear) source-filter theory is a very powerful first approximation for explaining many acoustic phenomena in human speech (Stevens 1998), singing (Miller 2008), and bioacoustics (Fitch 2000b; Fitch and Hauser 2002; Reby and McComb 2003), but it fails to describe faithfully all aspects of the physics of voice production. Nonlinear interactions between the vocal tract and the voice source, not predicted by the linear source-filter theory, were already described in early research (Flanagan 1968; Rothenberg 1981b). These interactions may make an important contribution to self-sustaining vocal fold vibration within the MEAD framework, as they constitute one option for establishing an asymmetric forcing function via vocal tract inertance, needed for the second oscillatory component described in Sect. 6.3.2.

The possible influences of the supra- and subglottal vocal tract have been classified as (Titze 2008) as follows:

- Level 1 interactions, in which the positive reactance of the vocal tract (caused by the inertance of the air column) influences the wave shape of the glottal air pulse. This effectively introduces additional harmonics into the wave shape of the glottal airflow (and as a consequence also into the acoustic output), which would not be present without an attached vocal tract. These changes can be facilitated by narrowing the epilarynx tube (the laryngeal vestibule), or by moving a vocal tract resonance (subglottal or supraglottal) just above the fundamental frequency (Rothenberg 1981a, b, Titze 2004a, b).
- Level 2 interactions, where changes of the vocal tract reactance (via the vocal tract geometry) directly influence the mechanics of vocal fold vibration and voice fundamental frequency. In this class of interactions, compliance (negative vocal tract reactance, regularly occurring just above each vocal tract resonance frequency) would add stiffness to the interactive vibrating system, thereby raising the fundamental frequency, whereas inertance (positive reactance, regularly occurring just below each vocal tract resonance frequency) adds mass, thereby lowering the fundamental frequency. This could have a possible effect on both fundamental frequency and the amplitude and mode of vocal fold vibration, introduce bifurcations and other nonlinear phenomena (Titze 2008), and also affect the onset of vocal fold vibration (Lucero et al. 2012).

[1] The terms vocal tract "resonance" and "format" are often used interchangeably, which may in some cases constitute a precarious simplification. Please refer to the discussion in Titze et al. (2015) for precise definitions and a historical perspective.

A comprehensive experimental proof for these two theoretical contributions is still outstanding, even for humans. This is particularly true for fundamental frequency shifts predicted in the context of level 2 interactions, which were found neither in human speech (Beil 1962; Hollien et al. 1977) nor in vocalizations of nonhuman mammals (Koda et al. 2012; Madsen et al. 2012). The occurrence of a strong *x*-11 vibratory mode of the vocal folds (see Sect. 6.3.3), as, for instance, predominantly present in human speech, may cause a weaker coupling between vocal tract and sound source, thus reducing interaction effects (Titze 2008).

In this context, the physical contribution of air sacs to mammalian sound production, which has been a topic of considerable scientific interest in the past decades (Fitch 2000a; Hewitt et al. 2002), should be mentioned. Based on anatomical and acoustical analysis and computer models, air sacs have been proposed to contribute to sound amplification and radiation (Gautier 1971; Riede et al. 2008), create acoustic coupling between vocal tract and air sac (Frey et al. 2007), potentially add and shift formant frequencies (Riede et al. 2008), or possibly affect the sound source by nonlinear coupling (Riede et al. 2008), thus potentially destabilizing the sound source (de Boer 2012). In contrast, Hilloowala and Lass (1978) report no contribution of air sacs to the vocal tract resonance properties of rhesus macaques (*Macaca mulatta*). Thus there are many theoretical options, and further empirical examination of these multiple hypotheses, preferably in a controlled laboratory situation, is vitally needed.

6.3.11 Glottal Efficiency

As stated in Sect. 6.2, from a physical perspective, sound production is the conversion of aerodynamic energy into acoustic energy. Having discussed some of the biophysical underpinnings of the subsystems involved in sound production within the MEAD theory, this final section is concerned with quantifying the effectiveness of that energy conversion process.

The efficacy of vocal energy conversion can be assessed by a parameter termed *glottal efficiency*, which is the ratio of aerodynamic power (i.e., the input to the system) to the radiated acoustic power (i.e., the system's output) (van den Berg 1956). The aerodynamic power P_{AIR} (expressed in watts) is defined as the product of time-averaged glottal airflow and subglottal pressure, and the radiated power P_{RAD} (also expressed in watts) is usually derived from the measured sound intensity (Titze 1988a). Finally, the glottal efficiency E_{GL}, expressed in dB (and sometimes in %), is calculated as

$$E_{GL}[\mathrm{dB}] = 10\log_{10}\frac{P_{RAD}}{P_{AIR}} \tag{6.4}$$

Glottal efficiency is surprisingly low, ranging from −40 to maximally −20 dB in human speech and singing (Bouhuys et al. 1968; Schutte 1980), so that only

0.0001% to maximally 1 % of the aerodynamic power is converted into radiated (acoutical) power. Even lower values in the range of −55 to −35 dB have been found in excised larynx experiments of rats (Welham et al. 2009), canines (Titze 1988a), tigers (Titze et al. 2010), and red deer (Herbst 2014), which can be explained by the lack of an impedance matching vocal tract, which affects the level of the radiated sound by about 5–10 dB (Titze 2006). A recent pilot study reported that glottal efficiency in red deer excised larynges was increased by about 2.5–3 dB when switching from periodic or subharmonic to irregular vocal fold vibration, suggesting that an irregular sound production mechanism at higher subglottal pressures could provide energetic advantage in animal vocal communication (Herbst 2014).

6.4 Summary

In this chapter, the basic physical (and some physiological) principles of the main sound production mechanism in mammals, MEAD, were discussed. As compared to humans, empirical data from other mammals is scarce, so current major theories mostly derive from experiments rooted in human voice science, which has a long tradition. In humans, investigative paradigms such as excised larynx experiments (Cooper 1986) and in vivo vocal fold imaging during phonation (Moore 1991) date back to the nineteenth century but have only recently been introduced for other species. Such approaches are capable of investigating the *dynamic* aspects of voice production, and they are to be preferred over "black box" strategies in which speculations about the physical underpinnings of the sound production organ are based solely on postmortem assessment of the anatomy (Negus 1932) or only on the analysis of the acoustic output.

Applying the knowledge and methods from human voice science to other mammals is a promising strategy, and a number of authors have successfully done that in the recent past (see, e.g., Brown and Cannito 1995). However, owing to the anatomical diversity in mammals, theories and hypotheses derived for humans should not be indiscriminately applied to a particular species, particularly if no corresponding empirical data are available. Because there may be variations to the human-based physical paradigms—some of which are already known (see, e.g., Charlton et al. 2013; Herbst et al. 2013b), and more are likely to be discovered—a challenge for future research is to conduct a solid cross-species comparisons of the physical aspects of sound production mechanisms in multiple mammalian species, across many orders.

Acknowledgments This publication was partly financed by the European Social Fund and the state budget of the Czech Republic within the project no. CZ.1.07/2.3.00/30.0004 "POST-UP" and partly by the ERC grant SOMACCA. I am very grateful to Aaron Johnson and Jan Svec for their comments and suggestions. I thank Megan Wyman for her aid in acquiring the data for Fig. 6.5 and Nadja Kavcik-Graumann for her help with the artwork of Figs. 6.1 and 6.2.

References

Alipour, F., Brücker, C., Cook, D. D., Gömmel, A., Kaltenbacher, M., Mattheus, W., Mongeau, L., Naumann, E., Schwarze, R., Tokuda, I., & Zörner, S. (2011). Mathematical models and numerical schemes for the simulation of human phonation. *Current Bioinformatics, 6*, 323–343.

Alipour, F., & Jaiswal, S. (2009). Glottal airflow resistance in excised pig, sheep, and cow larynges. *Journal of Voice, 23*(1), 40–50.

Alipour, F., Scherer, R. C., & Finnegan, E. (1997). Pressure-flow relationships during phonation as a function of adduction. *Journal of Voice, 11*(2), 187–194.

Alipour, F., & Titze, I. (1985). Viscoelastic modeling of canine vocalis muscle in relaxation. *Journal of Acoustical Society of America, 78*(6), 1939–1943.

Alipour, F., & Titze, I. (1991). Elastic models of vocal fold tissues. *Journal of the Acoustical Society of America, 90*(3), 1326–1331.

Alku, P., Vintturi, J., & Vilkman, E. (1999). On the linearity of the relationship between the sound pressure level and the negative peak amplitude of the differentiated glottal flow in vowel production. *Speech Communication, 28*, 269–281.

ANSI. (1960). USA Standard acoustical terminology (including mechanical shock and vibration). *Technical Report, S*1.1–1960 *(R*1976*)*.

Arai, T. (2001). The replication of Chiba and Kajiyama's mechanical models of the human vocal cavity. *Journal of the Phonetic Society of Japan, 5*(2), 31–38.

Arnold, G. E., & Pinto, S. (1960). Ventricular dysphonia: New interpretation of an old observation. *Laryngoscope, 70*, 1608–1627.

Attenborough, K. (2007). *Springer handbook of acoustics*. New York: Springer.

Au, W. W. L., & Suthers, R. A. (2014). Production of biosonar signals: Structure and form. In A. Surlykke, P. E. Nachtigall, R. R. Fay, & A. N. Popper (Eds.), *Biosonar* (pp. 61–105). New York: Springer.

Baer, T. (1981). Observation of vocal fold vibration: Measurements of excised larynges. In K. Stevens & M. Hirano (Eds.), *Vocal fold physiology* (pp. 119–133). Tokyo: University of Tokyo Press.

Baer, T., Gay, T., & Niimi, S. (1976). *Control of fundamental frequency, intensity, and register of phonation* (pp. 175–185). New Haven, CT: Haskins Laboratories, Status Report on Speech Research.

Bailly, L., Henrich, N., & Pelorson, X. (2010). Vocal fold and ventricular fold vibration in period-doubling phonation: Physiological description and aerodynamic modeling. *Journal of Acoustical Society of America, 127*(5), 3212–3222.

Baken, R. J. (1990). Irregularity of vocal period and amplitude: A first approach to the fractal analysis of voice. *Journal of Voice, 4*(3), 185–197.

Baken, R. J. (1992). Electroglottography. *Journal of Voice, 6*(2), 98–110.

Baken, R. J., & Orlikoff, R. F. (2000). *Clinical measurement of speech and voice* (2nd ed.). San Diego, CA: Singular Publishing, Thompson Learning.

Barney, A., Shadle, C. H., & Davies, P. O. A. L. (1999). Fluid flow in a dynamic mechanical model of the vocal folds and tract. I. Measurements and theory. *Journal of the Acoustical Society of America, 105*(1), 444–455.

Behrman, A., & Baken, R. (1997). Correlation dimension of electroglottographic data from healthy and pathologic subjects. *Journal of Acoustical Society of America, 102*(4), 2371–2379.

Beil, R. G. (1962). Frequency analysis of vowels produced in a helium-rich atmosphere. *Journal of Acoustical Society of America, 34*, 347–349.

Beranek, L. L. (1996). *Acoustics*. Melville, NY: American Institute of Physics for the Acoustical Society of America.

Bergé, P., Pomeau, Y., & Vidal, C. (1984). *Order within chaos: Towards a deterministic approach to turbulence*. Paris: Hermann and Wiley.

Berke, G. S., & Gerratt, B. R. (1993). Laryngeal biomechanics: An overview of mucosal wave mechanics. *Journal of Voice, 7*(2), 123–128.

Berke, G., Moore, D., Hanson, D., Hantke, D., Gerratt, B., & Burstein, F. (1987). Laryngeal modeling: Theoretical, in vitro, in vivo. *Laryngoscope, 97*, 871–881.

Berry, D. A., Herzel, H., Titze, I. R., & Krischer, K. (1994). Interpretation of biomechanical simulations of normal and chaotic vocal fold oscillations with empirical eigenfunctions. *Journal of Acoustical Society of America, 95*(6), 3595–3604.

Bouhuys, A., Mead, J., Proctor, D. F., & Stevens, K. N. (1968). Pressure-flow events during singing. *Annals of the New York Academy of Sciences, 155*, 165–176.

Brown, C. H., & Cannito, M. P. (1995). Modes of vocal variation in Syke's monkey (*Cercopithecus albogularis*) squeals. *Journal of Comparative Psychology, 109*(4), 398–415.

Carterette, E. C., Shipley, C., & Buchwald, J. S. (1979). Linear prediction theory of vocalization in cat and kitten. In B. Lindblom & S. Ohman (Eds.), *Frontiers in speech communication research* (pp. 245–255). London: Academic Press.

Charlton, B. D., Frey, R., McKinnon, A. J., Fritsch, G., Fitch, W. T., & Reby, D. (2013). Koalas use a novel vocal organ to produce unusually low-pitched mating calls. *Current Biology, 23*(23), R1035–R1036.

Chhetri, D. K., Neubauer, J., & Berry, D. A. (2012). Neuromuscular control of fundamental frequency and glottal posture at phonation onset. *Journal of Acoustical Society of America, 131*(2), 1401–1412.

Chiba, T., & Kajiyama, M. (1941). *The vowel: Its nature and structure*. Tokyo: Tokyo-Kaiseikan.

Childers, D. G., Hicks, D. M., Moore, G. P., & Alsaka, Y. A. (1986). A model for vocal fold vibratory motion, contact area, and the electroglottogram. *Journal of the Acoustical Society of America, 80*(5), 1309–1320.

Cooper, D. (1986). Research in laryngeal physiology with excised larynges. In C. W. Cummings, J. M. Fredrickson, L. A. Harker, D. E. Schuller, & C. J. Krause (Eds.), *Otolaryngology—Head and neck surgery* (Vol. 2, pp. 1728–1737). St. Louis, MO: C. V. Mosby.

de Boer, B. (2012). Air sacs and vocal fold vibration: Implications for evolution of speech. *Theoria et Historia Scientiarum, 9*, 13–28.

Dejonckere, P., Bradley, P., Clemente, P., Cornut, G., Crevier-Buchman, L., Friedrich, G., Van De Heyning, P., Remacle, M., & Woisard, V. (2001). A basic protocol for functional assessment of voice pathology, especially for investigating the efficacy of (phonosurgical) treatments and evaluating new assessment techniques Guideline elaborated by the Committee on Phoniatrics of the European Laryngological Society (ELS). *European Archives of Otorhinolaryngology, 258*, 77–82.

Döllinger, M., & Berry, D. (2006). Visualization and quantification of the medial surface dynamics of an excised human vocal fold during phonation. *Journal of Voice, 20*(3), 401–413.

Döllinger, M., Tayama, N., & Berry, D. A. (2005). Empirical eigenfunctions and medial surface dynamics of a human vocal fold. *Methods of Information in Medicine, 44*(3), 384–391.

Eckmann, J., Kamphorst, S. O., Ruelle, D., & Ciliberto, S. (1986). Liapunov exponents from time series. *Physical Review A, 34*(6), 4971–4979.

Eklund, R. (2008). Pulmonic ingressive phonation: Diachronic and synchronic characteristics, distribution and function in animal and human sound production and in human speech. *Journal of the International Phonetic Association, 38*, 235–324.

Fant, G. (1960). *Acoustic theory of speech production*. The Hague, The Netherlands: Mouton.

Fant, G. (1979). Glottal source and waveform analysis. *Speech, Music and Hearing Quarterly Progress and Status Report, 1*(1979), 85–107.

Finnegan, E. M., & Alipour, F. (2009). Phonatory effects of supraglottic structures in excised canine larynges. *Journal of Voice, 23*(1), 51–61.

Fitch, W. T. (1997). Vocal tract length and formant frequency dispersion correlate with body size in rhesus macaques. *Journal of the Acoustical Society of America, 102*(2), 1213–1222.

Fitch, W. T. (2000a). The evolution of speech: A comparative review. *Trends in Cognitive Sciences, 4*(7), 258–267.

Fitch, W. T. (2000b). The phonetic potential of nonhuman vocal tracts: Comparative cineradiographic observations of vocalizing animals. *Phonetica, 57*(2–4), 205–218.

Fitch, W. T., & Hauser, M. D. (1995). Vocal production in nunhuman primates: Acoustics, physiology, and functional constraints on “honest” advertisement. *American Journal of Primatology, 37*, 191–219.

Fitch, W. T., & Hauser, M. D. (2002). Unpacking “honesty”: Vertebrate vocal production and the evolution of acoustic signals. In A. M. Simmons, A. N. Popper, & R. R. Fay (Eds.), *Acoustic communication* (pp. 65–137). New York: Springer.

Fitch, W. T., Neubauer, J., & Herzel, H. (2002). Calls out of chaos: The adaptive significance of nonlinear phenomena in mammalian vocal production. *Animal Behaviour, 63*, 407–418.

Flanagan, J. (1968). Source-system interaction in the vocal tract. *Annals of the New York Academy of Sciences, 155*(1), 9–17.

Flanagan, J., & Landgraf, L. L. (1968). Self oscillating source for vocal tract synthesizers. *IEEE Transactions on Audio and Electroacoustics, AU-16*(1), 57–64.

Frey, R., Gebler, A., Fritsch, G., Nygren, K., & Weissengruber, G. E. (2007). Nordic rattle: The hoarse vocalization and the inflatable laryngeal air sac of reindeer (*Rangifer tarandus*). *Journal of Anatomy, 210*(2), 131–159.

Frey, R., & Riede, T. (2013). The anatomy of vocal divergence in North American elk and European red deer. *Journal of Morphology, 274*(3), 307–319.

Friedrich, G., & Dejonckere, P. H. (2005). Das Stimmdiagnostik—Protokoll der European Laryngological Society (ELS)—erste Erfahrungen im Rahmen einer Multizenterstudie. *Laryngo- Rhino- Otologie, 84*(10), 744–752.

Gautier, J.-P. (1971). Étude morphologique et fonctionelle des annexes extra-laryngées des Cercopithecinae: Liaison avec les cris d’espacement. *Biologia Gabonica, 7*, 230–267.

Glass, L., & Mackey, M. C. (1988). *From clocks to chaos*. Princeton, NJ: Princeton University Press.

Gleick, J. (1987). *Chaos: The amazing science of the unpredictable*. London: Vintage.

Gramming, P., Sundberg, J., Ternstrom, S., Leanderson, R., & Perkins, W. (1988). Relationship between changes in voice pitch and loudness. *Journal of Voice, 2*(2), 118–126.

Grassberger, P., & Procaccia, I. (1983). Measuring the strangeness of strange attractors. *Physica D, 9*, 189–208.

Hartley, D. J., & Suthers, R. A. (1988). The acoustics of the vocal tract in the horseshoe bat, *Rhinolophus hildebrandti. Journal of the Acoustical Society of America, 84*, 1201–1213.

Häusler, U. (2000). Vocalization-correlated respiratory movements in the squirrel monkey. *Journal of Acoustical Society of America, 108*(4), 1443–1450.

Herbst, C. T. (2014). Glottal efficiency of periodic and irregular in vitro red deer voice production. *Acta Acoustica united with Acoustica, 100*(4), 724–733.

Herbst, C. T., Herzel, H., Svec, J. G., Wyman, M. T., & Fitch, W. T. (2013a). Visualization of system dynamics using phasegrams. *Journal of the Royal Society Interface, 10*(85), 1–14.

Herbst, C. T., Hess, M., Müller, F., Svec, J. G., & Sundberg, J. (2015). Glottal adduction and subglottal pressure in singing. *Journal of Voice, 29*(4), 391–402.

Herbst, C. T., Howard, D. M., & Svec, J. G. (2015). The sound source in singing—basic principles and muscular adjustments for fine-tuning vocal timbre. In G. Welch, D. M. Howard, & J. Nix (Eds.), *The Oxford handbook of singing*. Oxford: Oxford University Press. Online Publication Date: Dec 2015. DOI: 10.1093/oxfordhb/9780199660773.013.011

Herbst, C. T., Qiu, Q., Schutte, H. K., & Švec, J. G. (2011). Membranous and cartilaginous vocal fold adduction in singing. *Journal of Acoustical Society of America, 129*(4), 2253–2262.

Herbst, C. T., Stoeger, A. S., Frey, R., Lohscheller, J., Titze, I. R., Gumpenberger, M., & Fitch, W. T. (2012). How low can you go? Physical production mechanism of elephant infrasonic vocalizations. *Science, 337*(6094), 595–599.

Herbst, C. T., & Svec, J. G. (2016). Basics of voice acoustics—A tutorial. In R. Sataloff (Ed.), *Sataloff’s comprehensive textbook of otolaryngology: Head and neck surgery: Laryngology* (Vol. 4, pp. 63–80). Philadelphia: Jaypee.

Herbst, C. T., Svec, J. G., Lohscheller, J., Frey, R., Gumpenberger, M., Stoeger, A. S., & Fitch, W. T. (2013b). Complex vibratory patterns in an elephant larynx. *Journal of Experimental Biology, 216*, 4054–4064.

Herbst, C. T., Ternström, S., & Švec, J. G. (2009). Investigation of four distinct glottal configurations in classical singing—A pilot study. *Journal of Acoustical Society of America, 125*(3), EL104–EL109.

Herzel, H. (1993). Bifurcations and chaos in voice signals. *Applied Mechanics Review, 46*(7), 399–413.

Herzel, H., Holzfuss, J., Kowalik, Z., Pompe, B., & Reuter, R. (1998). Detecting bifurcations in voice signals. In H. Kantz, J. Kurths, & G. Mayer-Kress (Eds.), *Nonlinear analysis of physiological data* (pp. 325–344). Berlin, Germany: Springer.

Hewitt, G., MacLarnon, A., & Jones, K. E. (2002). The functions of laryngeal air sacs in primates: A new hypothesis. *Folia Primatologica (Basel), 73*(2–3), 70–94.

Hilloowala, R. A., & Lass, N. J. (1978). Spectrographic analysis of laryngeal air sac resonance in rhesus monkey. *American Journal of Physical Anthropology, 49*(1), 129–131.

Hirano, M. (1974). Morphological structure of the vocal cord as a vibrator and its variations. *Folia Phoniatrica, 1974*(26), 89–94.

Hirano, M. (1981). *Clinical examination of voice*. New York: Springer.

Hirano, M., Hibi, S., & Hagino, S. (1995). Physiological aspects of vibrato. In P. H. Dejonckere, M. Hirano, & J. Sundberg (Eds.), *Vibrato* (pp. 9–33). San Diego, CA: Singular Publishing Group.

Hirano, M., Kurita, S., & Nakashima, T. (1983). Growth, development, and aging of human vocal folds. In D. Bless (Ed.), *Vocal fold physiology: Contemporary research and clinical issues* (pp. 22–43). San Diego, CA: College Hill Press.

Hixon, T. (1987). *Respiratory function in speech and song*. Boston: College-Hill.

Hollien, H., Shearer, W., & Hicks, J. W., Jr. (1977). Voice fundamental frequency levels of divers in helium-oxygen speaking environments. *Undersea Biomedical Research, 4*(2), 199–207.

Holmberg, E. B., Hillman, R. E., & Perkell, J. S. (1988). Glottal airflow and transglottal air pressure measurements for male and female speakers in soft, normal, and loud voice. *Journal of the Acoustical Society of America, 84*, 511–529.

Howard, D. M., & Angus, J. A. S. (2009). *Acoustics and psychoacoustics* (4th ed.). Oxford, England: Oxford University Press.

Husson, R. (1950). *Ètude des phénomènes physiologiques et acoustiques fondamentaux de la voix chantée.* Thesis, Université de Paris, Faculté des Sciences.

Ishizaka, K., & Flanagan, J. L. (1972). Synthesis of voiced sounds from a two-mass model of the vocal cords. *The Bell System Technical Journal, 51*(6), 1233–1268.

Jiang, J., O'Mara, T., Conley, D., & Hanson, D. (1999). Phonation threshold pressure measurements during phonation by airflow interruption. *Laryngoscope, 109*(3), 425–432.

Jiang, J., Zhang, Y., & McGilligan, C. (2006). Chaos in voice, from modeling to measurement. *Journal of Voice, 20*(1), 2–17.

Johnson, A. M., Ciucci, M. R., Russell, J. A., Hammer, M. J., & Connor, N. P. (2010). Ultrasonic output from the excised rat larynx. *Journal of Acoustical Society of America, 128*(2), EL75–EL79.

Kelleher, J. E., Siegmund, T., Du, M., Naseri, E., & Chan, R. W. (2013). Empirical measurements of biomechanical anisotropy of the human vocal fold lamina propria. *Biomechanics and Modeling in Mechanobiology, 12*(3), 555–567.

Khosla, S., Muruguppan, S., Gutmark, E., & Scherer, R. (2007). Vortical flow field during phonation in an excised canine larynx model. *The Annals of Otology, Rhinology, and Laryngology, 116*(3), 217–228.

Klatt, D., & Klatt, L. (1990). Analysis, synthesis, and perception of voice quality variations among female and male talkers. *Journal of the Acoustical Society of America, 87*(2), 820–857.

Koda, H., Nishimura, T., Tokuda, I. T., Oyakawa, C., Nihonmatsu, T., & Masataka, N. (2012). Soprano singing in gibbons. *American Journal of Physical Anthropology, 149*(3), 347–355.

Kurita, S., Nagata, K., & Hirano, M. (1983). A comparative study of the layer structure of the vocal fold. In D. M. Bless & J. H. Abbs (Eds.), *Vocal fold physiology: Contemporary research and clinical issues* (pp. 3–21). San Diego: College Hill.

Ladefoged, P., & McKinney, N. (1963). Loudness, sound pressure, and subglottal pressure in speech. *Journal of the Acoustical Society of America, 35*(4), 454–460.
Lauterborn, W., & Cramer, E. (1981). Subharmonic routes to chaos observed in acoustics. *Physical Review Letters, 47*, 1445–1448.
Lauterborn, W., & Parlitz, U. (1988). Methods of chaos physics and their application to acoustics. *Journal of Acoustical Society of America, 84*(6), 1975–1993.
Lindestad, P. A., Sodersten, M., Merker, B., & Granqvist, S. (2001). Voice source characteristics in Mongolian "throat singing" studied with high-speed imaging technique, acoustic spectra, and inverse filtering. *Journal of Voice, 15*(1), 78–85.
Lucero, J. C., Lourenco, K., Hermant, N., Van Hirtum, A., & Pelorson, X. (2012). Effect of source-tract acoustical coupling on the oscillation onset of the vocal folds. *Journal of Acoustical Society of America, 132*(1), 403–411.
Madsen, P. T., Jensen, F. H., Carder, D., & Ridgway, S. (2012). Dolphin whistles: A functional misnomer revealed by heliox breathing. *Biology Letters, 8*(2), 211–213.
McGlashan, J., Sadolin, C., & Kjelin, H. (2007). *Can vocal effects such as distortion, growling, rattle and grunting be produced without traumatising the vocal folds?* Paper presented at the 7th Pan-European Voice Conference (PEVOC), Groningen, The Netherlands.
Mende, W., & Herzel, H. (1990). Bifurcation and chaos in newborn infant cries. *Physics Letters A, 145*(8–9), 418–424.
Mergell, P., Herzel, H., & Titze, I. R. (2000). Irregular vocal-fold vibration—High-speed observation and modeling. *Journal of the Acoustical Society of America, 108*(6), 2996–3002.
Miller, D. G. (2008). *Resonance in singing*. Princeton, NJ: Inside View Press.
Miller, D. G., & Schutte, H. K. (1984). Characteristic patterns of sub- and supraglottal pressure variations within the glottal cycle. In *Transcripts of the XIIIth symposium: Care of the professional voice*. New York: The Voice Foundation.
Moore, P. (1991). Voice: A historical perspective. A short history of laryngeal investigation. *Journal of Voice, 5*(3), 166–281.
Negus, V. E. (1932). *The mechanism of the larynx* (pp. 220–227). London: Cassel.
Novick, A., & Griffin, D. R. (1961). Laryngeal mechanisms in bats for the production of orientation sounds. *Journal of Experimental Zoology, 148*, 125–145.
Oren, L., Khosla, S., & Gutmark, E. (2014). Intraglottal pressure distribution computed from empirical velocity data in canine larynx. *Journal of Biomechanics, 47*(6), 1287–1293.
Packard, N. H., Crutchfield, J. P., Farmer, J. D., & Shaw, R. S. (1980). Geometry from a time series. *Physical Review Letters, 45*(9), 712–716.
Powell, A. (1995). Lord Rayleigh's foundations of aeroacoustics. *Journal of Acoustical Society of America, 98*(4), 1839–1844.
Pye, J. D. (1967). Synthesizing the waveforms of bats' pulses. In R.-G. Busnel & J. F. Fish (Eds.), *Animal sonar systems: Biology and bionics* (pp. 43–67). Jouy-en-Josas, France: Laboratoire de Physiologie Acoustique, INRA-CNRS.
Reby, D., & McComb, K. (2003). Anatomical constraints generate honesty: Acoustic cues to age and weight in the roars of red deer stags. *Animal Behaviour, 65*, 519–530.
Remmers, J. E., & Gautier, H. (1972). Neural and mechanical mechanisms of feline purring. *Respiration Physiology, 16*, 351–361.
Riede, T. (2010). Elasticity and stress relaxation of rhesus monkey (*Macaca mulatta*) vocal folds. *Journal of Experimental Biology, 213*(Pt 17), 2924–2932.
Riede, T. (2011). Subglottal pressure, tracheal airflow, and intrinsic laryngeal muscle activity during rat ultrasound vocalization. *Journal Neurophysiology, 106*(5), 2580–2592.
Riede, T., Arcadi, A. C., & Owren, M. J. (2007). Nonlinear acoustics in the pant hoots of common chimpanzees (*Pan troglodytes*): Vocalizing at the edge. *Journal of Acoustical Society of America, 121*(3), 1758–1767.
Riede, T., & Titze, I. R. (2008). Vocal fold elasticity of the Rocky Mountain elk (*Cervus elaphus nelsoni*)—Producing high fundamental frequency vocalization with a very long vocal fold. *Journal of Experimental Biology, 211*(Pt 13), 2144–2154.

Riede, T., Tokuda, I. T., Munger, J. B., & Thomson, S. L. (2008). Mammalian laryngseal air sacs add variability to the vocal tract impedance: Physical and computational modeling. *Journal of Acoustical Society of America, 124*(1), 634–647.

Roark, R. M. (2006). Frequency and voice: Perspectives in the time domain. *Journal of Voice, 20*(3), 325–354.

Roberts, L. H. (1975). The rodent ultrasound production mechanism. *Ultrasonics, 13*(2), 83–88.

Rossing, T. (1990). *The science of sound*. Reading, MA: Addison-Wesley.

Rothenberg, M. (1973). A new inverse-filtering technique for deriving the glottal air flow waveform during voicing. *Journal of the Acoustical Society of America, 53*(6), 1632–1645.

Rothenberg, M. (1981a). The voice source in singing. In *Research aspects on singing* (pp. 15–33). Stockholm: Royal Swedish Academy of Music.

Rothenberg, M. (1981b). Acoustic interaction between the glottal source and the vocal tract. In K. N. Stevens & M. Hirano (Eds.), *Vocal fold physiology* (pp. 305–328). Tokyo: University of Tokyo Press.

Roux, J.-C., Simonyi, R. H., & Swinney, H. L. (1983). Observation of a strange attractor. *Physica D, 8*(1–2), 257–266.

Sanders, I., Weisz, D., Yang, B. Y., Fung, K., & Amirali, A. (2001). *The mechanism of ultrasonic vocalization in the rat*. Paper presented at the Society for Neuroscience Abstracts.

Scherer, R. C. (2014). Laryngeal function during phonation. In J. S. Rubin, R. Sataloff, & G. S. Korovin (Eds.), *Diagnosis and treatment of voice disorders* (4th ed., Vol. 117–144). San Diego, CA: Plural Publishing.

Schutte, H. (1980). *The efficiency of voice production*. Doctoral dissertation, University of Groningen.

Schutte, H., & Miller, D. G. (1988). Resonanzspiele der Gesangsstimme in ihren Beziehungen zu supra- und subglottalen Druckverläufen: Konsequenzen für die Stimmbildungstheorie. *Folia Phoniatrica, 40*, 65–73.

Serway, R. A. (1990). *Physics for scientists and engineers*. Philadelphia: Saunders College.

Sissom, D., Rice, D., & Peters, G. (1991). How cats purr. *The Zoological Society of London, 223*, 67–78.

Snelleman, J., Versnel, H., & Dejonckere, P. (2007). *Loud phonation: Subglottal pressures and closed quotients*. Paper presented at the 7th Pan-European Voice Conference (PEVOC), Groningen, The Netherlands.

Stevens, K. N. (1998). *Acoustic phonetics*. Cambridge, MA: The MIT Press.

Story, B. (2002). An overview of the physiology, physics and modeling of the sound source for vowels. *Acoustical Science and Technology, 23*(4), 195–206.

Story, B., & Titze, I. R. (1995). Voice simulation with a body cover model of the vocal folds. *Journal of the Acoustical Society of America, 97*(2), 1249–1260.

Strogatz, S. H. (2000). *Nonlinear dynamics and chaos: With applications to physics, biology, chemistry, and engineering*. Cambridge, MA: Westview Press.

Sundberg, J. (1981). *Research aspects on singing*. Stockholm: Royal Swedish Academy of Music.

Sundberg, J., Fahlstedt, E., & Morell, A. (2005). Effects on the glottal voice source of vocal loudness variation in untrained female and male voices. *Journal of Acoustical Society of America, 117*, 879–885.

Sundberg, J., Titze, I., & Scherer, R. (1993). Phonatory control in male singing: A study of the effects of subglottal pressure, fundamental frequency, and mode of phonation on the voice source. *Journal of Voice, 7*(1), 15–29.

Suthers, R. A., & Fattu, J. M. (1973). Mechanisms of sound production in echolocating bats. *American Zoologist, 13*, 1215–1226.

Suthers, R. A., & Fattu, J. M. (1982). Selective laryngeal neurotomy and control of phonation by the echolocating bat, *Eptesicus*. *Journal of Comparative Physiology, 145*, 529–537.

Svec, J. G. (2000). *On vibration properties of human vocal folds: Voice registers, bifurcations, resonance characteristics, development and application of videokymography*. Doctoral dissertation, University of Groningen.

Svec, J. G., Horacek, J., Sram, F., & Vesely, J. (2000). Resonance properties of the vocal folds: In vivo laryngoscopic investigation of the externally excited laryngeal vibrations. *Journal of the Acoustical Society of America, 108*(4), 1397–1407.

Svec, J. G., Schutte, H. K., & Miller, D. G. (1999). On pitch jumps between chest and falsetto registers in voice: Data from living and excised human larynges. *Journal of the Acoustical Society of America, 106*(3 Pt 1), 1523–1531.

Titze, I. R. (1988a). Regulation of vocal power and efficiency by subglottal pressure and glottal width. In O. Fujimura (Ed.), *Vocal fold physiology* (pp. 227–238). New York: Raven.

Titze, I. R. (1988b). The physics of small-amplitude oscillation of the vocal folds. *Journal of Acoustical Society of America, 83*(4), 1536–1552.

Titze, I. R. (1989). On the relation between subglottal pressure and fundamental frequency in phonation. *Journal of Acoustical Society of America, 85*(2), 901–906.

Titze, I. R. (1992). Phonation threshold pressure: A missing link in glottal aerodynamics. *Journal of Acoustical Society of America, 91*(5), 2926–2935.

Titze, I. R. (1995). *Workshop on acoustic voice analysis. Summary statement.* Iowa City, IA: National Center for Voice and Speech.

Titze, I. R. (2000). *Principles of voice production.* Iowa City, IA: National Center for Voice and Speech.

Titze, I. R. (2004a). A theoretical study of F0–F1 interaction with application to resonant speaking and singing voice. *Journal of Voice, 18*(3), 292–298.

Titze, I. R. (2004b). Theory of glottal airflow and source-filter interaction in speaking and singing. *Acta Acoustica united with Acoustica, 90*, 641–648.

Titze, I. R. (2006). *The myoelastic aerodynamic theory of phonation.* Denver, CO: National Center for Voice and Speech.

Titze, I. R. (2008). Nonlinear source-filter coupling in phonation: Theory. *Journal of the Acoustical Society of America, 123*(5), 2733–2749.

Titze, I. R. (2009). Phonation threshold pressure measurement with a semi-occluded vocal tract. *Journal of Speech, Language, and Hearing Research, 52*(4), 1062–1072.

Titze, I. R. (2011). Vocal fold mass is not a useful quantity for describing F0 in vocalization. *Journal of Speech, Language, and Hearing Research, 54*, 520–522.

Titze, I. R., Baken, R. J., Bozeman, K. W., Granqvist, S., Henrich, N., Herbst, C. T., Howard, D. M., Hunter, E. J., Kaelin, D., Kent, R. D., Kreiman, J., Kob, M., Löfqvist, A., McCoy, S., Miller, D. G., Noé, H., Scherer, R. C., Smith, J. R., Story, B. H., Svec, J. G., Ternström, S., & Wolfe, J. (2015). Toward a consensus on symbolic notation of harmonics, resonances, and formants in vocalization. *Journal of the Acoustical Society of America, 137*(5), 3005–3007.

Titze, I. R., Baken, R. J., & Herzel, H. (1993). Evidence of chaos in vocal fold vibration. In I. R. Titze (Ed.), *Vocal fold physiology: Frontiers in basic science* (pp. 143–188). San Diego, CA: Singular Publishing Group.

Titze, I. R., Fitch, W. T., Hunter, E. J., Alipour, F., Montequin, D., Armstrong, D. L., McGee, J. A., & Walsh, E. J. (2010). Vocal power and pressure-flow relationships in excised tiger larynges. *Journal of Experimental Biology, 213*(22), 3866–3873.

Titze, I. R., & Hunter, E. J. (2004). Normal vibration frequencies of the vocal ligament. *Journal of Acoustical Society of America, 115*(5 Pt 1), 2264–2269.

Titze, I. R., & Riede, T. (2010). A cervid vocal fold model suggests greater glottal efficiency in calling at high frequencies. *PLoS Computational Biology, 6*(8).

Titze, I. R., & Sundberg, J. (1992). Vocal intensity in speakers and singers. *Journal of the Acoustical Society of America, 91*(5), 2936–2946.

Tokuda, I., Horáček, J., Švec, J. G., & Herzel, H. (2007). Comparison of biomechanical modeling of register transitions and voice instabilities with excised larynx experiments. *Journal of Acoustical Society of America, 122*(1), 519–531.

Tokuda, I., Riede, T., Neubauer, J., Owren, M. J., & Herzel, H. (2002). Nonlinear analysis of irregular animal vocalizations. *Journal of Acoustical Society of America, 111*(6), 2908–2919.

van den Berg, J. (1956). Direct and indirect determination of the mean subglottic pressure. *Folia Phoniatrica, 8*, 1–24.
van den Berg, J. (1958). Myoelastic-aerodynamic theory of voice production. *Journal of Speech and Hearing Research, 3*, 227–244.
van den Berg, J. (1963). Vocal ligaments versus registers. *The NATS Bulletin, 16–31.*
van den Berg, J., Zantema, J. T., & Doornebal, P., Jr. (1957). On the air resistance and the Bernoulli effect of the human larynx. *Journal of the Acoustical Society of America, 29*(5), 626–631.
Vilkman, E. (1987). An apparatus for studying the role of the cricothyroid articulaion in the voice production of excised human larynges. *Folia Phoniatrica, 39*, 169–177.
Welham, N. V., Montequin, D. W., Tateya, I., Tateya, T., Choi, S. H., & Bless, D. M. (2009). A rat excised larynx model of vocal fold scar. *Journal of Speech, Language, and Hearing Research, 52*(4), 1008–1020.
Wilden, I., Herzel, H., Peters, G., & Tembrock, G. (1998). Subharmonics, biphonation, and deterministic chaos in mammal vocalization. *Bioacoustics, 9*, 171–196.
Zemlin, W. R. (1998). *Speech and hearing science: Anatomy and physiology* (4th ed.). Boston: Allyn & Bacon.
Zhang, Z. (2009). Characteristics of phonation onset in a two-layer vocal fold model. *Journal of Acoustical Society of America, 125*(2), 1091–1102.
Zhang, Y., Jiang, J. J., Tao, C., Bieging, E., & MacCallum, J. K. (2007). Quantifying the complexity of excised larynx vibrations from high-speed imaging using spatiotemporal and nonlinear dynamic analyses. *Chaos, 17*(4), 043114-1–043114-10.
Zhang, K., Siegmund, T., Chan, R. W., & Fu, M. (2009). Predictions of fundamental frequency changes during phonation based on a biomechanical model of the vocal fold lamina propria. *Journal of Voice, 23*(3), 277–282.

Chapter 7
Infrasonic and Seismic Communication in the Vertebrates with Special Emphasis on the Afrotheria: An Update and Future Directions

Peter M. Narins, Angela S. Stoeger, and Caitlin O'Connell-Rodwell

Abstract Infrasonic and seismic communication in terrestrial vertebrates is generally poorly known. Moreover, studies of these communication modalities have been restricted to relatively few vertebrate groups. In this chapter we begin with the non-Afrotherian vertebrates and review what is known about their infrasonic (including birds and mammals) and seismic (including amphibians, reptiles, birds, and mammals) communication. We then devote special sections to the Afrotherian vertebrates, concentrating on (1) infrasonic communication in elephants, (2) seismic communication in elephants, and (3) seismic communication in golden moles (Chrysocloridae). Motivated by the lack of detailed knowledge of vibration communication in chrysochlorids, we furnish a blueprint for a set of experiments that would provide novel and interesting data to fill the lacunae in our understanding of seismic signal detection and localization by these enigmatic animals.

P.M. Narins (✉)
Departments of Integrative Biology and Physiology, and Ecology and Evolutionary Biology, University of California Los Angeles,
621 Charles E. Young Drive South, Los Angeles, CA 90095, USA
e-mail: pnarins@ucla.edu

A.S. Stoeger
Department of Cognitive Biology, UZA 1, Biologiezentrum,
Althanstrasse 14, 1090 Vienna, Austria
e-mail: angela.stoeger-horwath@univie.ac.at

C. O'Connell-Rodwell
Department of Otolaryngology, Head and Neck Surgery, Stanford University School of Medicine, 801 Welch Road, Stanford, CA 94305-5739, USA
e-mail: ceoconnell@stanford.edu

R.A. Suthers et al. (eds.), *Vertebrate Sound Production and Acoustic Communication*, Springer Handbook of Auditory Research 53,
DOI 10.1007/978-3-319-27721-9_7

Keywords Acoustic camera • Acoustic communication • Amphibian papilla Basilar papilla • Golden moles • Infrasound • Mole-rats • Moles • Rayleigh waves Rumble vocalizations • Substrate-borne vibrations • Vibration communication

7.1 Introduction to Communication

Terrestrial acoustic communication (TAC) occurs when one organism exhibits a change in its behavior as a direct result of detecting a signal broadcast by a second organism via a channel between the two organisms. Thus, communication involves generation, transmission, and reception of signals. Infrasonic communication refers to TAC for which airborne signals fall in the infrasonic range (<20 Hz), the nominal limit of low-frequency human hearing. Seismic communication refers to TAC for which air is not the intervening channel, but rather signals propagate through a solid substrate such as the ground, a tree branch, or a blade of grass. Several quantitative reports have emerged of terrestrial vertebrates that *either* produce *or* detect infrasonic and/or seismic signals. Although these examples do not strictly qualify as "communication" (sensu the preceding definition), we are including some of them in this chapter because it is likely that further research will reveal infrasonic or seismic communication in these cases.

Infrasonic and seismic communication in terrestrial vertebrates is generally poorly known. Moreover, studies of these communication modalities have been restricted to relatively few vertebrate groups. In fact, the study of vibration communication in the invertebrates has a longer history (for recent reviews see Hill 2008; O'Connell-Rodwell 2010; Cocroft et al. 2014). This chapter attempts to review the current state of knowledge about both infrasonic and seismic communication in terrestrial vertebrates. It is organized in two sections as follows: In Sect. 7.2, a review of the current knowledge of infrasonic and seismic communication in the vertebrates is presented, including amphibians, reptiles, birds, and non-Afrotherian mammals; Sect. 7.3 focuses on these two communication modalities in the Afrotheria, a clade that appears to harbor several seismic specialists.

7.2 Section I

7.2.1 Review of Infrasonic Communication in the Non-Afrotherian Vertebrates

The definition of infrasound is clearly anthropocentric; that is, it is sound below 20 Hz, the nominal lower frequency limit of human hearing. In fact, the human audiogram is not as black and white as the infrasound definition would imply; sounds below 20 Hz may be audible to humans if the intensity is high enough (see human audiogram in Fig. 7.1).

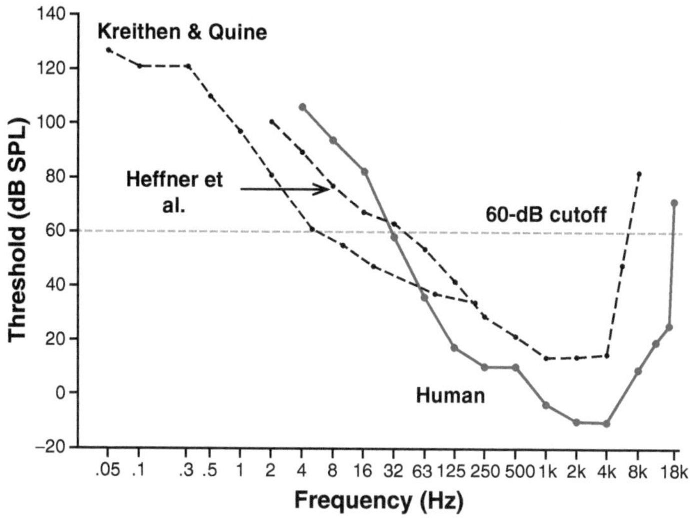

Fig. 7.1 Pigeon and human audiograms. Comparison of human audiogram (from Jackson et al. 1999) to two audiograms from the domestic pigeon, *Columba livia*, obtained from two different studies using two different methods (see text). The pigeon audiograms both indicate better sensitivity to low-frequency hearing than humans (Modified from Heffner et al. 2013)

7.2.1.1 Birds

Several bird species produce infrasound, but only those that either vocalize and/or have been documented as perceiving infrasound will be addressed here. Cassowaries (*Casuarius bennetti*) produce low-frequency vocalizations (Mack and Jones 2003) (ca. 23 Hz) that may extend into the infrasonic range. The sensitivity of chickens (*Gallus gallus domesticus*) to infrasound exceeds that of the homing pigeon (*Columba livia*) (Hill et al. 2014), which are famous for their sensitivity to infrasound, thought to facilitate orientation during migration (Yodlowski et al. 1977).

Rock Dove (Pigeons)

Rock doves (pigeons: *Columba livia*) show remarkable sensitivity to low-frequency sound (below 10 Hz), 50 dB more sensitive than humans; (Kreithen and Quine 1979). Figure 7.1 depicts the results from two different studies in which physiological (Kreithen and Quine 1979) and behavioral (Heffner et al. 2013) data were obtained. Electrophysiological recordings from the pigeon cochlear ganglion reveal sensitivity at frequencies of 1, 2, 5, and 10 Hz presented at 90 dB SPL (Schermuly and Klinke 1990).

Yodlowski et al. proposed that pigeons might detect thunderstorms, weather fronts, magnetic storms, earthquakes, and so forth, and use these low-frequency sounds for orientation, homing, and migration (Yodlowski et al. 1977). Others have suggested that pigeons might detect summer thermals and approach or avoid them (Schermuly and Klinke 1990). Hagstrum (2000) suggested that atmospheric processes can interfere with infrasonic map cues that cause homing pigeons to veer significantly off-course. Despite these intriguing hypotheses, the mechanisms underlying infrasound detection in pigeons, the best studied of all birds in this context, are still poorly understood.

Guinea Fowl

Using auditory evoked potentials (AEPs) and single-unit data recorded from the auditory midbrain nucleus (MLD) of unanesthetized Guinea fowl (*Numida meleagris*), Theurich et al. (1984) demonstrated that cells in the MLD of these animals exhibited phase-locked responses to extremely low-frequency sinusoids (2–10 Hz) at moderate intensities. The significance of infrasound detection in birds is likely to differ between species. Its use in navigation and homing is reasonable for the pigeon, but it may also be appropriate for the guinea fowl, which lead a primarily terrestrial lifestyle in which selection would presumably be strong for low-frequency acoustic signals that carried for considerable distances (Theurich et al. 1984) or the detection of distant thunderstorms and consequently rain in their natural arid habitat of savannah and grassland (Maier 1982). Follow-up behavioral studies of guinea fowl perception of infrasound are clearly needed.

Indian Peafowl

Freeman and Hare (2015) demonstrated that the male Indian peafowl (*Pavo cristatus*) produce bimodal displays consisting of a conspicuous visual component (tail or "train" erection and movement) and associated infrasonic signal production. In fact, the peacock vocalization (song) is almost completely infrasonic, as are the display components of wing-shaking, shiver train, and others. In acoustic playback experiments, they found that the male's concave train served as a radiator of acoustic signals, and that male wing-shaking displays produce infrasonic signals as much as 20–25 dB above the ambient noise level.

Quantitative morphological analyses of its inner ear confirms that like other galliforms, the Indian peafowl has an area of morphologically similar hair cells at the apical end of the basilar papilla (BP), indicative of a low-frequency specialization with most hair cells and more than half of the BP dedicated to frequencies below 1 kHz (Corfield et al. 2013). Single-unit recordings from the auditory nerve of the peafowl would confirm infrasonic sensitivity in the auditory periphery of these remarkable birds.

7.2.1.2 Mammals

To focus on a treatment of vocalizations made in only the range of 20 Hz and below, we chose to omit reports of "infrasonic" communication in large mammals such as the rhinoceros (*Diceros bicornis* spp.; O'Connell-Rodwell et al. 2001; Budde and Klump 2003), lion (*Panthera leo*; O'Connell-Rodwell et al. 2001), hippopotamus (*Hippopotamus amphibious*; Barklow 2004) and giraffes (*Giraffa camelopardalis* sp.; Baotic et al. 2015) , although these animals produce vocalizations containing low frequencies, published records thus far indicate that they are above 20 Hz.

Mountain Beaver

One of the more remarkable mammals with regard to its hearing range is the mountain beaver or sewellel (*Aplodontia rufa*). It appears to be the most primitive of all living rodents (Nowak 1999) and has a very large and unique cochlear nucleus complex; in fact, the dorsal cochlear nucleus (DCN) is four to seven times larger in volume than in any of 17 other rodent species examined (Merzenich et al. 1973). Single-unit recordings from more than one-third (78) of the 227 neurons studied in the specialized DCN of the mountain beaver responded to infrasonic frequencies below 10 Hz. Moreover, these units were also driven by much slower changes in air pressure. For a few neurons studied with pressure stimulation, the threshold at ca. 1 Hz was estimated to be in the range of 0.1–1.0 μbar. This prompted the suggestion that these DCN units may be specialized for the detection of slow changes in air pressure (Merzenich et al. 1973). The mountain beaver inhabits a large tunnel system in which the ability to detect and respond to pressure changes would be of obvious value. Field studies of this extraordinary animal could provide valuable insights into the function of these infrasonic cells unique within the rodentia.

7.2.2 *Review of Seismic Communication in the Non-Afrotherian Vertebrates*

7.2.2.1 Amphibians

The available evidence for seismic sensitivity in amphibians has been previously summarized in several reviews (Narins 1990, 2001; Narins et al. 2009; Gridi-Papp and Narins 2010). Some salient examples follow.

Caudate Amphibians: Salamanders

Acute seismic sensitivities (−90 to −130 dB rms re 1 g) have been reported in the salamanders: eastern newt (*Notophthalmus viridescens*), eastern red-backed salamander (*Plethodon cinereus*) (adults), and spotted salamander (*Ambystoma maculatum*) (larvae). Whereas such sensitivity measurements bypassed the natural coupling of the inner ears to the substrate (Ross and Smith 1979, 1980), terrestrial vertebrates exhibit several specializations for conducting vibrations of the substrate to the head and inner ear. Anatomical adaptations commonly rely on the skeleton, as the rigid structure of bones makes them suitable for faithfully transmitting vibrations with minimal loss. Amphibians possess an elaborate coupling solution, in which the opercularis muscle connects the scapula to the oval window (Wever 1973; Mason and Narins 2002). Seismic vibrations that reach the shoulders through the forelimbs are, this way, transmitted directly into the inner ear (Mason 2007a; Gridi-Papp and Narins 2010). This system appears to function quite efficiently, as the seismic sensitivity thresholds in *Notophthalmus viridescens* in the range from 100 to 300 Hz are the most sensitive of any vertebrate tested thus far (Gridi-Papp and Narins 2010). Whether salamanders are able to use low-level substrate vibrations as a source of information about their environment remains an open question (Hill 2009).

Anuran Amphibians: Frogs and Toads

American Bullfrog

Recordings from single axons in the VIIIth cranial nerve of the American bullfrog [*Rana (Lithobates) catesbeiana*] revealed the extraordinary sensitivity of this animal to substrate-borne vibrations (Koyama et al. 1982; Yu et al. 1991). The fibers with the lowest thresholds in this animal (in the frequency range from 15 to 200 Hz) exhibited clear responses to peak accelerations as low as 0.001 cm/s^2, making this the most sensitive quadraped vertebrate to substrate vibrations known at that time. Subsequently, other ranid species have been shown to exhibit remarkable seismic sensitivity: the common frog (*Rana temporaria*; Christensen-Dalsgaard and Jørgensen 1988, 1996; Christensen-Dalsgaard and Walkowiak 1999), and the northern leopard frog (*Rana pipiens*; Christensen-Dalsgaard and Narins 1993; and see "White-Lipped Frog" section).

White-Lipped Frog

The white-lipped frog of Puerto Rico (*Leptodactylus albilabris*) was the first vertebrate for which morphological, neurophysiological, and behavioral evidence has been garnered to support the notion that these animals communicate using seismic signals (Narins and Lewis 1984; Lewis and Narins 1985; Lewis et al. 2001).

Male white-lipped frogs are nocturnally active in the Puerto Rican rainforests, and often vocalize from cryptic calling sites in the moist substrate to attract females

(Lopez et al. 1988). These calls produce a conspicuous airborne component (peak energy at 2.4 kHz), but males also produce impulsive, low-frequency seismic vibrations ("thumps", peak energy <50 Hz) as their vocal sacs strike wet ground (Lewis and Narins 1985). Bimodal playback experiments reveal that males use these seismic thumps to adjust call timing, ensuring that their calls do not overlap temporally with those of neighboring frogs (Lewis et al. 2001). Although it has been suggested that thump vibrations may subserve the regulation of spacing between signaling males, this has yet to be experimentally demonstrated (Lewis and Narins 1985).

Single-unit recordings from the auditory nerve of males of the white-lipped frog revealed clear stimulus-evoked modulations of their resting discharge rates in response to sinusoidal seismic stimuli with peak accelerations less than 0.001 cm/s^2 (10^{-6} g). Thus, this animal exhibits the greatest sensitivity to substrate-borne vibrations for any known terrestrial vertebrate (Narins and Lewis 1984). Moreover, the vibration-sensitive units in this frog may be grouped into two classes: the first class consists of extremely sensitive fibers with best seismic frequencies (BSFs) between 20 and 160 Hz; the second class is made up of less-sensitive fibers with BSFs between 220 and 300 Hz (Lewis and Narins 1985). It is of note that the peak energy (<50 Hz) in the seismic "thump" signal generated during male calling falls in the low-frequency range of its most sensitive seismic units. This "matched filter" between an animal's seismic signal frequency and its best seismic sensitivity is thought to be one way in which animals improve the likelihood of successful communication in a noisy environment (Capranica and Moffat 1983; Smotherman and Narins 2004; Narins and Clark 2016).

Common Malaysian Treefrog

Frogs in the family Rhacophoridae, the Old World treefrogs, comprise 389 species in 18 genera (http://amphibiaweb.org/). One of these genera, *Polypedates*, contains 26 species found in Japan, eastern China, and throughout tropical southeast Asia (Narins 2001). Acoustic playback studies of the common Malaysian treefrog [*Polypedates leucomystax*; Narins et al. 1998 (non-striped morph raised to species status *P. discantus* sp. nov.; Rujirawan et al. 2013); Christensen-Dalsgaard et al. 2002] revealed that females initiate mating by producing a vibratory signal within the vegetation at night, by tapping their rear toes. The toe-tapping lasts for several minutes, only occasionally accompanied by vocalizations. Nearby males were observed to jump toward the toe-tapping female; amplexus ensued. Tapping may function as a vibrational signal advertising the female's presence to neighboring males, but experimental confirmation of this hypothesis remains lacking.

Red-Eyed Treefrog

In a study of the use of vibrational signals in agonistic interactions, experiments with red-eyed treefrogs (*Agalychnis callidryas*; Caldwell et al. 2010a, b) demonstrated that competing males produce chuckle calls and perform a tremulation display in which one male raises his body off of the substrate, rapidly contracts and extends his hind limbs and shakes his hind end (Fig. 7.2). The resulting vibrations

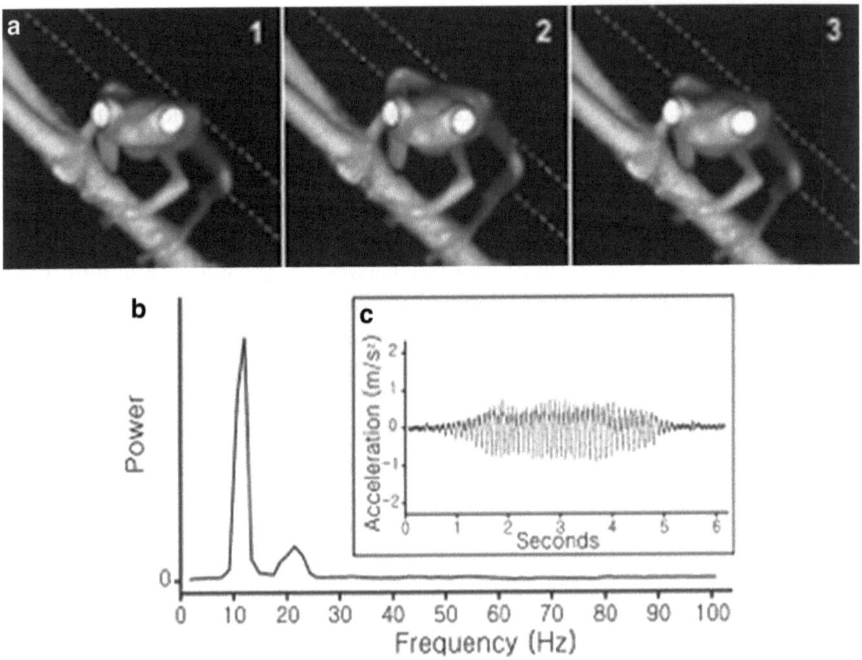

Fig. 7.2 Tremulation display. A1–A3: (**a**) tremulating male *A. callidryas*. (**b, c**) Power spectrum and waveform of a typical tremulation vibration, respectively (from Caldwell et al. 2010a)

are transmitted via the animal's legs to the substrate, often a tree branch or a sapling. A second male, often close by and on the same plant, can detect the substrate-borne vibrations from the first male and often responds to them by exhibiting submissive behavior—either fleeing or by remaining motionless. This is one of the few clear vertebrate examples of aggressive interactions mediated by vibrational signals.

In another series of experiments, Warkentin and her colleagues demonstrated that vibrational cues can trigger a predation response in red-eyed treefrog larvae. Tadpoles of red-eyed treefrogs respond to the vibrations produced during a predator (snake) attack by dropping out of their gelatinous egg mass into the stream below (Warkentin 2005; Warkentin et al. 2006, 2007; Warkentin and Caldwell 2009). Moreover, the vibrations caused by falling raindrops on the egg clutches do not have the same effect on the egg clutch. Thus the embryos are capable not only of detecting vibrations, but also of distinguishing between those produced by a benign stimulus (raindrops) and a potentially lethal source (snake).

7.2.2.2 Reptiles

Reptilian seismic communication has been summarized in a prior review of vertebrate vibration communication (Narins 2001). Some key examples from that review are reproduced here, in addition to several additional vertebrates that have been shown to produce and/or detect seismic signals.

Western Rattlesnake

Snakes are extremely sensitive to substrate vibrations: at the most sensitive frequencies (200–400 Hz), a 1 Å peak-to-peak amplitude is suprathreshold (Hartline 1971). Snakes possess two parallel sensory systems that respond to both airborne sound and substrate vibrations: one subserved by the VIIIth cranial nerve and inner ear (auditory), and the other mediated by the spinal cord and cutaneous mechanoreceptors (somatic). Multiunit evoked responses suggested that the auditory system of the western rattlesnake (*Crotalus viridus*) is not especially sensitive to airborne sound, unlike its remarkable sensitivity to substrate vibrations (see earlier in this section). Evoked potential recordings failed to demonstrate that either the somatic or auditory system can distinguish between airborne and substrate-borne vibrations. Although it has been postulated that vibration detection may subserve prey detection in snakes, this function of their remarkable seismic sensitivity has not been confirmed behaviorally in the western rattlesnake.

Sandfish Lizards

The sandfish lizard (*Scincus scincus*) is quite adept at detecting weak vibrations caused by the movements of insects on or below the surface of the sand (Hetherington 1989) at distances up to 15 cm. In addition, the lizard appears to be able to localize these vibrations while remaining submerged in the sand. The mechanism underlying the vibratory source localization in this lizard is unknown (but see "Localization of Seismic Stimuli: A Look Ahead" section for discussion of vibratory source localization in another desert vertebrate).

Veiled Chameleon

Barnett et al. (1999) provided the first example of reptiles using plant-borne vibrations for intraspecific communication. The veiled chameleon (*Chamaeleo calyptratus*) generates body vibrations just anterior to the front legs. Barnett et al. (1999) hypothesized that these vibratory signals served as vegetation-borne vibratory communication signals because they produced no detectable auditory component and they were produced in courtship and disturbance contexts where communication signals would be expected to occur. Nevertheless, conclusive evidence for these signals resulting in a change of behavior of the receiver in this species has not been forthcoming.

Saharan Sand Vipers

Sand vipers in the genus *Cerastes* are specialized semi-fossorial snakes that launch predatory strikes at lizards and rodents while partially buried in the soft sand of the Saharan desert (Young and Morain 2002). Presentation of chemosensory-neutral targets to the olfactory-denervated, temporarily blinded snakes of the species *Cerastes cerastes* resulted in predatory behaviors similar to those exhibited by intact animals, for both isothermic targets and those heated to mammalian body temperature. Remarkably, every temporarily blinded, anosmic sand viper tested succeeded in capturing free-ranging mice in every trial. These results provide the first experimental evidence for foraging by vibration detection in snakes (Young and Morain 2002).

Pig-Nosed Turtle

Doody et al. (2012), studying synchronous hatching in the nonvocalizing pig-nosed turtle (*Carettochelys insculpta*), demonstrated that this species has evolved rapid hatching in response to hypoxia during nest flooding (Webb et al. 1986). Latency to both hatching and emergence from experimental nests was significantly shorter in groups of eggs than in solitary eggs when subjected to hypoxic conditions, suggesting a group or "sibling effect." Although this study is suggestive of vibration-expedited hatching in the pig-nosed turtle, the recent discovery of underwater vocalizations in two species of fresh water turtles may implicate acoustic communication between embryos in at least some species of Chelonians (Giles et al. 2009; Ferrara et al. 2014).

Royal Python

Vibration and sound-pressure sensitivities were quantified in 11 royal pythons (*Python regius*) by measuring brainstem evoked potentials (Christensen et al. 2012). In this study, the auditory brainstem response to masked and unmasked click stimulations were compared, and forward masking was used to determine the thresholds of vibrational and acoustic sinusoidal stimuli (Berlin et al. 1991; Manley and Kraus 2010; Christensen-Dalsgaard et al. 2011). Their main result is that the pythons are very sensitive to low-frequency vibrations (best sensitivity: −54 dB re 1 m/s^2 at 80–120 Hz) and that the sensitivity to airborne sound is generated by sound-induced head vibrations. This was concluded because, in general, head vibrations induced by threshold-level sound pressure were equal to or greater than those induced by threshold-level vibrations, and therefore sound-pressure sensitivity can be explained by sound-induced head vibration. Thus it was postulated that pythons, and possibly all snakes, lost effective pressure hearing with the complete reduction of a functional outer and middle ear, but have an acute vibration sensitivity that may be used for communication and detection of predators and prey (Christensen et al. 2012).

7.2.2.3 Birds

Vibration communication per se has not been reported in birds. Nevertheless, sensitivity to low-level vibrations has been documented many times. For example, Herbst Corpuscles (HbCs) respond physically to small, uncalibrated displacements produced by a glass stylus attached to a piezo-electric element (Dorward and McIntyre 1980). HbCs are widely distributed in subcutaneous tissues in close association with the tibia and fibula of the legs of birds as well as close to the follicles of the large flight feathers (McIntyre 1980). Greatest sensitivity to vibration in the pigeon was found in the frequency range of 300–1000 Hz, with thresholds about 0.1 μm; the lowest threshold found was 0.04 μm at 500 Hz (Shen 1983). A subsequent study, also using heart-rate conditioning, determined the vibrational sensitivity of the pigeon wing (Hörster 1990). The highest sensitivity in this study was found at either 800 or 900 Hz, with amplitudes between 0.5 and 0.09 μm. Herbst corpuscles have been suggested to function as a warning device by detecting vibratory disturbances of the ground or other supporting surface (Dorward and McIntyre 1980), or because they respond to rapid oscillatory movements of the flight feathers, they could detect changes from laminar to turbulent air flow and thus act as stall indicators (McIntyre 1980) or as sensors in flight control (Hörster 1990). It appears that behavioral studies of birds' responses to either natural or artificial conspecific vibrations would be timely and would serve to demonstrate avian vibrational communication.

7.2.2.4 Non-Afrotherian Mammals

Mammalian seismic signaling has been the subject of several comprehensive reviews (Francescoli 2000; Mason and Narins 2010; Randall 2010). "Talpid Moles, Marsupial Moles, and Ctenomyid Rodents", "Spalacid Mole-Rats: Example—Blind Mole-Rat (*Nannospalax ehrenbergi*)", "Bathyergid Mole Rats: Example—Cape Mole-Rat", and "Gray Seal" sections review several salient examples that reveal general principles of this communication modality.

Talpid Moles, Marsupial Moles, and Ctenomyid Rodents

Only a subset of subterranean mammals appear to *generate* seismic signals for use in intraspecific communication, but the ability to *detect* substrate-borne vibrations is likely to be universal (Mason and Narins 2010). A striking example is the elaborate snout of the star-nosed mole (*Condylura*) containing about 30,000 Eimer's organs, possibly the most sensitive tactile organ yet discovered for its size (Catania 1995). Although one electrophysiological investigation of Eimer organ afferent sensitivity showed responses of one Pacinian Corpuscle (PC)-like unit to static displacements of 5 μm, the rapidly adapting fibers that responded best at frequencies between 250 and 300 Hz were considerably less sensitive (Marasco and Catania 2007). Perhaps, as these workers imply, it is best to consider that integration of the outputs of several Eimer's organs functions to detect small surface features during brief contact of the star to the surface.

Spalacid Mole-Rats: Example—Blind Mole-Rat (*Nannospalax ehrenbergi*)

There are examples of rodents in both the families Muridae (spalacine mole-rats) and Bathyergidae (bathyergid mole-rats) that have been shown to both generate and detect substrate-borne vibrations (Mason and Narins 2001). Probably the best-studied spalacine mole-rat is the blind mole-rat, a highly solitary fossorial form that rarely encounters conspecifics outside of the mating season (Nevo 1961). The unique morphology of the middle ear of the blind mole-rat and of the articulation between the lower jaw and the skull, coupled with its unusual "jaw-listening" behavior enable substrate-borne vibrations to be transmitted to the inner ear in this animal mainly by bone conduction (Rado et al. 1989). Moreover, it was shown that seismic communication signals are processed primarily by the auditory rather than the somatosensory system (Rado et al. 1998). More recently, evidence for a remarkable finding has emerged that *Nannospalax ehrenbergi* is capable of estimating the location and physical properties of underground obstacles using reflected self-generated seismic waves (seismic "echolocation") (Kimchi et al. 2005). Whether echolocation is considered true communication has been debated for years (Bradbury and Vehrencamp 2011), but regardless, the capacity of this animal to discern underground obstacles in the absence of visual cues certainly deserves further study.

Bathyergid Mole Rats: Example—Cape Mole-Rat

The Cape mole-rat (*Georychus capensis*) is a solitary fossorial animal that communicates with its conspecifics by alternately drumming its hind legs on the burrow floor (Narins et al. 1992). Signal production in this species is sexually dimorphic, and mate attraction is likely mediated primarily by seismic signaling between individuals in neighboring burrows. Foot-drumming signals consist of both auditory and seismic components and the seismic component alone is detectable at distances corresponding to natural interburrow distances (3–4 m); the amplitude of the acoustic component attenuates into the background noise level within 1 m of the source (Narins et al. 1992).

Gray Seal

A recent study of gray seals (*Halichoerus grypus*) showed that males perform body slap threat behaviors, in which a male slams his body onto the ground during male–male conflicts (Bishop et al. 2015), the vibrations of which reliably indicated male size measured more than 125 m from the source. These researchers also demonstrated that substrate-borne vibrations are robust across a range of environmental conditions (Bishop et al. 2015). Future studies will be needed to determine the gray seal's threshold for sensing substrate-borne vibrational cues, but the earlier work on elephant seals (Shipley et al. 1992) indicates that this mode of communication might be more prevalent than previously thought.

7.3 Section II

7.3.1 Infrasonic and Seismic Communication in the Afrotheria

Molecular evidence indicates the likelihood of a common African ancestry for several "odd" groups of mammals (http://www.afrotheria.net/information.php). This ancient radiation of African mammals, the Afrotheria, is a clade of mammals, the members of which belong to groups that are either currently living in Africa or of African origin (Springer et al. 1997; Stanhope et al. 1998). They include seven extant groups with little superficial resemblance to each other: the golden moles, sengis (or elephant-shrews), tenrecs, aardvarks, hyraxes, sea cows (manatees and dugongs), elephants, and the extinct Desmostylia. It is likely that all of the members of this clade have a high probability of communicating using infrasound and/or seismically; nevertheless, to date only two groups of Afrotherians have been studied in this regard. These are the elephants (family Elephantidae, genera *Loxodonta*, *Elephas*) and the golden moles (family Chrysochloridae, genera *Eremitalpa*, *Chrysochloris*). Consequently, the known studies of infrasonic and seismic communication in these two groups will be emphasized. It is hoped that this review will stimulate future work examining low-frequency communication in all Afrotheria.

7.3.1.1 Infrasonic Communication in Elephants

In terms of infrasonic communication, elephants [the African savannah elephant (*Loxodonta africana*), the African forest elephant (*Loxodonta cyclotis*), and the Asian elephant (*Elephas maximus*)] are especially noteworthy because they produce some of the loudest terrestrial animal sounds at frequencies between 10 and 35 Hz (Payne et al. 1986; Poole et al. 1988; de Silva 2010) (Fig. 7.3). These vocalizations with fundamental frequencies in the infrasonic range are commonly termed "rumbles" and can have amplitudes as high as 117 dB SPL at 1 m (Beranek 1988). The rumble is the most common (and also the most studied) vocalization of elephants, whereby most research has been conducted on the African savannah elephant (Langbauer 2000; Soltis 2010).

Rumble vocalizations seem to be multifunctional, being produced in almost all conceivable contexts, from close to long-distance communication within and between groups (Poole 2011). There is evidence that rumbles are used to coordinate the movement and spacing of social groups, helping affiliated individuals find one another as well as triggering defensive or exploratory behavior among those that are unaffiliated (McComb et al. 2000, 2003; Charif et al. 2005; Poole 2011). In addition, coordinated interactive rumble vocalizations generated within groups of bonded individuals result in longer calls that are repeated at a higher frequency than calls emitted in isolation (Fig. 7.4) (O'Connell-Rodwell et al. 2012). Longer repeated calls are more easily detected at long distances (see "Factors Enhancing Signal Propagation" section).

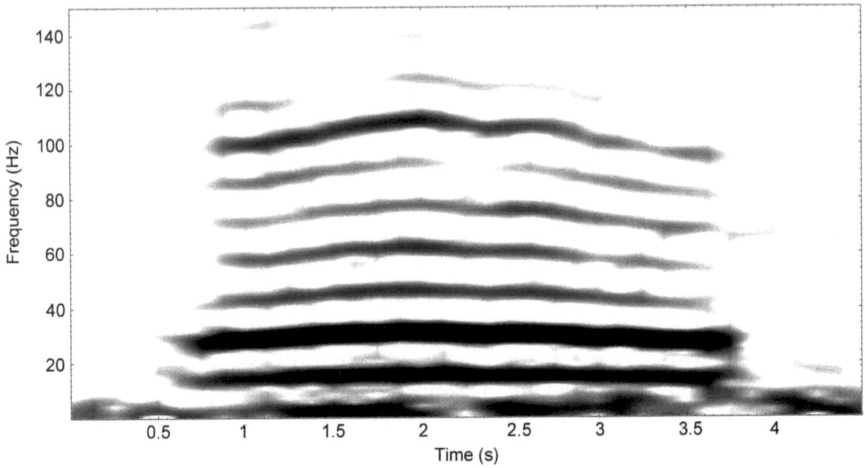

Fig. 7.3 Low-frequency rumble vocalization of a 19-year-old male African elephant (*L. africana*). The fundamental frequency is about 13–14 Hz

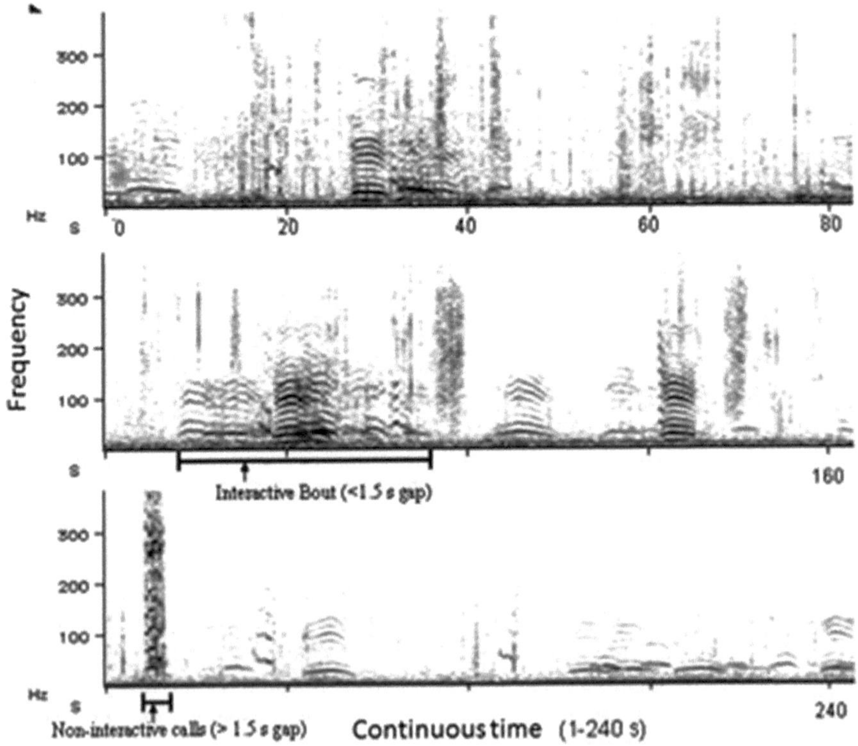

Fig. 7.4 Series of interactive antiphonal bouts of elephant rumble vocalizations (Modified from O'Connell-Rodwell et al. 2012)

One striking feature of the low-frequency rumble is the impressive information content apparently transmitted, where remarkable structural variability reflects all conceivable acoustic parameters including temporal, source, and filter related parameters (vocal tract resonances or formant frequencies). It is not surprisingly, therefore, that rumbles are individually distinctive and recognized as such by elephants (McComb et al. 2003; Soltis et al. 2005; O'Connell-Rodwell et al. 2007). They also differ according to age (Stoeger-Horwath et al. 2007; Stoeger et al. 2014), and structurally varying forms have been linked with reproductive (Poole et al. 1988; Poole 1989) and emotional states (Soltis et al. 2009).

Poole et al. (1988) proposed that rumbles could be used as long-distance mate attraction calls. Playback experiments showed that males respond to female "estrous calls" by orienting and walking 1 km or more toward the sound source (Langbauer et al. 1991). More recent playback experiments indicated that adult males in the hormonal state of musth (Poole and Moss 1981) and subadult males were more likely to respond to these calls than males that were not in musth (O'Connell-Rodwell et al. 2011).

Estrus rumbles are structurally distinct, including lower fundamental frequencies, lower first formant (vocal tract resonance) frequencies, and higher first formant amplitudes (Soltis et al. 2005). In addition, estrus calls are longer and repeated more often than other call types (Leong et al. 2003). These acoustic features are important in long-distance communication as high-amplitude, low-frequency long repeated calls facilitate better propagation.

Although elephant females are generally much more vocal than males, males do produce several distinctive rumble types; the best characterized is the "musth rumble" made in the context of the reproductive condition of musth (Poole et al. 1988), thought to advertise the animal's hormonal state over long distances to females as well as potential rivals (Poole 1989, 1999). Females respond to the rumbles of musth males by vocalizing, so males may indeed identify and locate estrous females over long distances (Poole 1999).

In elephant rumbles, formant variations have proven to be a highly relevant acoustic feature, being important in social context and in referential information coding (McComb et al. 2003; Soltis et al. 2014). Stoeger et al. (2012) demonstrated that African elephants shift between nasal and oral sound emission in rumbling vocalizations depending on social context. Nasal rumbles predominated during long-distance contact calling, whereas oral rumbles were mainly observed during close-distance social bonding. Nasal and oral rumbles varied considerably in their acoustic structure. In particular, the mean frequency spacing of the first two formants predicted the estimated lengths of the two vocal paths (Fig. 7.5). Formant frequency values are determined by the length and shape of the vocal tract, with longer vocal tracts producing lower, more closely spaced formants (Taylor et al., Chap. 9). The observed formant values in Stoeger et al. (2012) corresponded to a vocal tract length of about 2 m for nasal rumbles and about 0.7 m for oral rumbles in the investigated elephants (the study animals were younger than 17 years old and not yet fully grown). Thus, by using the nasal path, an elephant potentially lowers its formants by about threefold.

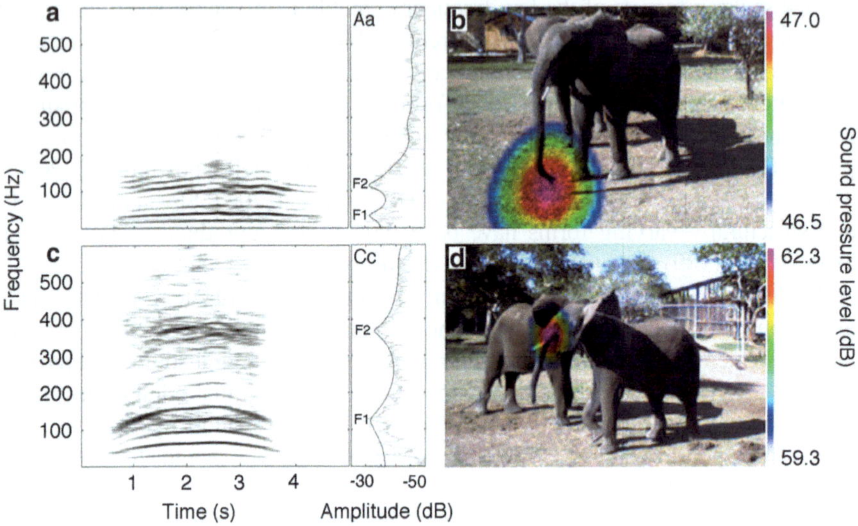

Fig. 7.5 Orally and nasally emitted rumble by a subadult female African elephant. Spectrogram and power spectra of a nasal (**a**) and oral (**c**) rumble revealing the differences in formant structure. (**b**) and (**d**) give the corresponding sound visualizations. *F* formant (Modified from Stoeger et al. 2012 for a nasal and oral rumble)

Formants provide the acoustic basis for discriminating vowels in human speech, transferring important information (Peterson and Barney 1952; Lieberman and Blumstein 1988). Several elephant studies have documented formant variation with context and/or arousal; specifically, an upward shift in the second formant seems to alert other elephants to potential danger (King et al. 2010). Also, females engaged in dominance interactions produce rumbles with lower formant dispersion (spacing) compared to rumbles produced in low-affect contexts (Soltis et al. 2009). Formants in elephants may well play a role in referential calling as elephants can discriminate distinct differences in formant frequencies between alarm rumbles made in response to bees and in response to human voices (Soltis et al. 2014).

In addition, formants are reliable cues to body size in several mammals, reflecting the close relationship between the caller's overall body size, vocal tract lengths, and the frequency spacing of the formants (Reby and McComb 2003; Fitch 2006). Morphological adaptations to elongate the vocal tract to lower formants are present in several species (Taylor et al., Chap. 9). Musth rumbles of male elephants, for example (or rumbles associated with hierarchical interactions), might be under similar selective pressure. Musth rumbles may also function to communicate the size of a musth male to listening males and females, in addition to advertising the hormonal state.

Lowering formants may also promote long-distance call propagation (McComb et al. 2003). The infrasonic fundamental frequency could be a by-product of the large size of the elephant's vocal folds (10.4 cm in an adult female African elephant; Herbst et al. 2012) rather than a specially evolved mechanism for long-distance vocal communication. The amplification of certain frequency regions by using the nasal vocal tract may have evolved as a result of selection pressures particularly relevant to social and reproductive long-distance communication. McComb et al. (2003) found that harmonics in the range of 115 Hz (which highlighted the second formant) decayed at a lower rate with increasing distance than frequency components below and above them. This finding suggests that harmonics in the second formant area (in nasal rumbles) may experience less interference from wind noise than the fundamental frequency contour itself, but more studies would have to be conducted to determine if this effect is true in the far field.

Low-frequency vocalizations are subject to atmospheric conditions, reducing an elephant's call range by an order of magnitude during the diurnal cycle (Garstang et al. 1995; Larom et al. 1997). Temperature inversions after sunset enhance propagation by channeling sound energy within the surface layer (spreading losses become effectively cylindrical rather than spherical) creating conditions where calls could be heard up to 10 km (Garstang et al. 1995). Elephants might adjust the timing and frequency of their low-frequency calls according to atmospheric conditions. Acoustic and seismic playback studies have been conducted between the hours of 4 P.M. and 2 A.M. because of the increase in family group movements to waterholes and vocalizations during this window (O'Connell-Rodwell et al. 2006, 2007), which supports the atmospheric correlation proposed in the preceding text.

Much less is known about elephant sound perception. To date, the hearing sensitivity of elephants has only been investigated in one 7-year-old female Asian elephant (Heffner and Heffner 1980, 1982) that was more sensitive to low frequencies than any other mammals previously tested. At an intensity of 60 dB SPL, the elephant was able to hear 17 Hz, nearly one octave below the comparable human threshold. Nonetheless, the elephant was still considerably less sensitive to frequencies below 100 Hz than to those between 100 Hz and 5 kHz; however, these measurements may underestimate acuity at the lower frequencies, as the longer waveforms of lower frequency sounds (16 Hz) presented in the Heffner and Heffner (1982) study may have required a longer window of time for detection (O'Connell-Rodwell et al. 2001).

The ability of animals to detect low frequencies is associated with the size of the body, head and distance between the ears, and the size of the anatomical hearing structures (Heffner et al. 1982; Rosowski 1994; Nummela 1995). African elephants have large pinna of about 0.5–1 m (Garstang 2004) and up to 1.8 m (Sikes 1971) which may act as a sound-gathering device and aid in sound localization (Heffner et al. 1982; Pye and Langbauer 1998). Listening elephants freeze, spread their ears, and scan their environment (Poole et al. 1988). In auditory experiments, Heffner and Heffner (1982) noted that their Asian elephant subject extended the ears only during sound localization tests, not during absolute frequency or frequency discrimination tests, locating low-frequency sounds to within an azimuth angle of 1°. Sound localization depends on the difference in waveform phases between the two ears, and such phase changes correspond to the interaural time delays (Hartman 1999). Lower frequencies produce longer interaural time differences. More definitive studies are needed to determine the frequency ranges of best sensitivity and the mechanisms of sound localization in individuals of both genera.

Recent investigations of the neuroanatomical structure of the elephant brain (mainly of the African savannah elephant) (Shoshani et al. 2006; Patzke et al. 2013a; Herculano-Houzel et al. 2014), including infrasonic vocal production and reception (Maseko et al. 2013), will no doubt lead to a better understanding of how elephants perceive low-frequency sound. New research has revealed that at least five regions of the combined vocalization production and auditory/seismic reception system are specialized in elephants (Maseko et al. 2012). Elephants possess a large and distinct nucleus ellipticus, which is otherwise found only in elephant seals and cetaceans (Patzke et al. 2013b). The nucleus ellipticus is a specialization of the periaqueductal gray matter and is suggested to be related to infrasonic vocalization production (Maseko et al. 2012).

In terms of reception and interpretation of infrasonic vocalizations, the enlarged lateral superior olivary nucleus and the unique transverse infrageniculate nucleus appear to be related to air-borne sound waves. The enlarged dorsal column nuclei and the ventral posterior inferior nucleus of the dorsal thalamus, in turn, seem to be related to the seismic aspects of the sound waves (Maseko et al. 2012). The investigations of the elephant diencephalon and the brainstem investigated in the latter study demonstrate that while much of the elephant neuroanatomy is typically mammalian, certain anatomical adaptations related to specialized behavior, including infrasonic and seismic communication, are present and highly instructive in understanding elephant behavior (Maseko et al. 2012).

7.3.1.2 Seismic Communication in Elephants

There have been several reviews on elephant seismic communication, but highlights are summarized in this section (see O'Connell-Rodwell 2007; O'Connell-Rodwell and Wood 2010). As discussed in the previous section, both African and Asian elephants emit (Payne et al. 1986; Poole et al. 1988) and detect (Heffner and Heffner 1982) low-frequency (~20 Hz), high-amplitude rumble vocalizations. These rumbles couple with the ground and propagate along the surface as Rayleigh waves (O'Connell-Rodwell et al. 2000; Gunther et al. 2004). The potential range of seismic rumbles based on average call intensities and different soil types is estimated between 2 and 16 km (O'Connell-Rodwell et al. 2000; Gunther et al. 2004).

Elephant family groups detect and respond to these ground-borne vocalizations by exhibiting defensive "bunching" behavior in response to seismic playbacks of antipredator calls (O'Connell-Rodwell et al. 2006), bunching being an indicator of vigilance (McComb et al. 2000). In addition, elephants exhibit more vigilant behaviors during presentations, orienting in the direction of the seismic signal, and spend significantly less time in the area when a seismic alarm is presented (O'Connell-Rodwell et al. 2006). Elephants are also able to discriminate subtle differences between familiar and unfamiliar callers through the ground (O'Connell-Rodwell et al. 2007). The sophistication with which elephants can detect vibrational cues indicates that the ground is an important medium for elephants in which to send and receive signals.

Seismic Signal Detection

Elephants have two possible pathways for detecting seismic signals, either through bone conduction and/or through vibration-sensitive mechanoreceptors in their feet and trunk (Reuter et al. 1998; O'Connell et al. 1999; Bouley et al. 2007). When vibrations transmit through bone, they first couple with the ground via the feet, then travel up through the legs, shoulders, and into the middle ear cavity (Rado et al. 1998) and detection would then be facilitated by the elephant's hypertrophied mallei (Reuter et al. 1998).

Elephants engage in "seismic listening," which is a freezing behavior that is distinctive from freezing in the context of airborne listening and appears to facilitate the detection of seismic information. Individuals lean forward with ears flat against their bodies, placing more weight on their larger front feet which, because of the unique graviportal structure of their forelimbs, are directly in line with the ear (O'Connell-Rodwell et al. 2006). Sometimes the foot is rolled forward onto the toenail, which would also facilitate bone conduction directly to the toe bones. The combination of this behavior and the presence of an enlarged malleus would suggest that elephants employ a bone-conduction pathway for seismic signal detection. The dense, fatty foot pad of the elephant appears similar to "acoustic fat," which would provide a mechanism for more efficient signal propagation and detection of seismic cues (O'Connell-Rodwell et al. 2001), or perhaps even serve as a "seismic lens" to improve sensitivity of the elephant to substrate-borne vibrations.

The role of "acoustic fat" is best known for dolphins, where it is found only in the mandibular channel and the melon (Varanasi and Malin 1971; Varanasi et al. 1975). The fat of the mandible causes a twofold increase in intensity of sound, serving as an impedance matching mechanism. The oil-rich lipid in the melon serves as an acoustic lens that efficiently couples acoustic energy to the water (Au 1993).

Elephants sometimes lift a front foot off the ground while freezing, which would facilitate localization through triangulation. This posture would not maximize bone conduction and suggests that there may be some situations where the somatosensory pathway of detection might be preferred to bone conduction. The amount of time an elephant spends with a portion of their trunk lying on the ground while freezing is additional evidence that the somatosensory pathway is important for detecting seismic cues.

Bone Conduction Enhancement

Fossil data (Barnes et al. 1985; Ketten 2000), immunological evidence (Gaeth et al. 1999), and the morphology of the fetal African elephant ear (Fischer 1990) indicate that Sirenia and elephants have a common aquatic ancestor. The cartilaginous, fat-filled lacunae of the manatee jaw, aerated skull sinuses and fatty deposits on the manatee skull are thought to play a role in coupling sound to the manatee's ear (Ketten et al. 1992; Gerstein et al. 1999). The structure of the manatee skull, incorporating unique fat deposits, may function to conduct sounds (Norris 1968). The African elephant's skull is also aerated by sinuses (van der Merwe et al. 1995) and except for the solid mandible, the cranium consists of inflated bones compartmentalized to form diploe (Shoshani 1998). The fatty deposits and aerated skull sinuses in the elephant may facilitate low-frequency seismic detection.

In addition, elephants have muscles surrounding the external auditory meatus that contract, occluding the opening of the ear canal (O'Connell-Rodwell et al. unpublished data), which would dampen acoustic signals and facilitate improved detection in the seismic environment. Pressure builds up in the ear canal upon closure, creating a "closed acoustic tube" that enhances bone conduction (Stenfelt et al. 2003). This anatomical feature, potentially a remnant of an aquatic ancestry, may facilitate acoustic reception of lower frequencies and/or a bone-conducted pathway for seismic detection.

Somatosensory Reception

Elephants have a second pathway for seismic detection through the somatosensory pathway. Pacinian corpuscles, or pressure receptors, are the largest peripheral mechanoreceptors in mammals (Saxod 1996). Pacinian corpuscles are deeply placed whereas the Meissner's corpuscles or touch receptors are superficial. In humans, the peak sensitivity of the Pacinian corpuscles is around 250 Hz with a frequency range of as low as 20 Hz and as high as 1000 Hz (Bolanowski and Zwislocki 1984), Meissner's corpuscles being equally sensitive between 10 and 65 Hz (Makous et al. 1995). The tip of the Asian elephant trunk contains both Pacinian and Meissner's corpuscles in extremely dense concentrations (Rasmussen and Munger 1996). Pacinian corpuscles have also been found in the elephant foot (Weissengruber et al. 2006; Bouley et al. 2007), mostly in the front and back of the dermal layer (Bouley et al. 2007).

Seismic Discrimination

Both pathways of detection would facilitate the discrimination of high-resolution frequency differences in seismic signals. The range of frequency modulation within an elephant acoustic antipredator call is approximately 15–19 Hz (O'Connell-Rodwell et al. 2007). The minimum perceptible frequency change (Δf) is related to the critical bandwidth (CBW) (Greenwood 1961), where $CBW = \Delta f * 20$. An estimated Δf of 0.75–0.95 Hz would allow elephants to detect very small changes in frequency modulation across these calls (O'Connell-Rodwell et al. 2007). If bone conduction to the ear is utilized, then frequency discrimination ability will be reliant on the elephant's ability to discriminate acoustic frequencies. Because the elephant's cochlea shows the sharpest resonance among seven species studied (von Békésy 1944/1960), these animals are capable of discriminating frequency changes within a narrow bandwidth.

If the pathway of detection is via vibration-sensitive corpuscles, then elephants should still be able to discriminate fine frequency differences. The frequency range of the second harmonic of the seismic antipredator calls that were played back to elephants varied from about 10 to 19 Hz, which should be within the range of vibrotactile frequency discrimination ability of elephants. As this measurement has not been made directly in the African elephant, estimates are based on work in other species, using similar sensory structures. The ability of touch receptors to discriminate very small changes in frequency (2 Hz) has been demonstrated in humans and other primates (Recanzone et al. 1992). It is likely that elephants have similar vibrotactile frequency discrimination abilities as primates, if not better given their ability to detect infrasound.

Seismic Signal Propagation

The vibration channel allows signals to propagate farther than acoustic signals owing to the outer limit on airborne signal propagation as defined by Snell's law, where sounds refract back into the atmosphere at 10 km. Airborne sound waves also attenuate more rapidly than Rayleigh waves as they spread spherically rather than cylindrically (losing 6 dB for every doubling of distance vs. 3 dB), and thus ground surface waves maintain integrity longer. Airborne waves are more susceptible to interference and alteration because of environmental factors such as wind and temperature fluctuations, whereas soil type and heterogeneity are factors influencing the propagation of a seismic signal (O'Connell-Rodwell et al. 2000, 2001; Gunther et al. 2004). Wind generates noise in the seismic channel, but that noise does not impart a directional dependent attenuation to the signal. An acoustic signal, however, is affected in a directional manner, the signal heavily attenuated when traveling upwind, while it travels slightly farther downwind.

Certain characteristics are needed for long-distance propagation of seismic stimuli. For percussive signals, large size is often associated with greater source amplitudes leading to a greater propagation range (e.g., Bishop et al. 2015). For vocal coupling, the low-frequency, high-amplitude nature of elephant vocalizations are important. The generation of such signals may be facilitated by a large diaphragm, a larynx with five rather than the nine bones present in most other mammals (Shoshani 1998), and an unusually large nasal cavity. Moreover, the weight of an elephant would facilitate the coupling of their vocalizations into the ground.

There are physical properties of seismic cues that, if detected on their own or in combination with acoustic cues, could enhance the elephant's ability to interpret signals. For example, localizing vocalizations centered around 20 Hz, with a wavelength of about 17 m, given an interear distance of approximately 0.5 m is challenging. Seismic signal localization may be facilitated when soil velocities are slower than air as is the case in some elephant habitats (210–250 m/s), creating a shorter wavelength of approximately 12.5 m. Thus the distance between an elephant's feet (2–2.5 m) would provide a greater phase difference to localize these shorter signals (O'Connell-Rodwell et al. 2000, 2001). Using the vibration-sensitive trunk would provide an additional advantage.

Factors Enhancing Signal Propagation

African elephant family groups vocalize within interactive bouts that result in multiple repetitions of a signal that is three or more times longer than one produced by a single individual (O'Connell-Rodwell et al. 2012). As auditory thresholds are based on temporal summation, longer signals would increase the signal-to-noise ratio, facilitating signal detection (Heil and Neubauer 2003). Repeated signals also facilitate detection (Hamilton 1957; Greenwood 1961). If temporal summation and repetition itself increase the detection probability, then it follows that signal detection and processing would be facilitated at greater distances.

Repetitive interactive calling behavior has also been documented among captive bonded individuals (Soltis et al. 2005). Furthermore, during departure from a resource, calling bouts are repeated at a greater rate (O'Connell-Rodwell et al. 2012). Elephants also increase their rate of calling during estrus, emitting calls that are longer in duration than other calls (Leong et al. 2003), adding further evidence that elephants may create longer repeated calls to facilitate their detection at greater distances. Since elephant cows have a very narrow window for ovulation (5 days every 4 years) (Moss 1983), it would be to their advantage to improve the advertising of their reproductive state.

Listening elephants at a distance would have an opportunity to optimize their physical orientation to better resolve multiple bouts of longer signals, which they appear to do by freezing for long periods and shifting positions, aligning themselves in the direction of acoustic or seismic signals. As the Asian elephant has the largest volume of cerebral cortex of all terrestrial mammals (Hart et al. 2001), they are presumably well equipped to integrate multimodal signals.

If a seismic and an acoustic signal are redundant in the near field, the ability to detect the same signal twice in different modalities would improve its chances of detection. Because of the different propagation velocities of the two modalities, the signal will arrive at different times, thus allowing the receiver to be alerted by the first arriving signal to concentrate on the second arriving signal to resolve any signal ambiguity and possibly estimate the distance of the signaler. Signal ambiguity can be mitigated by repeating the signal; however, if dual modalities are utilized, the signal is automatically repeated without any extra effort. Seismic communication could supplement airborne communication or be especially beneficial when airborne conditions are not ideal for transmission (O'Connell-Rodwell et al. 2000, 2001; Arnason et al. 2002). Elephants may also be able to distinguish less subtle seismic events such as an approaching vehicle, helicopters, airplanes, weather (thunder storms), or earthquakes.

Two main challenges remain in this research area. The first challenge is to determine the extent to which seismic correlates of elephant vocalizations propagate in the far field, and second, determining the sensitivity of an elephant's foot to vibrations produced from elephant vocalizations at a distance. Geophysicists normally collect data on body waves, such that high-amplitude surface waves produced by earthquakes (and elephant vocalizations) are considered noise and are filtered out of data sets. More research is needed to determine how ground borne waves behave in the far field and in soils of different compositions. These two remaining aspects of the sender–receiver process in seismic communication would solidify our understanding of how seismic communication enhances the elephant's already long-distance communication ability.

7.3.1.3 Seismic Communication in Golden Moles

Background and Review of Known Studies

Golden moles are nocturnal, surface-foraging mammals with rudimentary vision. Several species possess massively hypertrophied mallei that presumably confer low-frequency, substrate-vibration sensitivity through inertial bone conduction. The seismic sensitivity of golden moles has been studied anatomically (Mason 2003a, b, 2004, 2007b), behaviorally (Fielden et al. 1990; Narins et al. 1997; Lewis et al. 2006), and physiologically (Willi et al. 2006a, b). What is poorly understood is the mechanism by which these remarkable animals determine the source location of the seismic signals and how they are able to orient and move toward their source.

The quest for features in the golden mole's middle ear that might serve the localization of seismic disturbances is motivated by the known foraging behavior of the Namib Desert golden mole (*Eremitalpa granti namibensis*). In addition to other sand-dwelling invertebrates, the diet of this animal mainly consists of dune termites, *Psammotermes* (Fielden et al. 1990). It has been hypothesized that wind-blown dune grass sets the mounds into resonance, resulting in the emission of concentric Rayleigh (surface) waves, that can, in theory, provide the golden mole a homing vector to the source of this seismic beacon—the sand mounds—and hence to the food cache (Narins et al. 1997). It was demonstrated that in the absence of olfactory cues, golden moles are able to locate the food containing mounds at a distance, solely using vibrations generated by the wind blowing the dune grass (Lewis et al. 2006).

This golden mole navigates from one grassy tussock to another, punctuating its foraging trail with characteristic sand disturbances in which the animal "head-dips," presumably to obtain a vibrational "fix" on the next mound to be visited. It is thought that head-dipping serves to couple the animal's head to the sandy substrate, so that it can better detect the Rayleigh waves emanating from the mounds.

As the desert golden mole is currently protected under the Convention on International Trade in Endangered Species (CITES), another closely related species, the Cape golden mole (*Chrysochloris asiatica*), was the focus of preliminary studies of vibration localization in golden moles. *C. asiatica* and *Eremitalpa g. namibensis* belong to the same family, golden moles (Chrysochloridae), which inhabit sub-Saharan Africa. *C. asiatica* is not listed as a protected species and as a result is obtainable for study from the Republic of South Africa. The foraging behavior of *E.g. namibiensis* suggests that the animal is able to localize prey by detecting the seismic signals they emit (Fielden et al. 1990; Narins et al. 1997; Lewis et al. 2006). One anatomical feature both species (among others) exhibit, is a massively hypertrophied malleus (Mason 2003a). Hypertrophied mallei within the family Chrysochloridae have been known for some time (Forster Cooper 1928; von Mayer et al. 1995). Some of the species exhibit the heaviest ossicles relative to body mass of all mammals so far documented (Mason 2003a). Lombard and Hetherington (1993) proposed that the hypertrophied mallei of chrysochlorids are adapted to detect vibrations by means of inertial bone conduction, and Mason (2003b) developed a model, which is based on the middle ear anatomy of a golden mole exhibiting a hypertrophied malleus, and serves the detection of vibrations exploiting inertial bone conduction. Another prominent anatomical feature of some golden moles is the interbullar connection described for three genera of chrysochlorids, *Eremitalpa*, *Chrysochloris*, and *Chlorotalpa* (von Mayer et al. 1995). Its purpose is not known, but functionally it couples the two middle ear cavities. Interaural connections are described in reptiles, amphibians (Henson 1974), birds, and insects, but not in mammals other than moles and golden moles. Experiments in the European mole (*Talpa europaea*), a subterranean mammal that is not related to the golden mole but inhabits a similar environment, have shown that the trabeculated interaural connection enables acoustic coupling between the two ears (Coles et al. 1982). This study suggests that the ear of the European moles may act a pressure-gradient receiver (Mason 2014). The interbullar connection in the three golden mole genera is even more prominent than in the European mole, consisting of a wide and open tube. The perception of interaural time (ITD) and intensity (IID) differences becomes more difficult with smaller interaural distances (the size of the skull). Also, small animals perceive directional cues if the frequency of the source to be localized is high enough, but moles and golden moles inhabit a medium that favors the propagation of low frequencies over high frequencies (Heth et al. 1986). Therefore, some subterranean mammals might have been forced to find means other than detecting ITD and IID to localize a sound source. The interbullar connection might be such an adaptation.

Localization of Seismic Stimuli: A Look Ahead

Procedure for Measuring Directional Hearing in Golden Moles

One of us (P. M. N.), along with Urban Willi, established procedures and designed experiments that allowed us to begin to explore the mechanism of directional hearing in the Cape golden mole, *C. asiatica*. In our setup (Fig. 7.6), a precise and detailed yet preliminary description of the middle ear dynamics involved in the perception of directional cues from seismic and acoustic stimuli was obtained.

In one experiment, the response of one malleus head to vibration stimuli (thumps) presented at different azimuths (–90° to +90°, in 10° increments) at a distance of 20 cm from the animal's head was measured. The inflated epitympanic recess was opened to expose the distal portion of the malleus head of one ear. A thin layer of pure silver powder placed on the malleus head improved the reflection from the ossicle. The cavity was resealed with a piece of a glass cover slip applied with liquid

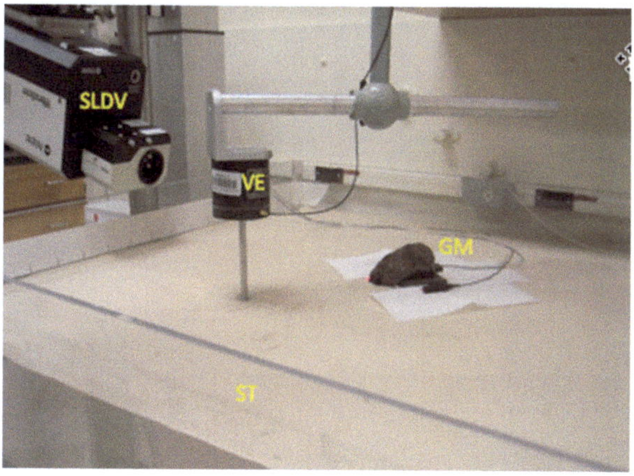

Fig. 7.6 Setup for measuring malleus velocity in response to seismic signals in the Cape golden mole. *GM* golden mole, *SLDV* scanning laser Doppler vibrometer, *VE* vibration exciter, *ST* sand tank

tissue adhesive, which reconstructed the middle ear volume but enabled access for scanning laser Doppler vibrometer (SLDV) measurements. The animal's head was placed exactly in the center of the sand tank. The head was partially buried in the sand in order to bring the long dimension of the malleus parallel with the ground and the head. The direction the animal faced was set as the 0° azimuth. A vibration exciter ("mini-shaker"; Bruel & Kjær, model 4810) was attached to a floor stand allowing us to position the vibration source at an arbitrary angle on a circle centered on the animal's head. Neither the SLDV nor the animal was moved during the measurements. During presentation of the seismic signal at 10° increments in azimuth, the horizontal velocity of the malleus head relative to that of the skull is measured in the time domain with the SLDV. The measurement was triggered by the signal output of the signal generator. Hence, differences in arrival time between seismic waves emanating from two different angles are due only to travel time in the substrate and the response of the malleus to the stimulus.

Although this vibration delivery system provided repeatable seismic stimuli, it was clear that tank reflections would result in secondary waves appearing at the geophone. Although the reflected waves arrived with a delay relative to the direct waves, and although the secondary wave amplitudes were attenuated relative to the direct waves, our stimulus system was nevertheless redesigned to eliminate or reduce secondary waves. Improved procedures have since been devised.

Improved Procedure for Measuring Directional Hearing in Golden Moles

To investigate directional hearing in *Chrysochloris asiatica* in the laboratory, an environment with physical properties similar to the free field needed to be simulated. In contrast to mechanical disturbances in fluids and gases, mechanical disturbances in elastic solids comprise not only compression-waves (P-waves), but also shear-waves (S-waves). Surface waves are a subclass of S-waves, which only exist

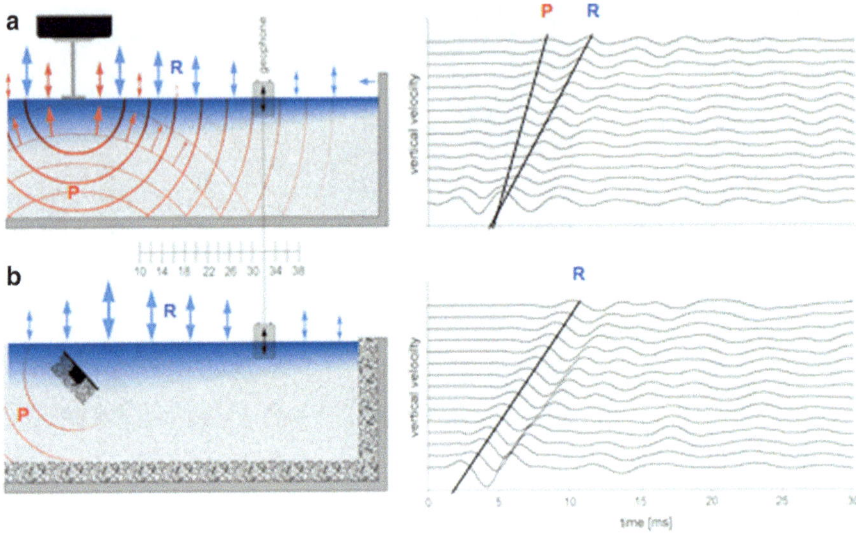

Fig. 7.7 Preliminary and improved setup to investigate directional hearing in the ground. The setup in the sand tank is shown on the *left illustrations*, the vertical velocities measured with a geophone at increasing distance (10–38 cm, 2-cm increments) from the source, is presented in the *right graphs*. (**a**) Interference patterns between the P-wave (P, *red*) and Rayleigh-waves (R, *blue*) are caused by the reflected P-wave in the preliminary setup. (**b**) These interferences were suppressed by reducing the induction of P-waves and reducing their reflection in the improved setup

at boundaries between two media. The two waveforms, P- and S-waves, can be discriminated by their propagation velocity and the orientation of particle motion. P-waves and S-waves in sandy soils travel at about 100 and 40 m/s, respectively (Ishimoto and Idia 1936). P-waves spread spherically into the ground and their attenuation is proportional to $1/r^2$ whereas S-waves propagate only along the surface and, therefore, and are attenuated less ($1/r$) (where 'r' is the distance from the source). In the free field it is relatively easy to separate the two waveforms, due to their difference in propagation velocity. In the lab, however, a medium of much smaller dimensions is a genuine constraint. Instead of the P-wave disappearing into the ground, it is reflected from the shallow tank bottom and interferes with the vertical S-wave (Rayleigh-wave) (Fig. 7.7a, left). The vertical motion measured with a geophone at increasing distance (10–38 cm, 2-cm increments) from the source reveals two components of different propagation speed (Fig. 7.7a, right). To interpret the directional cues reflected by the dynamic response of the moles' middle ear, it is critical to be able to determine the velocity of the stimulus approaching the animal. This was the impetus for instituting the improved properties of the sand tank. First, the entire tank was lined with open cell foam (25 mm thick) reducing the reflection of the P-wave. Second, the source was modified to decrease the generation of the P-wave component. The latter was achieved by using an electromagnetic

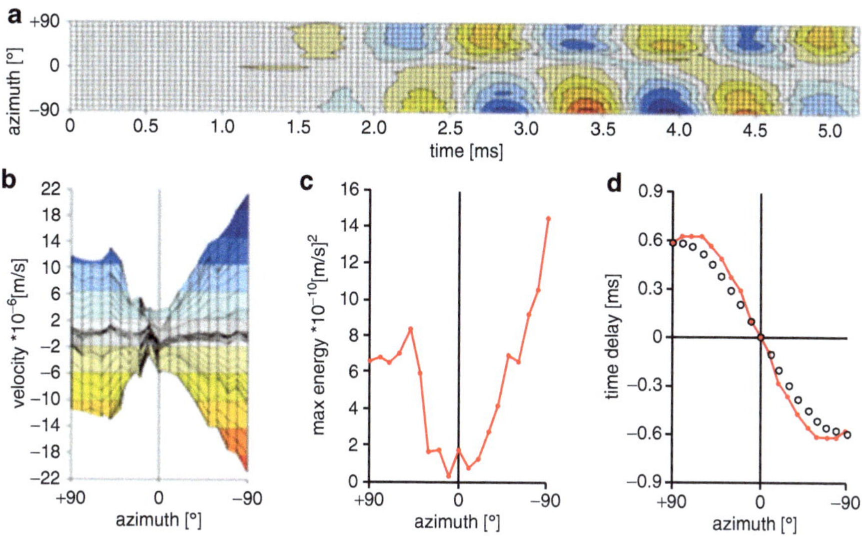

Fig. 7.8 Horizontal motion of the right malleus head as a response to a thump at different azimuths (–90° to +90°, 10° increments). Negative angles represent the ipsilateral side with respect to the right ear. (**a**) Response of the right malleus at all azimuths of the source. (**b**) Maximum response over the time signal shown in graph (**a**). (**c**) Power spectrum peaks at each azimuth over that same time signal. (**d**) *Open black circles*: Time delays calculated assuming a middle ear distance of 12 mm and a wave propagation speed of 40 m/s. *Red solid line*: Time delay calculated by an autocorrelation function, based on the time signals presented in graph (**a**). Hence, the positive and the negative values of the same azimuth were compared

transducer buried about 2–3 cm underneath the sand surface, facing the surface at an angle of 45°. In addition, the back of the transducer was embedded in a layer of open cell foam (25 mm thick). Using these procedures, the interference patterns and reflections disappeared, and the vertical motion generated in this new setup consists of a single slow surface-wave, a Rayleigh-wave (Fig. 7.7b). The peak attenuation over distance falls off as $1/r^{1/2}$ (Narins 1990). This setup allows us to simulate free field situations in the lab. Thus, it is expected that directional cues that might be exploited by the animal to localize seismic sources would be found in the temporal patterns of the malleus motion. Experiments are currently being planned to test these hypotheses.

Vibration Response of Malleus in the Golden Mole: A Look Ahead

In the Cape golden mole, peak horizontal velocities are greatest for ipsilateral stimulation and they gradually decay toward the zero azimuth, revealing a minimal response in front of the animal. In Fig. 7.8a, the time response of the relative malleus motion at each azimuth is shown and the maximal velocities of the same data are shown in Fig. 7.8b.

Contralaterally, the response increases again but does not reach the velocity value seen under ipsilateral stimulation. This phenomenon is most likely due to the anatomy of the ossicular chain: The firmest anchoring of the ossicular chain to the middle ear cavity is given by the ligament at the short process of the incus (LSPI). As the long axis of the malleus is parallel not only to the ground but also to the sagittal plane of the skull, the system is much more susceptible to sideways and vertical motions of the skull than to motion along the anteroposterior axis of the animal. An analogy to this is a pendulum moving relative to its suspension when latter moves sideways but not when it moves along the axis of its suspension (vertically). Figure 7.8c depicts the power spectral peaks at each azimuth and quantitatively confirms the findings in Fig. 7.8b. Figure 7.8d shows the time delay between stimuli reaching the same ear from the same azimuth on the ipsilateral and contralateral sides (e.g., +80°/–80°). The delays (red solid line) were evaluated by cross-correlating the ipsilateral to the contralateral time signal at equal azimuths. The open circles in the same graph represent an estimate of interaural time delay based on an interbullar distance of 12 mm and a measured wave propagation speed of 40 m/s. The estimate and the evaluation of the time delays show a good match, suggesting that the skull motion at each bulla reflects the motion of the ground and that time delays due to the propagation of the wave are accessible to the animal. The values reflect a relative motion of the malleus versus the skull, because the skull motion is already subtracted. Negative azimuth values refer to measurements using ipsilateral stimulation and positive azimuth values refer to measurements using contralateral stimulation. It is remarkable in the time domain that the polarity of the malleus motion switches when the source moves from the ipsilateral to the contralateral side. These data show clear directional cues present in the middle ear in response to the horizontal velocity component of the seismic disturbances and suggest a simple mechanism for localizing seismic cues in the substrate.

However, in this preliminary experiment only the horizontal component of the malleus motion was measured. Future experiments should explore the degree to which the middle ear is susceptible to both the vertical and horizontal motion components and the precise nature of the directional cues provided by them.

7.4 Conclusions

Infrasonic and vibrational forms of communication are still wide open fields of study. Although much is known about invertebrate vibration communication, it is not nearly as well understood in the vertebrates. Although some is known about the bird's ability to sense infrasound and vibrations, how they use this information needs further investigation. There are many cases of the use of vibrations as signals in reptiles and amphibians, and even small rodents, but the extent to which large mammals use vibrational cues either derived from infrasonic vocalizations or percussion still needs further exploration. Elephants can detect and distinguish low frequency acoustic callers and call types seismically. These seismic cues could supplement acoustic information, replace acoustic information under poor airborne conditions, or, under ideal seismic conditions, extend the elephant's range of communication within a complex multimodal communication repertoire. Remarkably, golden moles, the only member of the Afrotheria other than the elephant in which seismic behavior has been studied extensively, appears to rely on vibrational cues to both detect and localize prey. Examining the mechanisms underlying these fundamental behaviors needs additional study.

References

Arnason, B. T., Hart, L. A., & O'Connell-Rodwell, C. E. (2002). The properties of geophysical fields and their effects on elephants and other animals. *Journal of Comparative Psychology, 116*(2), 123–132.

Au, W. W. L. (1993). *The sonar of dolphins*. New York: Springer.

Baotic, A., Sicks, F., & Stoeger, A. S. (2015). Nocturnal "humming" vocalizations: Adding a piece to the puzzle of giraffe vocal communication. *BMC Research Notes, 8*, 425.

Barklow, W. E. (2004). Amphibious communication with sound in hippos, *Hippopotamus amphibious. Behaviour, 68*, 1125–1132.

Barnes, L. G., Domning, D. P., & Ray, C. E. (1985). Status of studies on fossil marine mammals. *Marine Mammal Science, 1*(1), 15–53.

Barnett, K. E., Cocroft, R. B., & Fleishman, L. J. (1999). Possible communication by substrate vibration in a chameleon. *Copeia, 1999*(1), 225–228.

Beranek, L. L. (1988). *Acoustical measurements*. Melville, NY: American Institute of Physics.

Berlin, C. I., Hood, L. J., Barlow, E. K., Morehouse, C. R., & Smith, E. G. (1991). Derived guinea pig compound VIIIth nerve action potentials to continuous pure tones. *Hearing Research, 52*, 271–280.

Bishop, A. M., Denton, P., Pomeroy, P., & Twiss, S. (2015). Good vibrations by the beach boys: Magnitude of substrate vibrations is a reliable indicator of male grey seal size. *Animal Behaviour, 100*, 74–82.

Bolanowski, S., & Zwislocki, J. J. (1984). Intensity and frequency characteristics of Pacinian corpuscle. II. Receptor potentials. *Journal of Neurophysiology, 51*, 812–830.

Bouley, D. M., Alarcon, C., Hildebrandt, T., & O'Connell-Rodwell, C. E. (2007). The distribution, density and three-dimensional histomorphology of Pacinian corpuscles in the foot of the Asian elephant (Elephas maximus) and their potential role in seismic communication. *Journal of Anatomy, 211*(4), 428–435.

Bradbury, J. W., & Vehrencamp, S. L. (2011). *Principles of animal communication* (2nd ed.). Sunderland, MA: Sinauer.

Budde, C., & Klump, G. M. (2003). Vocal repertoire of the black rhino *Diceros bicornis* ssp. And possibilities of individual identification. *Mammalian Biology, 68*, 42–47.

Caldwell, M. S., Johnston, G. R., McDaniel, J. G., & Warkentin, K. M. (2010a). Vibrational signaling in the agonistic interactions of red-eyed treefrogs. *Current Biology, 20*, 1012–1017.

Caldwell, M. S., McDaniel, J. G., & Warkentin, K. M. (2010b). Is it safe? Red-eyed treefrog embryos assessing predation risk use two features of rain vibrations to avoid false alarms. *Animal Behaviour, 79*, 255–260.

Capranica, R. R., & Moffat, A. J. M. (1983). Neurobehavioral correlates of sound communication in anurans. In J.-P. Ewert, R. R. Capranica, & D. J. Ingle (Eds.), *Advances in vertebrate neuroethology* (pp. 701–730). London: Plenum Press.

Catania, K. C. (1995). Structure and innervation of the sensory organs on the snout of the star-nosed mole. *Journal of Comparative Neurology, 351*, 536–548.

Charif, R. A., Ramey, R. R., Langbauer, W. R., Payne, K. B., Martin, R. B., & Brown, L. M. (2005). Spatial relationships and matrilineal kinship in African savanna elephant (*Loxodonta africana*) clans. *Behavioural Ecology and Sociobiology, 57*(4), 327–338.

Christensen, C. B., Christensen-Dalsgaard, J., Brandt, C., & Masden, P. T. (2012). Hearing with an atympanic ear: Good vibration and poor sound-pressure detection in the royal python, *Python regius. Journal of Experimental Biology, 215*, 331–342.

Christensen-Dalsgaard, J., Brandt, C., Wilson, M., Wahlberg, M., & Teglberg, P. M. (2011). Hearing in the African lungfish (*Protopterus annectens*): Pre-adaptation for pressure hearing in tetrapods? *Biology Letters, 7*, 139–141.

Christensen-Dalsgaard, J., & Jørgensen, M. B. (1988). The response characteristics of vibration-sensitive saccular fibers in the grassfrog, *Rana temporaria. Journal of Comparative Physiology, 162*, 633–638.

Christensen-Dalsgaard, J., & Jørgensen, M. B. (1996). Sound and vibration sensitivity of VIIIth nerve fibers in the grassfrog, *Rana temporaria. Journal of Comparative Physiology, 179*, 437–445.

Christensen-Dalsgaard, J., Ludwig, T. A., & Jørgensen, M. B. (2002). Call diversity in an old world treefrog: Level dependence and latency of acoustic responses. *Bioacoustics, 13*, 21–35.

Christensen-Dalsgaard, J., & Narins, P. M. (1993). Sound and vibration sensitivity of VIIIth nerve fibers in the frogs *Leptodactylus albilabris* and *Rana pipiens pipiens. Journal of Comparative Physiology, 172*, 653–662.

Christensen-Dalsgaard, J., & Walkowiak, W. (1999). In vitro and in vivo responses of saccular and caudal nucleus neurons in the grass frog (*Rana temporaria*). *European Journal of Morphology, 37*(2–3), 206–210.

Cocroft, R. B., Gogala, M., Hill, P. S. M., & Wessel, A. (2014). *Studying vibrational communication*. Heidelberg, Germany: Springer.

Coles, R. B., Gower, D. M., Boyd, P. J., & Lewis, D. B. (1982). Acoustic transmission through the head of the common mole, *Talpa europaea. Journal of Experimental Biology, 101*, 337–341.

Corfield, J. R., Krilow, J. M., Vande Ligt, M. N., & Iwaniuk, A. N. (2013). A quantitative morphological analysis of the inner ear of galliform birds. *Hearing Research, 304*, 111–127.

de Silva, S. (2010). Acoustic communication in the Asian elephant, *Elephas maximus. Behaviour, 147*, 825–852.

Doody, J. S., Stewart, B., Camacho, C., & Christian, K. (2012). Good vibrations? Sibling embryos expedite hatching in a turtle. *Animal Behaviour, 83*, 645–651.

Dorward, P. K., & McIntyre, A. K. (1980). Responses of vibration-sensitive receptors in the interosseous region of the duck's hind limb. *Journal of Physiology, 219*, 77–87.

Ferrara, C. R., Vogt, R. C., Sousa-Lima, R. S., Tardio, B. M. R., & Bernardes, V. C. D. (2014). Sound communication and social behavior in an Amazonian river turtle (*Podocnemis expansa*). *Herpetologica, 70*(2), 149–156.

Fielden, L. J., Perrin, M. R., & Hickman, G. C. (1990). Feeding ecology and foraging behaviour of the Namib Desert golden mole, *Eremitalpa granti namibensis* (Chrysochloridae). *Journal of Zoology (London), 220*, 367–389.

Fischer, M. S. (1990). Un trait unique de l'oreille des elephants et des sireniens (Mammalia): Un paradoxe phylogenetique. *Comptes Rendus de l'Académie des Sciences, 311*(4), 157–162.

Fitch, W. T. (2006). Production of vocalizations in mammals. In K. Brown (Ed.), *Encyclopedia of language and linguistics* (pp. 115–212). Oxford, England: Elsevier.

Forster Cooper, C. (1928). On the ear region of certain of the Chrysochloridae. *Philosophical Transactions of the Royal Society of London B: Biological Sciences, 216*, 265–283.

Francescoli, G. (2000). Sensory capabilities and communication in subterranean rodents. In E. A. Lacey, J. L. Patton, & G. N. Cameron (Eds.), *Life underground: The biology of subterranean rodents* (pp. 111–144). Chicago: The University of Chicago Press.

Freeman, A. R., & Hare, J. F. (2015). Infrasound in mating displays: A peacock's tale. *Animal Behaviour, 102*, 241–250.

Gaeth, A. P., Short, R. V., & Renfree, M. B. (1999). The developing renal, reproductive, and respiratory systems of the African elephant suggest an aquatic ancestry. *Proceedings of the National Academy of Sciences of the USA, 96*, 5555–5558.

Garstang, M. (2004). Long-distance, low-frequency elephant communication. *Journal of Comparative Physiology, 190*, 791–805.

Garstang, M., Larom, D., Raspet, R., & Lindeque, M. (1995). Atmospheric controls on elephant communication. *Journal of Experimental Biology, 198*, 939–951.

Gerstein, E. R., Gerstein, L., Forsythe, S. E., & Blue, J. E. (1999). The underwater audiogram of the West Indian manatee (*Trichechus manatus*). *Journal of the Acoustical Society of America, 105*, 3575–3583.

Giles, J. C., Davis, J. A., McCauley, X., & Kuchling, G. (2009). Voice of the turtle: The underwater acoustic repertoire of the long-necked freshwater turtle, *Chelodina oblonga. Journal of the Acoustical Society of America, 126*, 434–443.

Greenwood, D. (1961). Critical bandwidth and the frequency coordinates of the basilar membrane. *Journal of the Acoustical Society of America, 33*(484), 1344–1356.

Gridi-Papp, M., & Narins, P. M. (2010). Seismic detection and communication in amphibians. In C. E. O'Connell-Rodwell (Ed.), *The use of vibrations in communication: Properties, mechanisms and function across taxa* (pp. 69–83). Trivandrum, India: Transworld Research Network.

Gunther, R. H., O'Connell-Rodwell, C. E., & Klemperer, S. L. (2004). Seismic waves from elephant vocalizations: A possible communication mode? *Geophysical Research Letters, 31*(L11602), 1–4.

Hagstrum, J. T. (2000). Infrasound and the avian navigational map. *The Journal of Experimental Biology, 203*, 1103–1111.

Hamilton, P. M. (1957). Noise-masked thresholds as a function of tonal duration and masking noise bandwidth. *Journal of the Acoustical Society of America, 29*, 506–511.

Hart, B. L., Hart, L. A., McCoy, M., & Sarath, C. R. (2001). Cognitive behaviour in Asian elephants: Use and modification of branches for fly switching. *Animal Behaviour, 62*(5), 839–847.

Hartline, P. H. (1971). Physiological basis for detection of sound and vibration in snakes. *Journal of Experimental Biology, 54*, 349–371.

Hartman, W. M. (1999). How we localize sound. *Physics Today, 52*, 24.

Heffner, R. S., & Heffner, H. E. (1980). Hearing in the elephant (*Elephas maximus*). *Science, 208*(4443), 518–520.

Heffner, R. S., & Heffner, H. E. (1982). Hearing in the elephant (*Elephas maximus*): Absolute sensitivity, frequency discrimination, and sound localization. *Journal of Comparative Physiology and Psychology, 96*(6), 926–944.

Heffner, R. S., Heffner, H. E., & Stichman, N. (1982). Role of the elephant pinna in sound localization. *Animal Behaviour, 30*(2), 628–630.

Heffner, H. E., Koay, G., Hill, E. M., & Heffner, R. S. (2013). Conditioned suppression/avoidance as a procedure for testing hearing in birds: The domestic pigeon (*Columba livia*). *Behavioural Research, 45*, 383–392.

Heil, P., & Neubauer, H. (2003). A unifying basis of auditory thresholds based on temporal summation. *Proceedings of the National Academy of Sciences of the USA, 100*(10), 6151–6156.

Henson, O. W. (1974). Comparative anatomy of the middle ear. In W. D. Keidel & W. D. Neff (Eds.), *Handbook of sensory physiology* (Auditory system, Vol. V/1, pp. 39–110). Berlin, Germany: Springer.

Herbst, C. T., Stoeger, A. S., Frey, R., Lohscheller, J., Titze, I. R., Gumpenberger, M., et al. (2012). How low can you go? Physical production mechanism of elephant infrasonic vocalization. *Science, 337*(6094), 595–599.

Herculano-Houzel, S., Avelino-de-Souza, K., Neves, K., Porfírio, J., Messeder, D., Mattos Feijó, L., et al. (2014). The elephant brain in numbers. *Frontiers in Neuroanatomy, 8*, 46. doi:10.3389/fnana.2014.00046.

Heth, G., Frankenberg, E., & Nevo, E. (1986). Adaptive optimal sound for vocal communication in tunnels of a subterranean mammal (*Spalax ehrenbergi*). *Experientia, 42*, 1287–1289.

Hetherington, T. E. (1989). The use of vibratory cues for detection of insect prey by the sandswimming lizard *Scincus scincus. Animal Behaviour, 37*, 290–297.

Hill, P. S. M. (2008). *Vibration communication in animals*. Cambridge, MA: Harvard University Press.

Hill, P. S. M. (2009). How do animals use substrate-borne vibrations as an information source? *Naturwissenshaften, 96*, 1355–1371.

Hill, E. M., Koay, G., Heffner, R. S., & Heffner, H. E. (2014). Audiogram of the chicken (*Gallus gallus domesticus*) from 2 Hz to 9 kHz. *Journal of Comparative Physiology, 200*, 863–870.

Hörster, W. (1990). Vibrational sensitivity of the wing of the pigeon (*Columba livia*)—A study using heart rate conditioning. *Journal of Comparative Physiology, 167*, 545–549.

Ishimoto, M., & Idia, K. (1936). Determinations of elastic constants of soil by means of vibration methods. Part I. Young's modulus. *Bulletin of Earthquake Research of the Institute Tokyo, 14*, 632–657.

Jackson, L. L., Heffner, R. S., & Heffner, H. E. (1999). Free-field audiogram of the Japanese macaque (*Macaca fuscata*). *Journal of the Acoustical Society of America, 106*, 3017–3023.

Ketten, D. R. (2000). Cetacean ears. In W. W. L. Au, A. N. Popper, & R. R. Fay (Eds.), *Hearing by whales and dolphins* (pp. 43–108). New York: Springer.

Ketten, D. R., Odell, D. K., & Domning, D. P. (1992). Structure and adaptation of the manatee ear. In J. Thomas et al. (Eds.), *Marine mammal sensory systems* (pp. 77–95). New York: Plenum Press.

Kimchi, T., Reshef, M., & Terkel, J. (2005). Evidence for the use of reflected self-generated seismic waves for spatial orientation in a blind subterranean mammal. *Journal of Experimental Biology, 208*, 647–659.

King, L. E., Soltis, J., Douglas-Hamilton, I., Savage, A., & Vollrath, F. (2010). Bee threat elicits alarm call in African elephants. *PloS One, 5*(4), e10346. doi:10.1371/journal.pone.0010346.

Koyama, H., Lewis, E. R., Leverenz, E. L., & Baird, R. A. (1982). Acute seismic sensitivity in the bullfrog ear. *Brain Research, 250*, 168–172.

Kreithen, M. L., & Quine, D. B. (1979). Infrasound detection by the homing pigeon: A behavioral audiogram. *Journal of Comparative Physiology, 129*, 1–4.

Langbauer, W. R. (2000). Elephant communication. *Zoo Biology, 19*(5), 425–445.

Langbauer, W. R., Payne, K. B., Charif, R. A., Rapaport, L., & Osborn, F. (1991). African elephants respond to distant playbacks of low-frequency calls. *The Journal of Experimental Biology, 157*(1), 35–46.

Larom, D., Garstang, M., Payne, K., Raspet, R., & Lindeque, M. (1997). The influence of surface atmospheric conditions on the range and area reached by animal vocalizations. *The Journal of Experimental Biology, 200*(3), 421–434.

Leong, K. M., Ortolani, A., Burks, K. D., Mellen, J. D., & Savage, A. (2003). Quantifying acoustic and temporal characteristics of vocalizations for a group of captive African elephants (*Loxodonta africana*). *Bioacoustics, 13*(3), 213–232.

Lewis, E. R., & Narins, P. M. (1985). Do frogs communicate with seismic signals? *Science, 227*, 187–189.

Lewis, E. R., Narins, P. M., Cortopassi, K. A., Yamada, W. M., Poinar, E. H., Moore, S. W., et al. (2001). Do male white-lipped frogs use seismic signals for intraspecific communication? *American Zoologist, 41*, 1185–1199.

Lewis, E. R., Narins, P. M., Jarvis, J. U. M., Bronner, G., & Mason, M. J. (2006). Preliminary evidence for the use of microseismic cues for navigation by the Namib golden mole. *Journal of the Acoustical Society of America, 119*, 1260–1268.

Lieberman, P., & Blumstein, S. E. (1988). *Speech physiology, speech perception, and acoustic phonetics*. Cambridge, England: Cambridge University Press.

Lombard, R. E., & Hetherington, T. E. (1993). Structural basis of hearing and sound transmission. In J. Haken & B. K. Hall (Eds.), *The skull* (Vol. 3, pp. 241–302). London: University of Chicago Press.

Lopez, P. T., Narins, P. M., Lewis, E. R., & Moore, S. W. (1988). Acoustically-induced call modification in the white-lipped frog, *Leptodactylus albilabris*. *Animal Behaviour, 36*, 1295–1308.

Mack, A. L., & Jones, J. (2003). Low-frequency vocalizations by cassowaries (*Casuarius* spp.). *The Auk, 120*(4), 1062–1068.

Maier, V. (1982). Acoustic communication in the Guinea fowl (*Numida meleagris*): Structure and use of vocalizations, and the principles of message coding. *Zeitschrift für Tierpsychologie, 59*(1), 28–83.

Makous, J. C., Friedman, R. M., & Vierck, C. J., Jr. (1995). A critical band filter in touch. *Journal of Neuroscience, 15*(4), 2808–2818.

Manley, G. A., & Kraus, J. E. M. (2010). Exceptional high-frequency hearing and matched vocalizations in Australian pygopod geckos. *Journal of Experimental Biology, 213*, 1876–1885.

Marasco, P. D., & Catania, K. C. (2007). Response properties of primary afferents supplying Eimer's organ. *Journal of Experimental Biology, 213*, 765–780.

Maseko, B. C., Spocter, M. A., Haagensen, M., & Manger, P. R. (2012). Elephants have relatively the largest cerebellum size of mammals. *The Anatomical Record, 295*, 661–672.

Maseko, B. C., Patzke, N., Fuxe, K., & Manger, P. R. (2013). Architectural organization of the African elephant diencephalon and brainstem. *Brain, Behaviour and Evolution, 82*(2), 83–128.

Mason, M. J. (2003a). Morphology of the middle ear of golden moles (Chrysochloridae). *Journal of Zoology (London), 260*, 391–403.

Mason, M. J. (2003b). Bone conduction and seismic sensitivity in golden moles (Chrysochloridae). *Journal of Zoology (London), 260*, 405–413.

Mason, M. J. (2004). Functional morphology of the middle ear in *Chlorotalpa* golden moles (Mammalia, Chrysochloridae): Predictions from three models. *Journal of Morphology, 261*, 162–174.

Mason, M. J. (2007a). Pathways for sound transmission to the inner ear in amphibians. In P. M. Narins, A. S. Feng, R. R. Fay, & A. N. Popper (Eds.), *Hearing and sound communication in amphibians* (pp. 147–183). New York: Springer.

Mason, M. J. (2007b). Massive mallei in moles: Middle ear adaptations subserving seismic sensitivity. *Proceedings of the Institute of Acoustics, 29*, 69–76.

Mason, M. J. (2014). Internally coupled ears in mammals. *Abstract of presentation for the symposium internally coupled ears: Evolutionary origins, mechanisms, and neuronal processing from a biomimetic perspective, T.-U. Garching, Germany, 7.*

Mason, M. J., & Narins, P. M. (2001). Seismic signal use by fossorial mammals. *American Zoologist, 41*, 1171–1184.

Mason, M. J., & Narins, P. M. (2002). Vibrometric studies of the middle ear of the bullfrog (*Rana catesbeiana*). II: The operculum. *Journal of Experimental Biology, 205*, 3167–3176.

Mason, M. J., & Narins, P. M. (2010). Seismic sensitivity and communication in subterranean mammals. In C. E. O'Connell-Rodwell (Ed.), *The use of vibrations in communication: Properties, mechanisms and function across taxa* (pp. 121–139). Trivandrum, India: Transworld Research Network.

McComb, K., Moss, C., Sayialel, S., & Baker, L. (2000). Unusually extensive networks of vocal recognition in African elephants. *Animal Behaviour, 59*(6), 1103–1109.

McComb, K., Reby, D., Baker, L., Moss, C., & Sayialel, S. (2003). Long-distance communication of acoustic cues to social identity in African elephants. *Animal Behaviour, 65*(2), 317–329.

McIntyre, A. K. (1980). Biological seismography. *Trends in Neuroscience, 3*(9), 202–205.

Merzenich, M. M., Kitzes, L., & Aitkin, L. (1973). Anatomical and physiological evidence for auditory specialization in the mountain beaver (*Aplodontia rufa*). *Brain Research, 58*, 331–344.

Moss, C. J. (1983). Oestrous behaviour and female choice in the African elephant. *Behaviour, 86*, 167–196.

Narins, P. M. (1990). Seismic communication in anuran amphibians. *Bioscience, 40*, 268–274.

Narins, P. M. (2001). Vibration communication in vertebrates. In F. Barth & A. Schmidt (Eds.), *Ecology of sensing* (pp. 127–148). Berlin, Germany: Springer.

Narins, P. M. & Clark, G. A. (2016). Principles of matched filtering with auditory examples from selected vertebrates. In G. von der Emde, E. Warrant (Eds.).The Ecology of Animal Senses: Matched Filtering for Economical Sensing (pp. 111-140). Heidelberg: Springer-Verlag.

Narins, P. M., Feng, A. S., Yong, H.-S., & Christensen-Dalsgaard, J. (1998). Morphological, behavioral, and genetic divergence of sympatric morphotypes of the treefrog *Polypedates leucomystax* in Peninsula Malaysia. *Herpetologica, 54*, 129–142.
Narins, P. M., & Lewis, E. R. (1984). The vertebrate ear as an exquisite seismic sensor. *Journal of the Acoustical Society of America, 76*, 1384–1387.
Narins, P. M., Lewis, E. R., Jarvis, J. U. M., & O'Riain, J. (1997). The use of seismic signals by fossorial southern African mammals: A neuroethological gold mine. *Brain Research Bulletin, 44*, 641–646.
Narins, P. M., Losin, N., & O'Connell-Rodwell, C. E. (2009). Seismic and vibrational signals in animals. In L. R. Squire (Ed.), *Encyclopedia of neuroscience* (pp. 555–559). Amsterdam: Elsevier.
Narins, P. M., Reichman, O. J., Jarvis, J. U. M., & Lewis, E. R. (1992). Seismic signal transmission between burrows of the Cape mole-rat, *Georychus capensis. Journal of Comparative Physiology, 170*, 13–21.
Nevo, E. (1961). Observations on Israeli populations of the mole rat *Spalax E. ehrenbergi* Nehring 1898. *Mammalia, 25*, 127–144.
Norris, K. S. (1968). The evolution of acoustic mechanisms in odontocete cetaceans. In E. T. Drake (Ed.), *Evolution and environment* (pp. 297–324). New Haven, CT: Yale University Press.
Nowak, R. M. (1999). *Walker's mammals of the world* (6th ed.). Baltimore: The Johns Hopkins University Press.
Nummela, S. (1995). Scaling of the mammalian middle ear. *Hearing Research, 85*(1–2), 18–30.
O'Connell, C. E., Hart, L. A., & Arnason, B. (1999). Response to "Elephant hearing" [see comments]. *Journal of the Acoustical Society of America* 104, 1122–3 (1998). *Journal of the Acoustical Society of America, 105*, 2051–2052.
O'Connell-Rodwell, C. E. (2007). Keeping an "ear" to the ground: Seismic communication in elephants. *Physiology, 22*(4), 287–294.
O'Connell-Rodwell, C. E. (Ed.). (2010). *The use of vibrations in communication: Properties, mechanisms and function across taxa*. Trivandrum, India: Transworld Research Network.
O'Connell-Rodwell, C. E., Arnason, B., & Hart, L. A. (2000). Seismic properties of elephant vocalizations and locomotion. *Journal of the Acoustical Society of America, 108*(6), 3066–3072.
O'Connell-Rodwell, C. E., Erckie, R., Kilian, W., Wood, J. D., Kinzley, C., Rodwell, T. C., et al. (2011). Exploring the use of acoustics as a tool in male elephant/human conflict mitigation. *Journal of the Acoustical Society of America, 130*(4 Pt 2), 2459.
O'Connell-Rodwell, C. E., Hart, L. A., & Arnason, B. T. (2001). Exploring the potential use of seismic waves as a communication channel by elephants and other large mammals. *American Zoologist, 41*(5), 1157–1170.
O'Connell-Rodwell, C. E., & Wood, J. D. (2010). Vibration generation, propagation and detection in elephants. In C. E. O'Connell-Rodwell (Ed.), *The use of vibrations in communication: Properties, mechanisms and function across taxa* (pp. 121–139). Trivandrum, India: Transworld Research Network.
O'Connell-Rodwell, C. E., Wood, J. D., Kinzley, C., Rodwell, T. C., Poole, J. H., & Puria, S. (2007). Wild African elephants (Loxodonta Africana) discriminate between familiar and unfamiliar conspecific seismic alarm calls. *Journal of the Acoustical Society of America, 122*(2), 823–830.
O'Connell-Rodwell, C. E., Wood, J. D., Rodwell, T. C., Puria, S., Partan, S. R., Keefe, R., et al. (2006). Wild elephant (*Loxodonta africana*) breeding herds respond to artificially transmitted seismic stimuli. *Behavioural Ecology and Sociobiology, 59*(6), 842–850.
O'Connell-Rodwell, C. E., Wood, J. D., Wyman, M., Redfield, S., Hart, L. A., & Puria, S. (2012). Antiphonal vocal bouts associated with departures in free-ranging African elephant family groups (*Loxodonta africana*). *Bioacoustics, 21*(3), 215–224.
Patzke, N., Olaleye, O., Haagensen, M., Hof, P. R., Ihunwo, A. O., & Manger, P. R. (2013a). Organization and chemical neuroanatomy of the African elephant (*Loxodonta africana*) hippocampus. *Brain Structure and Function, 219*(5), 1587–601. doi:10.1007/s00429-013-0587-6.

Patzke, N., Spocter, M. A., Karlsson, K. Æ., Bertelsen, M. F., Haagensen, M., Chawana, R., et al. (2013b). In contrast to many other mammals, cetaceans have relatively small hippocampi that appear to lack adult neurogenesis. *Brain, Structure and Function, 220*(1), 361–83. doi:10.1007/s00429-013-0660-1.

Payne, K. B., Langbauer, W. R., & Thomas, E. M. (1986). Infrasonic calls of the Asian elephant (*Elephas maximus*). *Behavioural Ecology and Sociobiology, 18*(4), 297–301.

Peterson, G. E., & Barney, H. (1952). Control methods used in a study of the vowels. *Journal of the Acoustical Society of America, 24*(175), 175–184.

Poole, J. H. (1989). Mate guarding, reproductive success and female choice in African elephants. *Animal Behaviour, 37*(5), 842–849.

Poole, J. H. (1999). Signals and assessment in African elephants: Evidence from playback experiments. *Animal Behaviour, 58*(1), 185–193.

Poole, J. H. (2011). Behavioral contexts of elephant acoustic communication. In C. J. Moss, H. Croze, & P. C. Lee (Eds.), *The Amboseli elephants: A long-term perspective on a long-lived mammal* (pp. 125–161). Chicago: The University of Chicago Press.

Poole, J. H., & Moss, C. J. (1981). Musth in the African elephant, *Loxodonta africana. Nature, 292*(5826), 830–831.

Poole, J. H., Payne, K., Langbauer, W. R., & Moss, C. J. (1988). The social contexts of some very low frequency calls of African elephants. *Behavioural Ecology and Sociobiology, 22*(5), 385–392.

Pye, J. D., & Langbauer, W. R. (1998). Ultrasound and infrasound. In S. L. Hopp & M. J. Owren (Eds.), *Animal acoustic communication: Sound analysis and research methods* (pp. 221–245). Berlin: Springer.

Rado, R., Himelfarb, M., Arensburg, B., Terkel, J., & Wollberg, Z. (1989). Are seismic communication signals transmitted by bone conduction in the blind mole-rat? *Hearing Research, 41*, 23–30.

Rado, R., Terkel, J., & Wollberg, Z. (1998). Seismic communication signals in the blind mole rat (*Spalax ehrenbergi*): Electrophysiological and behavioral evidence for their processing by the auditory system. *Journal of Comparative Physiology. A, Neuroethology, Sensory, Neural, and Behavioral Physiology, 183*(4), 503–511.

Randall, J. A. (2010). Drummers and stompers: Vibrational communication in mammals. In C. E. O'Connell-Rodwell (Ed.), *The use of vibrations in communication: Properties, mechanisms and function across taxa* (pp. 99–120). Trivandrum, India: Transworld Research Network.

Rasmussen, L. E. L., & Munger, B. L. (1996). The sensorineural specialization of the trunk tip (finger) of the Asian elephant (*Elephas maximus*). *Anatomical Record, 246*, 127–134.

Reby, D., & McComb, K. (2003). Anatomical constraints generate honesty acoustic cues to age and weight in the roars of red deer stags. *Animal Behaviour, 65*(3), 519–530.

Recanzone, G. H., Jenkins, W. M., Hradek, G. T., & Merzenich, M. M. (1992). Progressive improvement in discriminative abilities in adult owl monkeys performing a tactile frequency discrimination task. *Journal of Neurophysiology, 67*(5), 1015–1030.

Reuter, T., Nummela, S., & Hemilea, S. (1998). Elephant hearing. *Journal of the Acoustical Society of America, 104*, 1122–1123.

Rosowski, J. J. (1994). Outer and middle ears. In R. R. Fay & A. N. Popper (Eds.), *Comparative hearing: Mammals* (pp. 172–247). New York: Springer.

Rujirawan, A., Stuart, B. L., & Aowphol, A. (2013). A new tree frog in the genus *Polypedates* (Anura: Rhacophoridae) from southern Thailand. *Zootaxa, 3702*(6), 545–565.

Saxod, R. (1996). Ontogeny of the cutaneous sensory organs. *Microscopy Research and Technique, 34*(4), 313–333.

Schermuly, L., & Klinke, R. (1990). Infrasound sensitive neurons in the pigeon cochlear ganglion. *Journal of Comparative Physiology, 166*, 355–363.

Shen, J.-X. (1983). A behavioral study of vibrational sensitivity in the pigeon (*Columba livia*). *Journal of Comparative Physiology, 152*, 251–255.

Shipley, C., Stewart, B. S., & Bass, J. (1992). Seismic communication in northern elephant seals. In J. A. Thomas, R. A. Kastelein, & A. Y. Supin (Eds.), *Marine mammal sensory systems* (pp. 553–562). New York: Plenum Press.

Shoshani, J. (1998). Understanding proboscidean evolution: A formidable task. *Trends in Ecology and Evolution, 13*, 480–487.

Shoshani, J., Kupsky, W. J., & Marchant, G. H. (2006). Elephant brain. 1. Gross morphology, functions, comparative anatomy, and evolution. *Brain Research Bulletin, 70*(2), 124–157.

Sikes, S. K. (1971). *The natural history of the African elephant*. New York: Elsevier.

Smotherman, M., & Narins, P. M. (2004). Evolution of the amphibian ear. In G. A. Manley, A. N. Popper, & R. R. Fay (Eds.), *Evolution of the vertebrate auditory system* (pp. 164–199). New York: Springer.

Soltis, J. (2010). Vocal communication in African elephants (*Loxodonta africana*). *Zoo Biology, 29*(2), 192–209.

Soltis, J., King, L. E., Douglas-Hamilton, I., Vollrath, F., & Savage, A. (2014). African elephant alarm calls distinguish between threats from humans and bees. *PLoS One, 9*(2), e89403. doi:10.1371/journal.pone.0089403.

Soltis, J., Leighty, K. A., Wesolek, C. M., & Savage, A. (2009). The expression of affect in African elephants (*Loxodonta africana*) rumble vocalizations. *Journal of Comparative Psychology, 123*(2), 222–225.

Soltis, J., Leong, K., & Savage, A. (2005). African elephant vocal communication II: Rumble variation reflects individual identity and emotional state of callers. *Animal Behaviour, 70*(3), 589–599.

Springer, M. S., Cleven, G. C., Madsen, O., de Jong, W. W., Waddell, V. G., Amrine, H. M., et al. (1997). Endemic African mammals shake the phylogenetic tree. *Nature, 388*, 61–64.

Stanhope, M. J., Waddell, V. G., Madsen, O., de Jong, W., Hedges, S. B., Cleven, G. C., et al. (1998). Molecular evidence for multiple origins of Insectivora and for a new order of endemic African insectivore mammals. *Proceedings of the National Academy of Sciences of the USA, 95*, 9967–9972.

Stenfelt, S., Wild, T., Hato, N., & Goode, R. L. (2003). Factors contributing to bone conduction: The outer ear. *Journal of the Acoustical Society of America, 113*(2), 902–913.

Stoeger, A. S., Heilmann, G., Zeppelzauer, M., Hensman, S., & Charlton, B. D. (2012). Visualizing sound emission of elephant vocalizations: Evidence for two rumble production types. *PloS One, 7*(11), e48907. doi:10.1371/journal.pone.0048907.

Stoeger, A., Zeppelzauer, M., & Baotic, A. (2014). Age-group estimation in free-ranging African elephants based on acoustic cues of low-frequency rumbles. *Bioacoustics, 23*(3), 231–46. doi: 10.1080/09524622.2014.888375.

Stoeger-Horwath, A. S., Stoeger, S., Schwammer, H. M., & Kratochvil, H. (2007). Call repertoire of infant African elephants: First insights into the early vocal ontogeny. *Journal of the Acoustical Society of America, 121*(6), 3922–3931.

Theurich, M., Langner, G., & Scheich, H. (1984). Infrasound responses in the midbrain of the guinea fowl. *Neuroscience Letters, 49*, 81–86.

van der Merwe, N. J., Bezuidenhout, A. J., & Seegers, C. D. (1995). The skull and mandible of the African elephant (*Loxodonta africana*). *Onderstepoort Journal of Veterinary Research, 62*, 245–260.

Varanasi, U., Feldman, H. R., & Malin, D. C. (1975). Molecular basis for formation of lipid sound lens in echolocating cetaceans. *Nature, 255*, 340–343.

Varanasi, U., & Malin, D. C. (1971). Unique lipids of the porpoise (*Tursiops gilli*): Differences in triacylglycerols and wax esters of acoustic (mandibular canal and melon) and blubber tissues. *Biochemica Biophysica Acta, 231*, 415–418.

von Békésy, G. (1944/1960). Frequency analysis in the cochleas of various animals. In E. G. Wever (Ed.), *Experiments in hearing* (pp. 500–510), New York: McGraw-Hill

von Mayer, A., O'Brien, G., & Sarmiento, E. E. (1995). Functional and systematic implications of the ear in golden moles (Chrysochloridae). *Journal of Zoology, 236*(3), 417–430. doi:10.1111/j.1469-7998.1995.tb02722.x/abstract.

Warkentin, K. M. (2005). How do embryos assess risk? Vibrational cues in predator-induced hatching in red-eyed treefrogs. *Animal Behaviour, 70*, 59–71.

Warkentin, K. M., & Caldwell, M. S. (2009). Assessing risk: Embryos, information, and escape hatching. In R. Dukas & J. Ratcliffe (Eds.), *Cognitive ecology II. The evolutionary ecology of learning, memory, and information use* (pp. 177–200). Chicago: University of Chicago Press.

Warkentin, K. M., Caldwell, M. S., & McDaniel, J. G. (2006). Temporal pattern cues in vibrational risk assessment by embryos of the red-eyed treefrog, *Agalychnis callidryas. Journal of Experimental Biology, 209*, 1376–1384.

Warkentin, K. M., Caldwell, M. S., Siok, T. D., D'Amato, A. T., & McDaniel, J. G. (2007). Flexible information sampling in vibrational assessment of predation risk by red-eyed treefrog embryos. *Journal of Experimental Biology, 210*, 614–619.

Webb, G. J. W., Choquenot, D., & Whitehead, P. J. (1986). Nests, eggs, and embryonic development of *Carettochelys insculpta* (Chelonia: Carettochelidae) from Northern Australia. *Journal of Zoology, 1*, 521–550.

Weissengruber, G. E., Egger, G. F., Hutchinson, J. R., Groenewald, H. B., Elsasser, L., Famini, D., et al. (2006). The structure of the cushion in the feet of African elephants (Loxodonta africana). *Journal of Anatomy, 209*, 181–192.

Wever, E. G. (1973). The ear and hearing in the frog, *Rana pipiens. Journal of Morphology, 141*, 461–478.

Willi, U. B., Bronner, G. N., & Narins, P. M. (2006a). Ossicular differentiation of airborne and seismic stimuli in the Cape golden mole (*Chrysochloris asiatica). Journal of Comparative Physiology, 192*, 267–277.

Willi, U. B., Bronner, G. N., & Narins, P. M. (2006b). Middle ear dynamics in response to seismic stimuli in the Cape golden mole (*Chrysochloris asiatica). Journal of Experimental Biology, 209*, 302–313.

Yodlowski, M. L., Kreithen, M. L., & Keeton, W. T. (1977). Detection of atmospheric infrasound by homing pigeons. *Nature, 265*, 725–726.

Young, B. A., & Morain, M. (2002). The use of ground-borne vibrations for prey localization in the Sahara sand vipers (*Cerastes*). *The Journal of Experimental Biology, 205*, 661–665.

Yu, X.-L., Lewis, E. R., & Feld, D. (1991). Seismic and auditory tuning curves from bullfrog saccular and amphibian papillar axons. *Journal of Comparative Physiology, 169*, 241–248.

Chapter 8
Vocal Production by Terrestrial Mammals: Source, Filter, and Function

Anna M. Taylor, Benjamin D. Charlton, and David Reby

Abstract In little over two decades, researchers have moved from a situation in which most studies of terrestrial mammal vocal signals focused on conspicuous characteristics, such as their rate of occurrence, and where the spectral acoustic variation was largely ignored or poorly quantified, to a field of study in which there is a much better understanding of the nature and function of the acoustic parameters that compose vocalizations. The source-filter theory, originally developed for the analysis of speech signals, has played a large role in this progress. Understanding how the acoustic variability of vocalizations is grounded within their mechanism of production has enabled researchers to predict the type of information that vocal signals are likely to contain, and to predict their co-variation with morphological and/or physiological attributes of callers. Moreover, the powerful theoretical platform derived from the source-filter theory not just conceptually supports the formulation of multilevel hypotheses, but also paves the way to develop the corresponding methodologies needed to address them. Although the full range of acoustic diversity of terrestrial mammal signals has yet to be explored, this chapter draws together a wealth of research conducted over the last two decades, and describes how source- and filter-related acoustic components encode functionally relevant information in the vocal communication systems of terrestrial mammal and how selection pressures have led to the evolution of anatomical innovations that enable animals to produce exaggerated vocal traits.

Keywords Acoustic signals • Formants • Larynx • Nonlinear phenomena pitch • Sexual communication • Source-filter theory • Vocal apparatus • Vocal communication

---

A.M. Taylor • D. Reby (✉)
School of Psychology, University of Sussex, Falmer BN1 9QH, UK
e-mail: anna.magdalena.taylor@gmail.com; d.reby@sussex.ac.uk

B.D. Charlton
School of Biology and Environmental Science, Science Centre West, University College Dublin (UCD), Belfield, Dublin 4, Ireland
e-mail: benjamin.charlton@ucd.ie

R.A. Suthers et al. (eds.), *Vertebrate Sound Production and Acoustic Communication*, Springer Handbook of Auditory Research 53,
DOI 10.1007/978-3-319-27721-9_8

8.1 Introduction

Animal vocalizations, like all acoustic signals, are intrinsically difficult to study because they are transient phenomena whose complexity cannot be reliably quantified by ear. Spectrographic representations originally developed in the 1940s, and now readily available as freeware on affordable personal computers, have greatly facilitated the visualization and quantification of sound (see Read et al. 1992 for a history and evaluation of speech analysis systems). However, this technological advance raises a new problem: with the high number of redundant acoustic variables that can be derived from a spectrographic analysis, how can researchers efficiently determine which are important for communication and which are not? Because mammalian acoustic perception is typically nonlinear (Kuhl and Miller 1975), and likely to vary between species as a consequence of their ecology (Gibson 1966), it is, in essence, difficult to objectively and efficiently determine where to focus investigations. Understanding the anatomy of vocal production alongside the biomechanical processes that dictate the form and structure of vocal signals can help researchers predict the type of information that they are likely to contain, and in doing so, determine which acoustic variables are likely to be functionally relevant.

A well-established theoretical and methodological framework for examining the vocalizations of terrestrial mammals already exists. The "source-filter theory of speech production," originally developed for the analysis of speech signals (Chiba and Kajiyama 1941; Fant 1960), posits that the production of voiced signals follows a two-stage process involving independent contributions from distinct parts of the vocal apparatus. The "source" signal is generated by the vibration of the laryngeal vocal folds and then modified in the supralaryngeal vocal tract, which acts as a "filter." In the 1990s researchers investigating the acoustic structure of nonhuman primate vocalizations realized that because the vocal apparatus is fundamentally similar across mammalian species (Fitch 1997; Owren et al. 1997), the source-filter theory could be naturally generalized to nonhuman vocalizations (Owren and Bernacki 1998; Riede and Fitch 1999). Following these breakthrough studies, the source-filter theory was to become central to integrative investigations of mammalian vocal communication over the following two decades, incorporating every stage of vocal communication from the microanatomical aspects of signal production (Herbst, Chap. 7) to the behavioral consequences of signal perception (see Taylor and Reby 2010). Although this chapter focuses specifically on vocal communication in terrestrial mammals, it should be noted that the source-filter theory has also been extended to avian species (Elemans et al. 2008; Ohms et al. 2010): the avian syrinx acts as the "source" and the suprasyringeal cavities (including, but not limited to, the trachea and the oral cavity) functions as the "filter."

The generalization of the source-filter theory has given researchers a highly useful framework that provides a theoretical platform for the formulation of multilevel hypotheses, as well as the corresponding methodologies to address them. These methodologies include dedicated methods for analyzing specific production-related components of the vocal signals, as well as resynthesis techniques [e.g., Pitch Synchronous Overlap and Add (PSOLA) resynthesis] for independently manipulating these components along their natural variation, to test the functional relevance of acoustic variation in playback experiments (in the laboratory or in the field). Although this chapter opens with a succinct and simplified introduction to the biological and physical bases of the source-filter theory, a detailed description of the mammalian vocal apparatus is not provided here, and readers are referred to Herbst (Chap. 7) for a more detailed account of the biomechanical processes involved in mammalian phonation and their effects on the acoustics of the produced signals.

8.2 The Source-Filter Theory of Vocal Production

According to the source-filter theory, most speech signals can be interpreted as resulting from two independent production processes: a source signal produced by the larynx, which is subsequently filtered in the cavities of the vocal tract (Fig. 8.1; Chiba and Kajiyama 1941; Fant 1960).

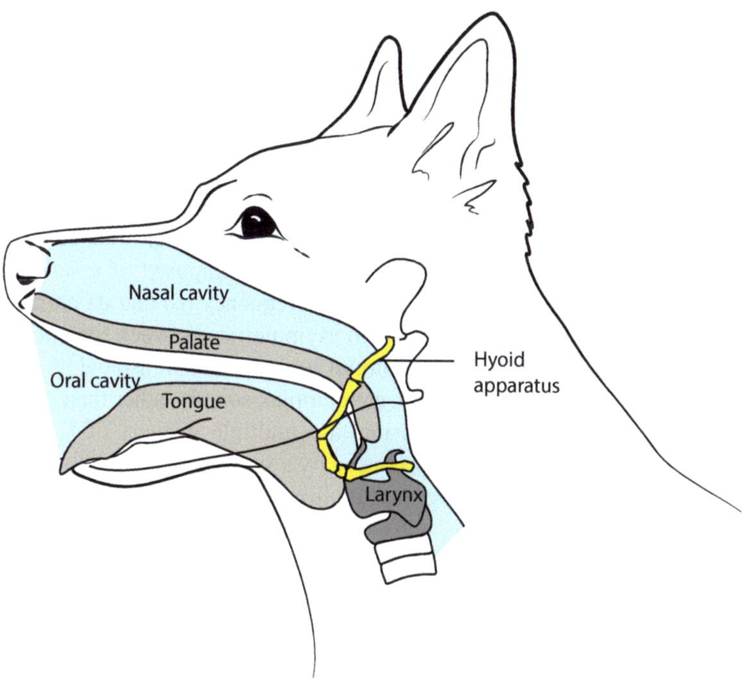

Fig. 8.1 Main components of the mammalian vocal apparatus. The source signal is produced at the level of the larynx and subsequently filtered in the supralaryngeal vocal tract (the oral cavity and the nasal cavity, separated by the palate)

8.2.1 The Source

The initial production of voiced sounds takes place inside the larynx (shaded in red on Fig. 8.1), a cartilaginous structure located where the pharyngeal cavity splits into the trachea and the esophagus. Although the human larynx is comparatively low and externally visible in adult males (the "Adam's apple"), most nonhuman mammals have a more elevated, laryngeal position, with the larynx tightly attached to the base of the skull and protruding into the nasal cavity. This intranarial larynx position allows animals to breathe through the nose while they feed (Fitch 2006). The mammalian larynx is composed of several cartilages, including the thyroid cartilage, the cricoid cartilage, and a pair of arytenoid cartilages. Voiced sounds are produced by vibration of the vocal folds, a pair of mucous membranes that are anteriorly attached to the thyroid cartilage (at the level of the laryngeal prominence) and posteriorly to the arytenoid cartilages. Vocal folds are fleshy, lip-like structures that consist of three layers: the thyro-arythenoid muscle, the vocal ligament or "lamina propria" (itself composed of the deep, intermediate, and superficial layers), and the epithelium. Although they are sometimes colloquially referred to as "vocal cords," the term "vocal folds" is anatomically more correct (vocal folds are not strings, albeit they can behave in a "string-like" manner under certain models; e.g. Woods 1893) and thus preferred in the scientific literature (Titze 1994; Fitch 2006).

Voiced sounds are generated as air expelled from the lungs passes through the space between the vocal folds (termed the glottis), causing them to open and close. The effect of air pressure differences across the glottis (Bernoulli forces) combined with the biomechanical properties (elastic recoil) of vocal fold tissue make the folds rapidly snap shut again after they have been forced open. This sequence of vocal fold opening and closing generates a cyclic variation in air pressure across the larynx (Titze 1994; Chan et al. 2009). The vibration of the vocal folds generates the source signal, or "glottal wave," which is typically composed of a series of frequency components known as the fundamental frequency (f_0) and its harmonically related overtones or "harmonics." Owing to asymmetries between the opening and closing phases of the glottal cycle, the glottal source is not sinusoidal. As a consequence, most mammalian vocalizations are complex sounds rather than pure tones, with energy contained at frequencies that are multiple integers of f_0 (Fig. 8.2). Because most vertebrate signals are produced by a single source, overtones are usually integer multiples of the f_0. However, variations on this structure are described in Sect. 8.3.8 of this chapter.

The rate of vibration of the vocal folds can be predicted using the string model:

$$F_0 = \frac{1}{2L}\sqrt{\frac{\sigma}{\rho}} \qquad (8.1)$$

where is the longitudinal stress, L is the length, and the tissue density of the vocal folds (Titze 1994). The f_0 is determined primarily by the anatomical length of the vocal folds, with longer vocal folds vibrating at a slower rate than smaller vocal folds (Hollien 1960), but it can also be varied dynamically by controlling the

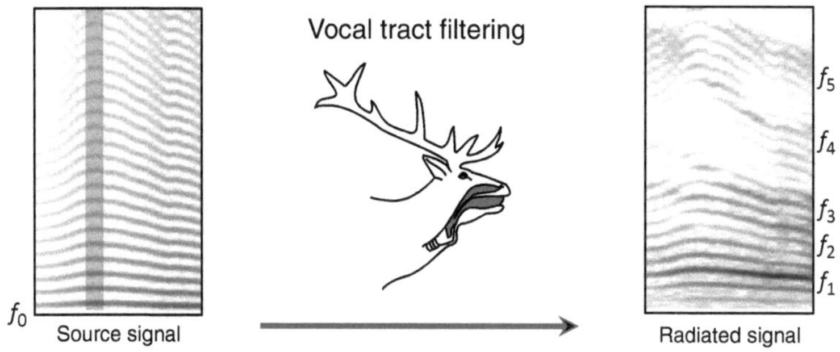

Fig. 8.2 Application of the source-filter theory to the production of a red deer roar. Air forced out of lungs through the glottis causes the vocal folds to vibrate, generating a source wave composed of a fundamental frequency (f_0) and its series of harmonic overtones. Supralaryngeal cavities that compose the vocal tract filter have natural resonances that shape the spectral envelope of the glottal source, creating broad frequency peaks called *formants* (f_1, f_2, etc.)

subglottal pressure (the air pressure in the trachea, below the larynx) as well as the length, stiffness, and tension of the vocal folds (Hirano et al. 1969). Other characteristics of the source signal, such as its intensity contour, can also be controlled by adjusting lung pressure and glottal opening, and contribute to the temporal structure of the vocalization (Titze 1994).

The f_0 is the main factor determining the perceived pitch of a voice: vocalizations with a high mean fundamental frequency tend to be perceived as high pitched and vice versa. It should be noted, however, that the "pitch" of a vocalization is a perceptual attribute that can also be affected by other dimensions of the sound, such as its spectral envelope or its amplitude. Consequently, although the perceived pitch of a vocalization is largely determined by f_0, it is preferable not to use this term as a synonym of f_0 in the scientific literature (Fitch 2002). Finally, the modulation of f_0 over the course of a vocalization constitutes its f_0 contour (intonation in speech signals; Titze 1994), an important parameter that often contributes to the inter- and intraspecific diversity of mammalian vocal repertoires by defining different call types (domestic dog: Cohen and Fox 1976; Corsican deer, *Cervus elaphus corsicanus*: Kidjo et al. 2008).

8.2.2 *The Filter*

The source signal must travel through the supralaryngeal vocal tract (Fig. 8.1) before it is eventually radiated into the environment. The vocal tract consists of the cavities that connect the glottis to the lips and nostrils: the (oro- and naso-) pharyngeal cavities, the oral cavity, and the nasal cavity. The resonance properties of these cavities selectively dampen or enhance specific frequencies of the source signal, resulting in a heterogeneous spectrum, with bands of relatively high amplitude separated by bands of relatively low amplitude (Fig. 8.2). The areas of enhanced frequencies are referred to as vocal tract resonances or "formants" (Fant 1960), and their distribution affects the perceived "timbre" of the vocal signal (Childers and Lee 1991).

In humans, the production of speech sounds involves complex movements of the flexible components of the vocal tract, or "articulators" (mainly the tongue, velum, lips, and lower jaw) that dynamically alter its shape, and thereby alter the resonance properties of the vocal tract. Articulation particularly affects the respective position of the lower formants of the voice spectrum (F_1, F_2, and to a lesser extent F_3) and results in the production of the phonetic elements of speech: vowels and consonants (Fant 1960; Ladefoged 2001). In contrast, most nonhuman mammals have limited control over the shape and dimensions of the vocal tract, which makes the formants in their vocalizations typically more static and predictable (Fitch 1994, 2002). One of the key factors affecting the frequency distribution of formants in mammal vocal signals is the existence of an inverse relationship between formant frequencies and vocal tract length: if the vocal tract is approximated as a straight uniform tube closed at one end (the glottis) and open at the other end (the lips), the centre frequencies of the successive formants (F_1, F_2, ... F_N) generated by such a resonator are related to the length of the vocal tract by the equation

$$F_i = \frac{(2i-1)c}{4VTL} \text{ for } i = 1,2,\ldots,N \tag{8.2}$$

where c is the speed of sound in air (ca. 350 m/s in the warm, humid air of a mammalian vocal tract) and VTL is the length of the vocal tract. As a consequence, the spacing between any two consecutive formants in the frequency spectrum is constant and given by

$$\Delta F = F_{i+1} - F_i = \frac{c}{2\text{VTL}} \tag{8.3}$$

Therefore, the longer the vocal tract, the lower the formant frequencies, and the narrower their overall frequency spacing. Larger animals, with longer vocal tracts, are therefore expected to produce vocalizations with lower formant frequencies (Fitch 1997, 2000a, b).

Finally, a key assumption of the source-filter theory of voice production is that the functioning of the source is independent and decoupled from the functioning of the filter: f_0 can be varied independently from formant frequencies and vocal tract filtering does not affect vocal fold vibration. In other words, there is no feedback of the vocal tract filter on the glottal source, and vice versa (Titze 1994). The important consequences of source filter independence as well as some possible exceptions to this rule are discussed in Sect. 8.5.

8.2.3 On the Origin of Information and Function: Production Constraints as Cues to Physical Attributes

Vocal signals are used by animals to mediate many social interactions such as sexual competition, territorial maintenance, partner or parent/young recognition, and coordination of defense against predators (Clutton-Brock and Albon 1979;

Owings and Morton 1998). Because the outcome of such interactions often depends on physical attributes of individuals (Schmidt-Nielsen 1975), receivers benefit from being able to perceive and assess any variable that can provide indexical information about the caller such as their body size, physical condition, age, or sex. The source-filter theory predicts that vocal signals should contain indexical cues that arise directly from the biomechanical constraints affecting their production: signals encode "honest" information, independently of the cost associated with their production, because they obey simple principles of acoustics and biomechanics (Fitch 1997; Reby and McComb 2003). Understanding the origin, nature, and function of such indexical cues has become one of the most active areas of vocal communication research (see Taylor and Reby 2010). Although the role of indexical cues usually originates in the physical relationship that ties them to the dimension they express, selection pressures may lead to the evolution of mechanisms enabling callers to partially escape anatomical constraints to minimize or exaggerate the impression of the primary dimension expressed in the acoustic cue (Morton 1977; Maynard-Smith and Harper 2003). In some species, selection processes have even led to the evolution of anatomical adaptations enabling the production of "permanently exaggerated" traits, which may still obey allometric principles and therefore contain honest and reliable information (Fitch and Reby 2001; Sanvito et al. 2007). Following the key assumption of source and filter independence, cues generated at the level of the source and those shaped in the filter are reviewed in turn.

8.3 Vocal Cues Encoded at the Source

8.3.1 Mean f_0 *as a Static Cue to Interspecific Size Variation*

Initially, studies of animal vocal signals focused principally on understanding the variability and indexical content of f_0 (Morton 1977; August and Anderson 1987). Because the size of the larynx is expected to grow in proportion with other body dimensions, larger species can reasonably be assumed to have longer vocal folds. As a result, f_0 is expected to obey general allometric principles, with larger animals producing lower f_0 calls. This assumption is broadly verified across species: the fundamental frequency of vocalizations is minimal in the heaviest species (e.g., African elephants, *Loxodonta* sp.: 16.8 Hz), and highest in the lightest (bats, suborder Yinpterochiroptera: 63.8 kHz), as illustrated in Fig. 8.3 (see Herbst, Chap. 7).

8.3.2 Mean f_0 *as a Static Cue to Intraspecific Size Variation*

As a consequence of this allometric relationship, f_0 also reflects size-related differences between breeds in species possessing an unusually large range of morphological variations (domestic dogs: Taylor et al. 2008), as well as between age categories (Seyfarth and Cheney 1986; Briefer and McElligott 2011) and between sexes in

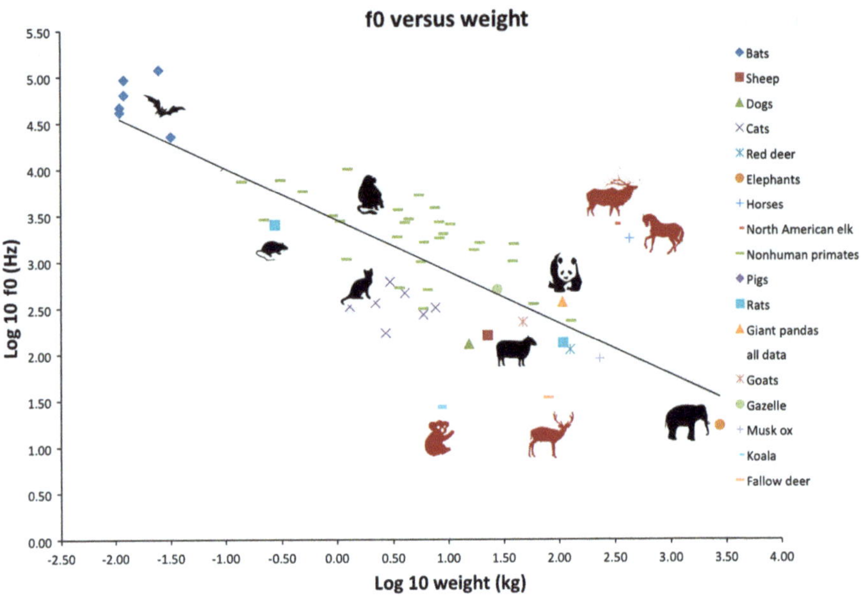

Fig. 8.3 Correlation between body mass and f_0 across mammalian species. Some species have evolved specialized vocal folds or alternative voice production mechanisms to deviate from the expected acoustic allometry

sexually dimorphic species (Rendall et al. 2004). In some species, such as red deer (*Cervus elaphus*), the vocal folds of males continue to grow in length after the animal itself has stopped growing, leading to a partially size-independent association between f_0 and age throughout the lifetime of the individual (Reby and McComb 2003). Despite this broad relationship between body size and f_0, the dimensions of the vocal folds (and consequently the range of f_0 they can produce) are relatively unconstrained by the anatomical structures surrounding the larynx (Fitch 2006). In addition, most mammals can adjust the length and tension of the vocal folds to modulate f_0 within and between call types. Consequently, the relationship between f_0 and body size often breaks down within species (Smith et al. 2005; Pisanski et al. 2014), making f_0 a comparatively poor indicator of interindividual variation in body size.

8.3.3 f$_0$ *as a Cue to Mate Quality in Sexual Communication*

In many species, males have lower f_0 than females, even after accounting for size differences between the sexes (humans: Huber et al. 1999; red deer: Reby and McComb 2003; Corsican deer: Kidjo et al. 2008). The sexual dimorphism of f_0 suggests a role of sex hormones in vocal fold development. Indeed, it is well documented that in humans the vocal folds start to lengthen and thicken disproportionately during male puberty, presumably as a consequence of the maturation of the testes and increased levels of circulating testosterone (Harries et al. 1997; Fitch and Giedd 1999). Although the exact mechanism by which testosterone influences such vocal fold growth has not yet been identified (some studies have failed to identify androgen receptors on the vocal folds themselves; e.g., Schneider et al. 2007; Nacci et al. 2011), a controlled experiment demonstrated a clear positive dose–response relationship between doses of testosterone and dihydrotestosterone administered to castrated lambs and the resultant laryngeal changes (Beckford et al. 1985). In humans, the puberty-related growth results in a decrease in f_0 of 50 % in comparison to same-aged women (comparatively, the body size variation between adult men and women is approximately 20 %; Fitch and Giedd 1999). Although Harries et al. (1997) suggest that after puberty, short-term variation in circulating levels of testosterone no longer affect f_0, a more recent study showed that adult men's voices do in fact show diurnal variations in f_0 that track saliva testosterone levels (Evans et al. 2008). Moreover, higher testosterone levels are known to be predictive of lower voice pitch in adult men (Dabbs and Mallinger 1999; Cartei et al. 2014). Mean f_0 may thus act as an index of androgen levels in humans (Bruckert et al. 2006), although to date no empirical evidence is known to support this claim in nonhuman mammals. Another aspect of f_0 has, however, been directly linked to sex hormone levels in giant pandas (*Ailruopoda melanoleuca*): the rate of f_0 modulation in male bleat vocalizations is a reliable cue to interindividual variation in androgen levels, and the duration of the vocalization signals short-term changes in the caller's androgen levels (Charlton et al. 2011c). In addition, male and female giant pandas attend to this information in male bleats, and respond in a functionally meaningful manner to bleats representing high- versus low-testosterone males (Charlton et al. 2012b).

Adult men with lower f_0 report more sexual partners (Apicella et al. 2007) and are judged by female listeners as sounding more masculine (Cartei et al. 2014) and more attractive (Puts et al. 2007, Apicella and Feinberg 2009; although interdependence with other acoustic variables should be considered; see Pisanski and Rendall 2011). However, despite this consistent result, Simmons et al. (2011) found that men with lower f_0 did not have better sperm quality than men with higher f_0, suggesting that there may be a functional trade-off between attractiveness and sperm production (see also Puts et al. 2012). Fallow deer (*Dama dama*) provide an interesting parallel with humans, with their highly sexually dimorphic larynx and f_0. Male fallow deer that produce lower f_0 groan vocalizations are in better physical condition and less fatigued than those producing higher f_0 groans and are more competitive both on an intra- and intersexual level (Vannoni and McElligott 2008; Pitcher et al. 2014). In direct contrast, male chacma baboons (*Papio cynocephalus ursinus*) with high dominance status produce calls with a higher f_0 than lower ranked males (Fischer et al. 2004). Similarly, red deer stags with a higher minimum f_0 hold hinds for longer during the breeding season (a proxy for reproductive success in this species) than stags with lower minimum f_0 (Reby and McComb 2003). These findings in chacma baboons and red deer are likely to indicate greater arousal levels in dominant individuals, leading to higher subglottal pressures and/or increased vocal fold tension during sound production. Playback experiments using resynthesized calls indicate that estrous red deer hinds prefer males that produce high f_0 roars in mate choice contexts (Reby et al. 2010), but further work had found that red deer stags do not respond differently to high versus low f_0 roars from unfamiliar intruders (Garcia et al. 2013). These observations suggest that divergent sexual selection pressures affect the functional relevance of f_0 in red deer, and this is something that should be kept in mind when working on other mammalian species.

8.3.4 *Evading the Size/f_0 Allometry: Permanent Adaptations for the Production of Abnormally High or Abnormally Low f_0s*

Several terrestrial mammals have evolved specializations of vocal anatomy that allow them to produce unexpectedly wide-ranging f_0 (Fitch 2006). For instance, some bats have thin vocal membranes on the edges of their vocal folds that enable the production of very high f_0 ultrasonic calls for echolocation purposes (Griffin 1958). When these membranes are cut experimentally, the bats are no longer able to produce ultrasound (Novick and Griffin 1961; see Au and Suthers 2014, for a review of the production of biosonar signals). Similarly, the very high f_0 of some nonhuman primates is thought to be produced using thin membranes on the edge of the vocal folds that, because of their low mass, can oscillate periodically at very high frequencies when exposed to airflow from the lungs (Schön-Ybarra 1995; Mergell et al. 1999). Other adaptations allow mammals to produce disproportionately low f_0 for their size. For instance, hammerhead bats (*Hypsignathus monstrosus*) and howler monkeys (*Alouatta* sp.) have hypertrophied larynges (Kelemen and Sade 1960; Bradbury 1977), whereas roaring cats (*Panthera* sp.) and Mongolian gazelles (*Procapra gutturosa*) have thick fleshy pads on their vocal folds (Frey and Gebler 2003; Titze et al. 2010). These specializations may reflect selection pressures to efficiently produce low f_0, either to enhance the salience of vocal tract resonances (as discussed in this chapter) or because low f_0 in itself signals an aspect of the caller that is important in the species' communication system.

An extreme case of adaptation for low-frequency sound production is found in the koala (*Phascolarctos cinereus*: Charlton et al. 2013b), which has evolved an additional set of vocal folds (termed "velar vocal folds") outside the larynx that allow it to produce extremely low-pitched vocalizations (ca. 30 Hz; see Fig. 8.3). Indeed, the koala's velar folds are much larger than its laryngeal vocal folds, and hence can oscillate at much lower frequencies. These velar folds are the only known example of an anatomical specialization for the production of sound outside the larynx in a terrestrial mammal (Charlton et al. 2013b). Future studies should investigate similar allometric anomalies, where animals produce unexpectedly high- or low-pitched vocalizations. For example, it is unclear how Diana monkeys (*Cercopithecus diana*) produce vocalizations with a disproportionately low f_0 for their size, despite not possessing an enlarged larynx (Riede and Zuberbühler 2003a).

8.3.5 *Intraindividual* f_0 *Modulation as a Dynamic Cue to Motivations and Emotions*

Beyond the encoding of static caller attributes, vocal signals can function to communicate the current motivational or emotional state of individuals, especially when information about dynamic attributes of callers is central to the complex social interactions that characterize species living in large groups. Following a comparative review of vocalizations used in agonistic displays in a range of mammalian and avian species, Morton (1977) predicted that animals with aggressive motivation produced low-pitched, broadband vocalizations (such as growls and hisses), while friendly or submissive animals produced higher pitched vocalizations (such as whimpers and whines). Morton's hypothesis (also known as Morton's motivation-structural rules) is based on the observation that aggressive and dominant individuals of many species seek to project, both visually and acoustically, an impression of a larger body size, whereas friendly or submissive individuals seek to project a smaller body size (Morton 1977; Owings and Morton 1998). Interestingly, in human speech voices with lower f_0 are perceived as coming from a larger speakers even though f_0 is a very poor cue to body size variation between individuals (Rendall et al. 2007), accounting for less than 2 % of size variance within sexes (Pisanski et al. 2014). The vocal repertoires of several species suggest that f_0 variation generally follows the framework of Morton's motivation-structural rules; for example, barks of domestic dogs recorded in an aggressive context have a significantly lower f_0 than those recorded in a playful setting (Yin 2002; Pongrácz et al. 2005). Similarly, bugle calls emitted by males of North American elks (*Cervus canadensis*) in aggressive contexts are lower in frequency than bugle calls emitted during nonaggressive interactions (Feighny et al. 2006).

Dynamic variations in f_0 that provide cues to affective state are most likely mediated by changes in physiological arousal, such as rate of respiration or muscular tension in the vocal folds (Scherer 1986). Indeed, as previously mentioned, f_0 can be modulated by controlling subglottal pressure or by adjusting the vibrating properties of the vocal folds. Contraction or relaxation of the cricothyroid or thyroarytenoid muscles enables considerable modulation of vocal fold length, stiffness, and tension, thereby widening the range of f_0 that can be produced by any given caller (Fitch 2006). As a consequence, source characteristics typically vary between vocalizations from the same caller. This can be due to intentional vocal control (such as the control of intonation in human speech: Ohala 1984; Banse and Scherer 1996) or because of the uncontrolled effects of emotional arousal or tension (Briefer 2012). Several characteristics of f_0 (such as mean f_0, peak f_0, and f_0 modulation) have been linked to the context in which calls are emitted. Classification methods such as discriminant function analysis are useful in confirming the acoustic categorization of vocalizations emitted in different contexts. For example, Yin (2002) found that domestic dog barks occurred on a scale, showing a continuum of acoustic gradations on several frequency parameters depending on the context in which they were emitted. Statistical divisions based on the co-variation of maximum and mean f_0, duration, and interbark interval enables the reliable classification of barks into different context-specific subsets (Yin 2002; Yin and McCowan 2004). These parameters are salient to human listeners, who are able to categorize barks according to their recording context (Pongrácz et al. 2005). More recent investigations have shown that human listeners use rules derived from the assessment of emotional content in speech when evaluating the emotional valence of dog barks (Faragó et al. 2014), suggesting that some mechanisms for assessing the emotional content of vocalizations may be shared across mammals.

8.3.6 *Temporal Characteristics of the Source*

A further dimension of the source implicated in the communication of motivational state is calling rate. As calling rate is linked to the rate of respiration, it can provide immediate information about an individual's current level of physiological arousal (Rendall et al. 1999), physical condition (Pitcher et al. 2014), and motivational or emotional state (Taylor et al. 2009). A communicative role for calling rate has been identified in several species, with overall higher calling rates and/or longer duration vocalizations tending to indicate high arousal contexts, whereas slower calling rates and/or shorter duration calls are more typical of low arousal contexts. For example, calling rate has been found to advertise stamina and fitness in red deer (Clutton-Brock and Albon 1979) and baboons (Fischer et al. 2004), and the very high vocalization rates of fallow deer bucks (ca. 3000 groans per hour) appear to signal the caller's motivational state/willingness to engage in direct competition for females (McElligott and Hayden 1999, 2001). Indeed, Vannoni et al. (2005) confirmed that fallow deer bucks appear to prioritize the maintenance of a high groaning rate rather than favoring other indexical cues available to them, such as the maximization of body size via laryngeal retraction. Longer calls are also more commonly produced in contexts of high arousal. For example, in domestic dogs, both barks and growls are longer in aggressive contexts (Yin 2002; Taylor et al. 2009), and meerkats (*Suricata suricatta*) produce longer duration calls when the threat of predation is more imminent (Manser 2001).

8.3.7 f_0 *and Individual Recognition*

The ability to discriminate between individuals based on their vocalizations is important for many terrestrial mammals. Within the source-filter framework, several studies have demonstrated that identity can be coded within the source-related characteristics of vocalizations; however, the way in which this occurs depends on the ecological needs of the species. For example, in wolves, mean and maximum f_0 as well as frequency modulation are individually distinctive and can be used for individual recognition (Palacios et al. 2007). In hyena (*Crocuta crocuta*) giggles, caller identity is coded primarily by the range of f_0 in the giggle, but in hyena whoop calls, it is the pronounced intracall modulation of f_0 that encodes caller identity (Mathevon et al. 2010). Both f_0 range and frequency modulation have been found to encode caller identity in a number of species including yellow-bellied marmots (*Marmota flaviventris*: Blumstein and Munos 2005) and fallow deer (Vannoni and McElligott 2007). In some cases, highly distinctive f_0 contours are said to constitute a uniquely recognizable "vocal signature." Although this has been most extensively studied in nonterrestrial mammals such as the bottlenose dolphin (Tursiops truncates: Janik et al. 2006), there is evidence that some bats may also use similar acoustic cues to identify specific individuals (Carter et al. 2008; Melendez and Feng 2010).

One social context where the acoustic signaling of individual identity is of fundamental importance is that of mother–young recognition. The frequency modulation of high-pitched pup calls are used by white-winged vampire bat (*Diaemus youngi*) mothers to identify their young (Carter et al. 2008). In fur seals (*Arctocephalus tropicalis*), f_0 and its first two harmonics are sufficient to enable mothers to recognize their own pup from among several hundreds of others (Charrier et al. 2002). Finally, vocalizations that are characterized by rapid f_0 and amplitude modulation, often termed "bleats," appear to be particularly suited to individual recognition in a number of mammalian species, both in the mother–young context (sheep: Sébe et al., 2010; Australian sea lions, *Neophoca cinerea*: Pitcher et al. 2012) and in adults (giant panda, *Ailuropoda melanoleuca*: Charlton et al. 2009b). Playback experiments confirmed that the highly individualized amplitude modulation of giant panda bleats enables individual recognition by conspecifics (Charlton et al. 2009b).

8.3.8 Nonlinear Phenomena

Nonlinear phenomena (NLP) are ubiquitous and form part of the normal vocal repertoire of most mammalian species (Fitch et al. 2002; Tokuda et al. 2002; Riede et al. 2007). The presence of additional harmonics (subharmonics) visible in the spectrum beneath f_0 and/or between harmonics is called double (one subharmonic) or triple (two subharmonics) vibration, and adds to the perceived roughness of the vocalization. The presence of two (or more) independent, nonharmonically related glottal sources is called biphonation (or triphonation, etc.). Both phenomena have been reported in African wild dogs (*Lycaon pictus*) where they may contribute to signaling motivation and status (Wilden et al. 1998) and chimpanzees (*Pan troglodytes*) where their relative prevalence may indicate a measure of physical condition (Riede et al. 2004a). A third type of NLP termed "deterministic chaos" is characterized by the presence of broadband frequency "noise" that partially masks the signals' periodicity and increases the harsh-sounding quality of vocalizations (Titze 2008). Deterministic chaos has been documented in the calls of several species, including the sexually selected vocalizations of polygynous deer, which produce distinct "harsh calls" in mating contexts. Recent playback experiments have shown that although harsh roars are not more attractive than common roars to estrous female red deer (Charlton et al. 2014), they may function as "attention grabbing" signals in this species' sexual communication (Reby and Charlton 2002). An additional study suggests that harsh roars may signal motivation during male contests, as red deer stags react less strongly to sequences of harsh versus common roars from unfamiliar intruders, possibly to avoid escalating a contest with a highly motivated competitor in the absence of visual confirmation (Garcia et al. 2014).

It has also been suggested that NLP in terrestrial mammal vocalizations may function to generate unpredictability to prevent listeners from habituating to alarm calls (see discussions by Fitch et al. 2002; Blumstein and Recapet 2009; Townsend and Manser 2011). In yet another hypothesis, recent experiments using excised female deer larynges suggest that chaotic vibrational regimes may also enhance glottal efficiency, thereby enabling the more efficient production of intense vocalizations (Herbst 2014). Although nonlinearities are ubiquitous in vocal repertoires, and undoubtedly play a key role in mammal vocal communication systems, their production and function remain poorly understood, and reliable methods for identifying and quantifying them are clearly needed.

8.3.9 Source Amplitude

Although male terrestrial mammals often produce loud vocalizations during competitive interactions (Fitch and Reby 2001; Sanvito et al. 2007), very few studies have directly investigated the functional relevance of call amplitude (source intensity) in terrestrial mammals. This is partly due to the inherent technical difficulty of obtaining reliable measures of amplitude in the field, as these vary with the caller's distance, direction, atmospheric conditions, and the local environment's vegetation and topography. Furthermore, these same limitations mean that amplitude is less likely to provide consistent and reliable information about callers to receivers (Pfefferle and Fischer 2006). Despite these limitations some studies have managed to quantify amplitude: Sanvito and Galimberti (2003) report that source intensity was reliably correlated with age, body size, and breeding status in elephant seals (*Mirounga* sp.), although the researchers concluded that source intensity was unlikely to be a relevant assessment cue. Wyman et al. (2008) found a positive correlation between the amplitude of North American plains bison (*Bison bison*) bellows and the physical condition and motivation of the caller, whereas other measures of quality, such as mating and reproductive success, were negatively correlated with bellow amplitude.

These findings indicate a possible trade-off between different acoustic properties. Specifically, it is worth noting that higher call amplitude may affect f_0, at any given vocal fold dimension, because lower f_0 calls are typically produced at lower amplitudes using lower subglottal pressures. In contrast, high subglottal pressures will tend to produce high f_0 calls with higher intensities that can probably be perceived over greater distances. Future studies should investigate the impact of these potential trade-offs on the diversification of vocal signal repertoires in terrestrial mammals.

8.4 Vocal Cues Encoded in the Filter

8.4.1 Static Formants as Honest Cues to Body Size

Because the development of the vocal tract is anatomically constrained by surrounding skeletal structures, vocal tract length is expected to correlate with body size (Fitch 2000a). Anatomical investigations have confirmed the existence of significant correlations between vocal tract length and body size (Fitch 1997; Riede and Fitch 1999; Plotsky et al. 2013). As a consequence, formant frequencies (which were demonstrated to be inversely related to the length of the vocal tract in Sect. 8.2.3 of this chapter) can provide reliable indices of body size. Several indices have been proposed to quantify size-related variation in formants; this topic has been recently reviewed in a meta-analysis of speaker size estimations from formant frequencies in human speech by Pisanski et al. (2014). A useful predictor of body size is "formant frequency spacing," or ΔF (Reby and McComb 2003). Assuming that the vocal tract is a quarter wavelength resonator fully open at the level of the lips and closed at the level of the glottis, formants are expected at $Fn = kn * \Delta F$, where $kn = (2n - 1)/2$. ΔF can therefore be obtained by plotting observed formant frequencies against kn and estimating the slope of the line of best fit (setting the intercept to 0). This linear regression method provides a reliable estimate of the overall frequency spacing between consecutive formants (see Reby and McComb 2003 for details of this method). Formant frequency spacing can also be used to estimate the "apparent vocal tract length" (aVTL) of the animal that produces the vocalization, by applying (8.3) given earlier in this chapter.

Figure 8.4 illustrates how formant frequency spacing is inversely correlated with body mass across mammalian species: unsurprisingly, smaller species produce vocalizations with higher, more widely spaced formants and vice versa. Within species, smaller (often younger) animals produce calls with higher formant spacing (rhesus macaques, *Macaca mulatta*: Fitch 1997). In heterogeneous species such as domestic dogs, smaller breeds give calls with higher formants (Taylor et al. 2008). Similarly, in sexually dimorphic species, including humans (Fitch and Giedd 1999), males tend to produce vocalizations with proportionally and even disproportionally lower vocal tract resonances (red deer: Reby and McComb 2003; fallow deer: McElligott et al. 2006). The relationship between formants and body size also holds within same sex and age class in several species (Sanvito et al. 2007; Charlton et al. 2009a).

The observation that formant frequency spacing has the potential to function as an index of caller body size has led to a series of studies investigating whether receivers can perceive, and use, this acoustic variation. In some studies, human listeners have been shown to be able to use formant frequencies to correctly estimate speaker height (Griesbach 1999; Pisanski and Rendall 2011), and spontaneous discrimination of size-related formant variation has been demonstrated in several species using habituation–discrimination paradigms (Fitch and Fritz 2006; Charlton et al. 2007a, 2012a) and behavioural observations (Taylor et al. 2010).

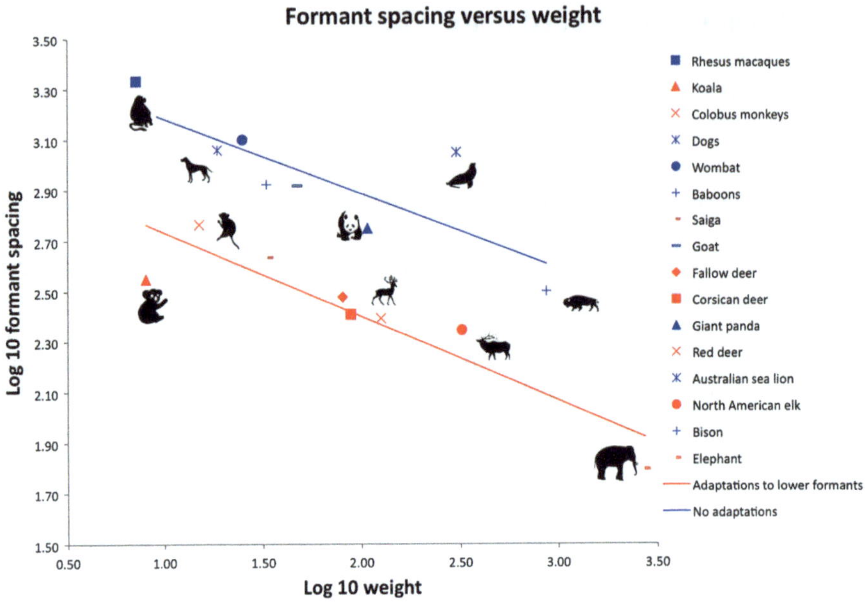

Fig. 8.4 The relationship between body mass and formant frequency spacing across 16 mammalian species. The *blue line* shows the relationship in species that lack vocal tract extensions, and the *red line* in species with proboscises, descended larynges, or air sacs

Domestic dogs have also been shown to spontaneously match acoustic size information with corresponding visual stimuli (Taylor et al. 2011). Rhesus macaques associate smaller formant dispersions with pictures of mature conspecifics, and wider formant dispersions with pictures of immature individuals (Ghazanfar et al. 2007). Together these studies show that formant frequencies encode information on body size at inter- and intraspecific levels, that receivers are capable of perceiving this size related variation in formant frequencies, and that they spontaneously match acoustic size with visual size.

8.4.2 *Function of Formants in Sexual Communication*

In the context of male competition, size-related variation in formants can provide receivers with cues to the resource holding potential of competitors, and inform decisions about whether or not to escalate an agonistic interaction with another individual. Males of several species have been shown to adjust their behavioral responses when they are played resynthesized male sexual calls in which the formant frequency spacing has been scaled to mimic intruders of different sizes (red deer: Reby et al. 2005; koala: Charlton et al. 2013c). In humans, men with lowered vocal tract resonances are perceived by other men as more socially and physically dominant (Puts et al. 2007). Interestingly, men are better than female listeners at perceiving very small size-related differences in synthetic voice-like signals (Charlton et al. 2013a), suggesting a particularly strong role for formant frequencies as size cues during male–male competition.

Playback experiments using similar stimuli have also shown that formant frequencies can play a role in female mate choice. In red deer and koalas, estrous females preferentially approach speakers broadcasting resynthesized male calls with lower formants that simulate larger callers (Charlton et al. 2007b, 2012c). In humans, male voices with lower formant frequencies are rated as sounding more attractive than voices with higher formant frequencies (Feinberg et al. 2005; Puts et al. 2007). Interestingly, women rate voices with low formant dispersion as sounding especially attractive when they are most likely to conceive (Hodges-Simeon et al. 2010). Together these studies suggest that formant frequency spacing functions as an index of body size in both inter- and intrasexual selection in several terrestrial mammals.

8.4.3 Anatomical Adaptations for the Exaggeration of Acoustic Size

In several species of terrestrial polygynous mammals, intense sexual selection has led to the evolution of anatomical adaptations that enable callers to permanently or temporarily extend their vocal tract, thereby lowering the formant frequency spacing of their vocalizations, and maximizing the impression of their body size conveyed to receivers. Male red deer (Fitch and Reby 2001), fallow deer (McElligott et al. 2006), Mongolian and goitred gazelles (Frey and Gebler 2003; Frey et al. 2011), koalas (Charlton et al. 2011b), and several species of big cats (Weissengruber et al. 2002) possess a "descended larynx," an anatomical peculiarity that was previously believed to be unique to humans (Lieberman et al. 1972; Lieberman 1984). Instead of resting in an intranarial, elevated position at the back of the oral cavity (as seen in most nonhuman mammals) the larynx of these species resides in an unusually low position in the throat (Fitch and Reby 2001; McElligott et al. 2006). As a consequence, males of these species have a longer vocal tract than would normally be expected for their body size. Accordingly, vocalizations from species that possess a descended larynx may partially escape the "normally" expected acoustic allometry, and are characterized by a lower formant frequency spacing and concomitantly exaggerated impression of the caller's body size relative to other species lacking this anatomical innovation. In several deer and gazelle species males also have the ability to retract their larynx toward the sternum during the production of mating calls, allowing them to further elongate the vocal tract and lower formants (McElligott et al. 2006; Frey et al. 2011). In red deer, laryngeal retraction is made possible by the presence of a highly elastic thyrohyoid membrane linking the larynx to the hyoid apparatus, and extremely well-developed sternothyroid and sternohyoid muscles that pull the larynx down toward the sternum during vocalization (Fitch and Reby 2001). Because these muscles originate at the sternum, the larynx cannot be pulled lower than the sternum, thereby placing an anatomical limitation on laryngeal retraction, and maintaining the proximate honesty of this signal via a "reconfigured" acoustic allometry (Reby and McComb 2003).

In the koala, the larynx and hyoid bone have both descended, mirroring the structural arrangement seen in humans (Charlton et al. 2011b). Interestingly, while adult humans have a descended larynx, adult males also have a disproportionately longer vocal tract than adult human females, due to a secondary descent of the larynx that only affects adolescent males during puberty (Fitch and Giedd 1999). Given the importance of formant frequency spacing for the communication of size (Pisanski and Rendall 2011), dominance (Puts et al. 2007; Wolff and Puts 2010), androgen levels (Bruckert et al. 2006), and attractiveness (Feinberg et al. 2005) in the male human voice, it is reasonable to hypothesize that the lowered resting position of the larynx in humans may also have evolved due to sexual selection pressures, predating the development of speech.

Other anatomical adaptations or vocal gestures may also be involved in the exaggeration of acoustic size. For example, male saiga antelopes (*Saiga tatarica*) are able to increase the length of their vocal tract while producing mating calls by performing a specific vocal posture involving a strongly tensed and extended trunk (Volodin et al. 2009). Furthermore, some species possess a pronounced proboscis, elongating the nasal region of the vocal tract and potentially influencing the spacing of formant frequencies (elephant seals: Sanvito et al. 2007). The very closely spaced formant frequencies of African elephant rumbles (McComb et al. 2003) indicates that the trunk acts as a resonator, although some rumbles appear to be radiated via the oral cavity only (Stoeger et al. 2012). Black-and-white colobus monkeys (*Colobus guereza*) have evolved a subhyoid air sac that is inflated to act as an additional resonator during roars, thereby lowering their formants in comparison to the values that would normally be expected for an animal of its size (Harris et al. 2006). Finally, some animals may be able to temporarily use body parts that are not part of the vocal apparatus as behavioral adaptations to modify their acoustic output (vocal gestures). This has been documented in the Bornean orangutan (*Pongo pygmaeus wurmbii*) hand kiss–squeak call, in which individuals position the hand (and sometimes detached leaves) in front of the lips during call production (Hardus et al. 2009). Hand kiss–squeak calls have clearer resonances at a lower frequency than unaided kiss-squeak calls, as would be the theoretically predicted acoustic outcome of lengthening the vocal tract (de Boer et al. 2015).

8.4.4 Formants as Cues to Motivational State

According to the "frequency code" or "size code" hypothesis (Ohala 1984, 1994), signals that initially provide a reliable index of static attributes can evolve a secondary function to also encode dynamic information. Ohala (1984) identifies the conditions that must be met for the evolution of such signals: there needs to be a physical and predictable relationship between some acoustic parameter and the body size of the caller (such as the relationship between formants and vocal tract length) and limited variation of this acoustic parameter must be possible within a physically constrained range to manipulate the impression of the physical feature with which it is associated. The caller can thus use acoustic variation in a way that maximizes or minimizes their apparent body size depending on the context of an interaction. A useful visual analogy is the pilomotor reflex observed in many mammals, where the raising of fur (or feathers, as in the case of birds: see Wilson 1972) makes the animal look larger than it really is, thus exploiting the relationship between visual and actual body size when reacting to threat or during ritualized dominance displays.

Owing to the reliability of formant frequencies as an acoustic correlate of body size, small variations in formant dispersion may thus gain a secondary function of signaling motivational state. Effectively, the signaling of body size can become a ritualized advertisement of emotional or motivational state (Ohala 1984, 1996; also see Morton's 1977 motivation-structural rules). In several species, callers have been observed to retract the lips in positive situations or in encounters where it is beneficial for them to appease another individual (such as a smile or fear grin; Fox 1970; Drahota et al. 2008) and to protrude the lips during socially stressful or agonistic encounters when it is beneficial to appear larger or more dominant (Fox 1970; Harris et al. 2006). Moreover, in domestic dogs, growls emitted when guarding a valuable food item have been found to contain lower formants than those emitted in situations with lower emotional valence (Faragó et al. 2010). It seems likely that such ritualized communication is widespread across mammals, and this would be a fruitful area for future research.

8.4.5 Formant Modulation at the Basis of Referential Signals?

Going beyond the size code, it has been suggested that animals can use relatively complex control of filter components for the encoding of context-specific information. Indeed, recent studies on Diana monkeys (Riede and Zuberbühler 2003b; Riede et al. 2005a) and meerkats (Townsend et al. 2014) have shown that active modulation of the first two formants during vocalizations appears to play the greatest role in referential communication. This is perhaps not surprising as F_1 and F_2 are dependent on mouth and tongue configurations, which have the most potential for volitional manipulation. Rudimentary modulation of the first two formants, as illustrated in Fig. 8.5, may hint at the origins of human speech articulation, where volitional changes in vocal tract shape are performed to encode information about external events or objects in the formant frequencies (Fant 1960; Lieberman and Blumstein 1988) and may reflect the transition from essentially affective calls (present in most mammals) to functionally referential calls (present in some nonhuman primates), and ultimately intentionally referential signals as used by humans (see Evans 1997).

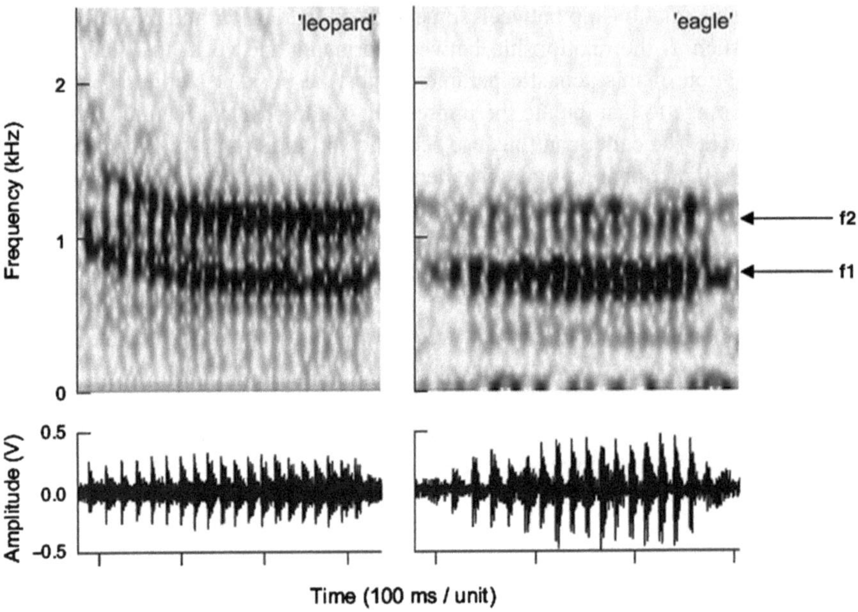

Fig. 8.5 Spectrogram and time series of a leopard and eagle alarm call uttered by a male Diana monkey. Note the downward modulation of the first (f_1) and second (f_2) formants at the beginning of the leopard call, which is absent in the eagle alarm call (Reproduced with permission from Riede et al. 2005a)

8.4.6 Formants and Individual Recognition

The center frequencies, bandwidth, and spacing of formants are linked to the shape, tissue (absorption) properties, and size of the vocal tract (Titze 1994). Because the morphology of the vocal tract is likely to vary between individuals, it has been hypothesized that formant frequencies should contain cues to identity in the vocal signals of terrestrial mammals. In particular, formants are likely to be a reliable source of identity cues in mammals that do not perform dynamic vocal tract modifications during call production (so that the formants are comparatively stable within individuals across calls) and in which the excitation source adequately highlights the caller's distinctive formant pattern (Charlton et al. 2011a; Owren and Rendall 2003). Acoustic analyses of the vocalizations of several species have confirmed this hypothesis (Rendall et al. 1998; McComb et al. 2003). In red-bellied lemurs (*Eulemur rubriventer*), formants have been found to be the most individually distinctive acoustic feature: using formants more than 80 % of vocalizations could be attributed to the correct caller, whereas using f_0 alone lowered correct classification to 25 % of the vocalizations (Gamba et al. 2011). Formant structure may also reliably cue identity in formant-modulated calls. A good example of this is the meerkat noisy bark vocalization, in which statistical analyses have found significant interindividual variation in formant patterns (Townsend et al. 2014). These findings suggest that the communicative potential of formants is not limited to a single function in nonhuman mammals, and may, as in humans, encode a variety of information both within and between calls.

8.5 Interplay of Source and Filter, Source-Filter Coupling and Nonlinear Propagation

So far it has been discussed how source and filter components can independently provide indexical, affective, and functionally referential information in the vocal communication systems of terrestrial mammals. Although source and filter characteristics can broadly be assumed to be independently produced (with the exception of second-order interactions; see Titze 2008 for a discussion), the combined variation of source- and filter-related features is expected to have an effect on the availability and perception of information. For example, the density of the source spectrum has a direct influence on the presence, and perceptibility of formant frequencies (Ryalls and Lieberman 1982; Lieberman and Blumstein 1988). Several studies of human perception confirm that vowel perception is improved when f_0 is lowered (Kewley-Port et al. 1996; Smith et al. 2005). In addition, a recent experiment has shown that lowering f_0 in synthetic voice-like signals improves the perception of size-related formant information by human listeners (Charlton et al. 2013a). In nonhuman mammals, however, the potential interplay and communicative effects of interactions between source and filter is less well understood (cf. Charlton et al. 2008, 2010). Nevertheless, there are likely to be several important interactions taking place between source and filter acoustic components of vocalizations. Studies on birds have shown that male ringdoves (*Streptopelia risoria*) and northern cardinals (*Cardinalis cardinalis*) adjust their formants to track f_0 in order to increase signal amplitude (Riede et al. 2004b, 2006). This same technique of shifting a formant to track f_0, or vice versa, appears to be used by soprano opera singers (Sundberg 1975) and gibbons (Hylobatidae sp.: Koda et al. 2012). It is reasonable to expect that other nonhuman mammals might also align f_0 with formants to increase the intensity of vocalizations to increase signal propagation distances in their natural environments.

It is thus apparent that increasing spectral density may function to highlight the vocal tract resonances in some species. A similar effect could also be achieved by modulating, rather than lowering f_0, so that the harmonics have overlapping ranges that fully scan and excite vocal tract resonances. Such f_0 modulation could be expected to improve formant perception, particularly at high fundamental frequencies when the spectral envelope is sampled more sparsely. Indeed, it has been shown that vibrato-like frequency modulation of the fundamental frequency helps human listeners to better discriminate and identify spectral envelopes with different formant patterns (McAdams and Rodet 1988). It should be noted, however, that this study by McAdams and Rodet was conducted on four human subjects, and ideally should be replicated using a larger sample. Future studies of rapid f_0 modulated calls (bleats, twitters, geckers, etc.) in nonhuman animals should also investigate whether f_0 modulation increases the salience of formant-related information (Charlton et al. 2009a). This could be especially relevant for species with relatively high f_0 for their size, and in which an otherwise unmodulated f_0 would be expected to sample the formants very sparsely. Future studies designed to determine the effect of f_0 modulation on formant perception in nonhuman animals are clearly warranted.

Finally, although the source-filter theory of voice production generally assumes that sound propagation in the vocal tract filter is linear, this may not always be the case in high-intensity vocalizations forced through a long vocal tract resonator. For example, the unusual brassiness of elephant trumpet calls is consistent with the occurrence of shockwaves (nonlinear propagation) in the trunk (Gilbert et al. 2014). Whether similar mechanisms are involved in the production of vocal signals by other species with unusually long vocal tracts should be investigated.

8.6 Conclusion

In little over two decades, researchers have moved from a situation where most studies of mammal vocal signals focused on conspicuous, easily measured characteristics, such as their rate of occurrence, and where the spectral acoustic variation was largely ignored or poorly quantified, to a field of study in which there is a much better understanding of the nature and function of the acoustic parameters that compose vocalizations. The multilevel framework derived from the source-filter theory outlined in this chapter enables investigators to understand the acoustic components of mammal vocal signals according to their mode of production, and to predict their covariation with morphological and/or physiological attributes of callers. With sufficient knowledge of vocal anatomy, the acoustic structure of calls can thus be decomposed into production-related components, and specific hypotheses can be formulated regarding the static and dynamic information that these components encode. Importantly, the perceptual and functional relevance of these components can be tested in playback experiments using resynthesized signals that are systematically varied to mimic naturally occurring variation in the parameters of interest.

While the application of this multilevel approach has clearly enabled considerable advances, much work remains to be done, as only a fraction of the acoustic diversity of terrestrial mammal signals has been described and understood. Recent discoveries of anatomical innovations illustrate how little is still known about the diversity of the functional anatomy of mammalian vocal production. Although the dissection of dead specimens is a useful first step, it is insufficient for inferring the full range and complexity of the mechanisms of vocal production observed in live animals (see Fitch 2000b). The availability of sophisticated imaging techniques, such as cine-MRI, will undoubtedly assist in understanding how animals use their vocal apparatus, and how this affects the signals they produce. Although most of the research described here has focused on periodic oscillation and propagation, and assumes independence of source and filter, the contribution of nonlinear phenomena to the diversity of mammalian vocal repertoires remains largely underinvestigated, and should be the focus of future studies.

8.7 Summary

Understanding animal vocal communication requires investigating the information content of acoustic signals to establish their function. The source-filter theory, originally developed for the investigation of human speech production, provides a powerful and versatile multilevel framework that grounds the acoustic variability of vocalizations into their mechanism of production. Indeed, selection pressures operate at the level of signal production, rather than directly on the signals themselves, as the responses of receivers to information encoded in vocalizations ultimately affects the control and anatomy of the vocal apparatus of the animal that produces the signal. As a consequence, a better understanding of the morphology and biometry of a given species' vocal anatomy will enable researchers to make specific predictions about the information encoded in vocal signals.

Dedicated sound analyses and sound manipulation tools are also now available to test any predictions about the potential communicative function of specific acoustic features of mammal vocal signals, by presenting subjects with resynthesized (or synthetic) call stimuli in different contexts and observing and quantifying any behavioral responses. Throughout this chapter, numerous examples of how recent research within the source-filter theory framework have been presented to show that source- and filter-related acoustic components encode functionally relevant information in the vocal communication systems of terrestrial mammals. The selection pressures leading to the evolution of anatomical innovations that enable animals to produce exaggerated vocal traits have also been described. Finally potentially fruitful areas for future studies are discussed, including the need for more detailed studies of nonlinear phenomena and source-filter interactions in mammalian vocal signals.

References

Apicella, C. L., & Feinberg, D. R. (2009). Voice pitch alters mate-choice-relevant perception in hunter-gatherers. *Proceedings of the Royal Society B: Biological Sciences, 276*, 1077–1082.

Apicella, C. L., Feinberg, D. R., & Marlowe, F. W. (2007). Voice pitch predicts reproductive success in male hunter-gatherers. *Biology Letters, 3*, 682–684.

Au, W. W., & Suthers, R. A. (2014). Production of biosonar signals: Structure and form. In A. Surlykke, P. Nachtigall, R. R. Fay, & A. N. Popper (Eds.), *Biosonar* (pp. 61–105). New York: Springer.

August, P., & Anderson, J. (1987). Mammal sounds and motivation-structural rules: A test of the hypothesis. *Journal of Mammalogy, 68*, 1–9.

Banse, R., & Scherer, K. R. (1996). Acoustic profiles in vocal emotion expression. *Journal of Personality and Social Psychology, 70*, 614–636.

Beckford, N. S., Rood, S. R., Schaid, D., & Schanbacher, B. (1985). Androgen stimulation and laryngeal development. *The Annals of Ontology, Rhinology, and Laryngology, 94*, 634–640.

Blumstein, D. T., & Munos, O. (2005). Individual, age and sex-specific information is contained in yellow-bellied marmot alarm calls. *Animal Behaviour, 69*, 353–361.

Blumstein, D. T., & Recapet, C. (2009). The sound of arousal: The addition of novel non-linearities increases responsiveness in marmot alarm calls. *Ethology, 115*, 1074–1081.

Bradbury, J. (1977). Lek mating behavior in the hammer-headed bat. *Zeitschrift für Tierpsychologie, 45*, 225–255.

Briefer, E. F. (2012). Vocal expression of emotions in mammals: Mechanisms of production and evidence. *Journal of Zoology, 288*, 1–20.

Briefer, E. F., & McElligott, A. G. (2011). Indicators of age, body size and sex in goat kid calls revealed using the source-filter theory. *Applied Animal Behaviour Science, 133*, 175–185.

Bruckert, L., Liénard, J., Lacroix, A., Kreutzer, M., & Leboucher, G. (2006). Women use voice parameters to assess men's characteristics. *Proceedings of the Royal Society B: Biological Sciences, 273*, 83–89.

Cartei, V., Bond, R., & Reby, D. (2014). What makes a voice masculine: Physiological and acoustical correlates of women's ratings of men's vocal masculinity. *Hormones and Behavior, 66*, 569–576.

Carter, G. G., Skowronki, M. D., Faure, P. A., & Fenton, B. (2008). Antiphonal calling allows individual discrimination in white-winged vampire bats. *Animal Behaviour, 76*, 1343–1355.

Chan, R. W., Siegmund, T., & Zhang, K. (2009). Biomechanics of fundamental frequency regulation: Constitutive modeling of the vocal fold lamina propria. *Logopedics, Phoniatrics, Vocology, 34*, 181–189.

Charlton, B. D., Ellis, W. A. H., Brumm, J., Nilsson, K., & Fitch, W. T. (2012a). Female koalas prefer bellows in which lower formants indicate larger males. *Animal Behaviour, 84*, 1565–1571.

Charlton, B. D., Ellis, W. A. H., Larkin, R., & Fitch, W. T. (2012b). Perception of size-related formant information in male koalas (*Phascolarctos cinereus*). *Animal Cognition, 15*, 999–1006.

Charlton, B. D., Ellis, W. A. H., McKinnon, A. J., Brumm, J., Nilsson, K., & Fitch, W. T. (2011a). Perception of male caller identity in koalas (*Phascolarctos cinereus*): Acoustic analysis and playback experiments. *PLoS One, 6*, e20329.

Charlton, B. D., Ellis, W. A. H., McKinnon, A., Cowin, G. J., Brumm, J., Nilsson, K., et al. (2011b). Cues to body size in the formant spacing of male koala (*Phascolarctos cinereus*) bellows: Honesty in an exaggerated trait. *Journal of Experimental Biology, 214*, 3414–3422.

Charlton, B. D., Frey, R., McKinnon, A., Fritsch, G., Fitch, W. T., & Reby, D. (2013a). Koalas use a novel vocal organ to produce unusually low-pitched mating calls. *Current Biology, 23*, R1035–R1036.

Charlton, B. D., Keating, J. L., Kersey, D., Rengui, L., Huang, Y., & Swaisgood, R. R. (2011c). Vocal cues to male androgen levels in giant pandas. *Biology Letters, 7*, 71–74.

Charlton, B. D., Reby, D., & McComb, K. (2007a). Female perception of size-related formant shifts in red deer (*Cervus elaphus*). *Animal Behaviour, 74*, 707–714.

Charlton, B. D., Reby, D., & McComb, K. (2007b). Female red deer prefer the roars of larger males. *Biology Letters, 3*, 382–385.

Charlton, B. D., Reby, D., & McComb, K. (2008). Effect of combined source (F0) and filter (formant) variation on red deer hind responses to male roars. *Journal of the Acoustical Society of America, 123*, 2936–2943.

Charlton, B. D., Swaisgood, R. R., Zhihe, Z., & Snyder, R. J. (2012c). Giant pandas attend to androgen-related variation in male bleats. *Behavioral Ecology and Sociobiology, 66*, 969–974.

Charlton, B. D., Taylor, A. M., & Reby, D. (2013b). Are men better than women at acoustic size judgements? *Biology Letters, 9*(4), 20130270.

Charlton, B. D., Whisson, D. A., & Reby, D. (2013c). Free-ranging male koalas use size-related variation in formant frequencies to assess rival males. *PLoS One, 8*(7), e70279.

Charlton, B. D., Wyman, M., Locatelli, Y., Fitch, W. T., & Reby, D. (2014). Do red deer hinds prefer stags that produce harsh roars in mate choice contexts? *Journal of Zoology, 293*, 57–62.

Charlton, B. D., Zhang, Z. H., & Snyder, R. J. (2010). Giant pandas perceive and attend to formant frequency variation in male bleats. *Animal Behaviour, 79*, 1221–1227.

Charlton, B. D., Zhihe, Z., & Snyder, R. J. (2009a). The information content of giant panda, *Ailuropoda melanoleuca*, bleats: Acoustic cues to sex, age and size. *Animal Behaviour, 78*, 893–898.

Charlton, B. D., Zhihe, Z., & Snyder, R. J. (2009b). Vocal cues to identity and relatedness in giant pandas (*Ailuropoda melanoleuca*). *Journal of the Acoustical Society of America, 126*, 2721–2732.

Charrier, I., Mathevon, N., & Jouventin, P. (2002). How does a fur seal mother recognize the voice of her pup? An experimental study of *Arctocephalus tropicalis*. *Journal of Experimental Biology, 205*, 603–612.

Chiba, T., & Kajiyama, M. (1941). *The vowel: Its nature and structure*. Tokyo: Tokyo-Keiseikan.

Childers, D. G., & Lee, C. K. (1991). Vocal quality factors: Analysis, synthesis, and perception. *Journal of the Acoustical Society of America, 90*, 2394–2410.

Clutton-Brock, T., & Albon, S. (1979). The roaring of red deer and the evolution of honest advertising. *Behavior, 69*, 145–170.

Cohen, J. A., & Fox, M. W. (1976). Vocalizations in wild canids and possible effects of domestication. *Behavioral Proceedings, 1*, 77–92.

Dabbs, J. M., & Mallinger, A. (1999). High testosterone levels predict low voice pitch among men. *Personality and Individual Differences, 27*, 801–804.

de Boer, B., Wich, S. A., Hardus, M. E., & Lameira, A. R. (2015). Acoustic models of orangutan hand-assisted alarm calls. *Journal of Experimental Biology, 218*, 907–914.

Drahota, A., Costall, A., & Reddy, V. (2008). The vocal communication of different kinds of smile. *Speech Communication, 50*, 278–287.

Elemans, C., Zaccarelli, R., & Herzel, H. (2008). Biomechanics and control of vocalizations in a non-songbird. *Journal of the Royal Society Interface, 5*, 691–703.

Evans, C. S. (1997). Referential signals. *Perpectives in Ethology, 12*, 99–143.

Evans, S., Neave, N., Wakelin, D., & Hamilton, C. (2008). The relationship between testosterone and vocal frequencies in human males. *Physiology and Behavior, 93*, 783–788.

Fant, G. (1960). *Acoustic theory of speech production*. The Hague, The Netherlands: Mouton.

Faragó, T., Andics, A., Devecseri, V., Kis, A., Gács, M., & Miklósi, A. (2014). Humans rely on the same rules to assess emotional valence and intensity in conspecific and dog vocalizations. *Biology Letters, 10*, 20130926.

Faragó, T., Pongrácz, P., Range, F., Virányi, Z., & Miklósi, A. (2010). 'The bone is mine': Affective and referential aspects of dog growls. *Animal Behaviour, 79*, 917–925.

Feighny, J. J., Williamson, K. E., & Clarke, J. A. (2006). North American elk bugle vocalizations: Male and female bugle call structure and context. *Journal of Mammalogy, 87*, 1072–1077.

Feinberg, D. R., Jones, B. C., Little, A. C., Burt, D. M., & Perrett, D. I. (2005). Manipulations of fundamental and formant frequencies affect the attractiveness of human male voices. *Animal Behaviour, 69*, 561–568.

Fischer, J., Kitchen, D. M., Seyfarth, R. M., & Cheney, D. L. (2004). Baboon loud calls advertise male quality: Acoustic features and relation to rank, age, and exhaustion. *Behavioral Ecology and Sociobiology, 56*, 140–148.

Fitch, W. T. (1994). *Vocal tract length perception and the evolution of language*. PhD thesis, Brown University.

Fitch, W. T. (1997). Vocal tract length and formant frequency dispersion correlate with body size in rhesus macaques. *Journal of the Acoustical Society of America, 102*, 1213–1222.

Fitch, W. T. (2000a). Skull dimensions in relation to body size in non-human primates: The causal bases for acoustic allometry. *Zoology, 103*, 40–58.

Fitch, W. T. (2000b). The phonetic potential of nonhuman vocal tracts: Comparative cineradiographic observations of vocalizing animals. *Phonetica, 57*, 205–218.
Fitch, W. T. (2002). Comparative vocal production and the evolution of speech: Reinterpreting the descent of the larynx. In A. Wray (Ed.), *The transition to language* (pp. 21–45). Oxford, England: Oxford University Press.
Fitch, W. T. (2006). Production of vocalizations in mammals. In K. Brown (Ed.), *Encyclopedia of language and linguistics* (pp. 115–121). Oxford, England: Elsevier.
Fitch, W. T., & Fritz, J. (2006). Rhesus macaques spontaneously perceive formants in conspecific vocalizations. *Journal of the Acoustical Society of America, 120*, 2132–2141.
Fitch, W. T., & Giedd, J. (1999). Morphology and development of the human vocal tract: A study using magnetic resonance imaging. *Journal of the Acoustical Society of America, 106*, 1511–1522.
Fitch, W. T., Neubauer, J., & Herzel, H. (2002). Calls out of chaos: The adaptive significance of nonlinear phenomena in mammalian vocal production. *Animal Behaviour, 63*, 407–418.
Fitch, W. T., & Reby, D. (2001). The descended larynx is not uniquely human. *Proceedings of the Royal Society B: Biological Sciences, 268*, 1669–1675.
Fox, M. (1970). A comparative study of the development of facial expressions in canids: Wolf, coyote and foxes. *Behavior, 35*, 49–73.
Frey, R., & Gebler, A. (2003). The highly specialized vocal tract of the male Mongolian gazelle (*Procapra gutturosa*). *Journal of Anatomy, 203*, 451–471.
Frey, R., Volodin, I., Volodina, E., Soldatova, N. V., & Juldaschev, E. T. (2011). Descended and mobile larynx, vocal tract elongation and rutting roars in male goitred gazelles (*Gazella subgutturosa*). *Journal of Anatomy, 218*, 566–585.
Gamba, M., Colombo, C., & Giacoma, C. (2011). Acoustic cues to caller identity in lemurs: A case study. *Journal of Ethology, 30*, 191–196.
Garcia, M., Charlton, B. D., Wyman, M. T., Fitch, T. W., & Reby, D. (2013). Do red deer stags (*Cervus elaphus*) use roar fundamental frequency (f0) to assess rivals? *PLoS One, 8*(12), e83946.
Garcia, M., Wyman, M. T., Charlton, B. D., Fitch, T. W., & Reby, D. (2014). Response of red deer stags (*Cervus elaphus*) to playback of harsh versus common roars. *NaturwissenschaftenI, 101*, 851–854.
Ghazanfar, A., Turesson, H., Maier, J., van Dinther, R., Patterson, R., & Logothetis, N. (2007). Vocal tract resonances as indexical cues in the rhesus monkeys. *Current Biology, 17*, 425–430.
Gibson, J. J. (1966). *The senses considered as perceptual systems*. Boston: Houghton Mifflin.
Gilbert, J., Dalmont, J., Potier, R., & Reby, D. (2014). Is nonlinear propagation responsible for the brassiness of elephant trumpet calls? *Acta Acustica united with Acustica, 100*, 734–738.
Griesbach, R. (1999). Estimation of speaker height from formant frequencies. *International Journal of Speech, Language, and the Law, 6*, 265–277.
Griffin, D. R. (1958). *Listening in the dark*. New Haven, CT: Yale University Press.
Hardus, M. E., Lameira, A. R., van Schaik, C. P., & Wich, S. A. (2009). Tool use in wild orang-utan modified sound production: A functionally deceptive innovation? *Proceedings of the Royal Society of London B: Biological Sciences, 276*, 3689–3694.
Harries, M. L. L., Walker, J. M., Williams, D. M., Hawkins, S., & Hughes, I. A. (1997). Changes in the male voice at puberty. *Archives of Disease in Childhood, 7*, 445–447.
Harris, T. R., Fitch, W. T., Goldstein, L. M., & Fashing, P. J. (2006). Black and white colobus monkey (*Colobus guereza*) roars as a source of both honest and exaggerated information about body mass. *Ethology, 112*, 911–920.
Herbst, C. T. (2014). Glottal efficiency of periodic and irregular in vitro red deer voice production. *Acta Acustica united with Acustica, 100*, 724–733.
Hirano, M., Ohala, J., & Vennard, W. (1969). The function of laryngeal muscles in regulating fundamental frequency and intensity of phonation. *Journal of Speech and Hearing Research, 12*, 616–628.
Hodges-Simeon, C. R., Gaulin, S. J. C., & Puts, D. A. (2010). Different vocal parameters predict perceptions of dominance and attractiveness. *Human Nature, 21*, 406–427.

Hollien, H. (1960). Vocal pitch variation related to changes in vocal fold length. *Journal of Speech and Hearing Research, 3*, 150–156.

Huber, J. E., Stathopoulos, E. T., Curione, G. M., Ash, T. A., & Johnson, K. (1999). Formants of children, women, and men: The effects of vocal intensity variation. *Journal of the Acoustical Society of America, 106*, 1532–1542.

Janik, V. M., Sayigh, L., & Wells, R. (2006). Signature whistle shape conveys identity information to bottlenose dolphins. *Proceedings of the National Academy of Sciences of the USA, 103*, 8293–8297.

Kelemen, G., & Sade, J. (1960). The vocal organ of the howling monkey (*Alouatta palliata*). *Journal of Morphology, 107*, 123–140.

Kewley-Port, D., Li, X., Zheng, Y., & Neel, A. (1996). Fundamental frequency effects on thresholds for vowel formant discrimination. *Journal of the Acoustical Society of America, 100*, 2462–2470.

Kidjo, N., Cargnelutti, B., Charlton, B. D., Wilson, C., & Reby, D. (2008). Vocal behavior in the endangered Corsican deer, description and phylogenetic implications. *Bioacoustics, 18*, 159–181.

Koda, H., Nishimura, T., Tokuda, I. T., Oyakawa, C., Nihonmatsu, T., & Masatka, N. (2012). Soprano singing in gibbons. *American Journal of Physical Anthropology, 149*, 347–355.

Kuhl, P. K., & Miller, J. D. (1975). Speech-perception by chinchilla—voiced voiceless distinction in alveolar plosive consonants. *Science, 190*, 69–72.

Ladefoged, P. (2001). *Vowels and consonants: An introduction to the sounds of languages*. Oxford, England: Blackwell.

Lieberman, P. (1984). *The biology and evolution of language*. Cambridge, MA: Harvard University Press.

Lieberman, P., & Blumstein, S. E. (1988). *Speech physiology, speech perception, and acoustic phonetics*. Cambridge, England: Cambridge University Press.

Lieberman, P., Crelin, E. S., & Klatt, D. H. (1972). Phonetic ability and related anatomy of the newborn and adult human, Neanderthal man, and the chimpanzee. *American Anthropologist, 74*, 287–307.

Manser, M. (2001). The acoustic structure of suricates' alarm calls varies with predator type and the level of response urgency. *Proceedings of the Royal Society of London B: Biological Sciences, 268*, 2315–2324.

Mathevon, N., Koralek, A., Weldele, M., & Teunissen, F. E. (2010). What the hyena's laugh tells: Sex, age, dominance and individual signature in the giggling call of *Crocuta crocuta*. *BMC Ecology, 10*, 9. doi:10.1186/1472-6785-10-9.

Maynard-Smith, J., & Harper, D. (2003). *Animal signals*. Oxford, England: Oxford University Press.

McAdams, S., & Rodet, X. (1988). The role of FM-induced AM in dynamic spectral profile analysis. In H. Duifhuis, J. W. Jorst, & H. P. Witt (Eds.), *Basic issues in hearing* (pp. 359–369). London: Academic Press.

McComb, K., Reby, D., Baker, L., Moss, C., & Sayialel, S. (2003). Long-distance communication of acoustic cues to social identity in African elephants. *Animal Behaviour, 66*, 317–329.

McElligott, A. G., Birrer, M., & Vannoni, E. (2006). Retraction of the mobile descended larynx during groaning enables fallow deer bucks (*Dama dama*) to lower their formant frequencies. *Journal of Zoology, 270*, 340–345.

McElligott, A. G., & Hayden, T. (1999). Context-related vocalization rates of fallow bucks, *Dama dama. Animal Behaviour, 58*, 1095–1104.

McElligott, A. G., & Hayden, T. (2001). Postcopulatory vocalizations of fallow bucks: Who is listening? *Behavioral Ecology, 12*, 41–46.

Melendez, K. V., & Feng, A. S. (2010). Communication calls of little brown bats display individual-specific characteristics. *Journal of the Acoustical Society of America, 128*, 919–923.

Mergell, P., Fitch, W. T., & Herzel, H. (1999). Modeling the role of non-human vocal membranes in phonation. *Journal of the Acoustical Society of America, 105*, 2020–2028.

Morton, E. S. (1977). On the occurrence and significance of motivation-structural rules in some birds and mammal sounds. *American Naturalist, 111*, 855–869.

Nacci, A., Fattori, B., Basolo, F., Filice, M. E., De Jeso, K., Giovannini, L., et al. (2011). Sex hormone receptors in vocal fold tissue: A theory about the influence of sex hormones in the larynx. *Folia Phoniatrica et Logopaedica, 63*, 77–82.

Novick, A., & Griffin, D. R. (1961). Laryngeal mechanisms in bats for the production of orientation sounds. *Journal of Experimental Zoology, 148*, 125–145.

Ohala, J. J. (1984). An ethological perspective on common cross-language utilization of F0 of voice. *Phonetica, 41*, 1–16.

Ohala, J. J. (1994). The frequency codes underlies the sound symbolic use of voice pitch. In L. Hinton, J. Nichols, & J. J. Ohala (Eds.), *Sound symbolism* (pp. 325–347). Cambridge, England: Cambridge University Press.

Ohala, J. J. (1996). Ethological theory and the expression of emotion in the voice. In *Proceedings of ICSLP 96*, October 3–6, 1996. [*4th International Conference on Spoken Language Processing*, Philadelphia] 3, 1812–1815. Wilmington: University of Delaware.

Ohms, V. R., Snelderwaard, P. C., ten Cate, C., & Beckers, G. J. L. (2010). Vocal tract articulation in zebra finches. *PLoS One, 5*(7), e11923.

Owings, D., & Morton, E. (1998). *Animal vocal communication: A new approach*. Cambridge, England: Cambridge University Press.

Owren, M. J., & Bernacki, R. H. (1998). Applying linear predictive coding (LPC) to frequency-spectrum analysis of animal acoustic signals. In S. L. Hopp, M. J. Owren, & C. S. Evans (Eds.), *Animal acoustic communication: Sound analysis and research methods* (pp. 130–162). New York: Springer.

Owren, M. J., & Rendall, D. (2003). Salience of caller identity in rhesus monkey (*Macaca mulatta*) coos and screams: Perceptual experiments with human (*Homo sapiens*) listeners. *Journal of Comparative Psychology, 117*, 380–390.

Owren, M. J., Seyfarth, R. M., & Cheney, D. L. (1997). The acoustic features of vowel-like grunt calls in chacma baboons (*Papio cyncephalus ursinus*): Implications for production processes and functions. *Journal of the Acoustical Society of America, 101*, 2951–2963.

Palacios, V., Font, E., & Márquez, R. (2007). Iberian wolf howls: Acoustic structure, individual variation, and a comparison with North American populations. *Journal of Mammalogy, 88*, 606–613.

Pfefferle, D., & Fischer, J. (2006). Sounds and size: Identification of acoustic variables that reflect body size in hamadryas baboons, Papio hamadryas. *Animal Behaviour, 72*, 43–51.

Pisanski, K., Fraccaro, P. J., Tigue, C. C., O'Connor, J. M. M., Röder, S., Andrews, P. W., et al. (2014). Vocal indicators of body size in men and women: A meta-analysis. *Animal Behaviour, 95*, 89–99.

Pisanski, K., & Rendall, D. (2011). The prioritization of voice fundamental frequency or formants in listeners' assessments of speaker size, masculinity, and attractiveness. *Journal of the Acoustical Society of America, 129*, 2201–2212.

Pitcher, B. J., Briefer, E. F., Vannoni, E., & McElligott, A. G. (2014). Fallow bucks attend to vocal cues of motivation and fatigue. *Behavioral Ecology, 25*, 392–401.

Pitcher, B. J., Harcourt, R. G., & Charrier, I. (2012). Individual identity encoding and environmental constraints in vocal recognition of pups by Australian sea lion mothers. *Animal Behaviour, 83*, 681–690.

Plotsky, K., Rendall, D., Riede, T., & Chase, K. (2013). Radiographic analysis of vocal tract length and its relation to overall body size in two canid species. *Journal of Zoology, 291*(1). doi:10.1111/jzo.12048.

Pongrácz, P., Molnár, C., Miklósi, A., & Csányi, V. (2005). Human listeners are able to classify dog (*Canis familiaris*) barks recorded in different situations. *Journal of Comparative Psychology, 119*, 136–144.

Puts, D., Hodges, C., Cardenas, R., & Gaulin, S. (2007). Men's voices as dominance signals: Vocal fundamental and formant frequencies influence dominance attributions among men. *Evolution and Human Behavior, 28*, 340–344.

Puts, D. A., Jones, B. C., & DeBruine, L. M. (2012). Sexual selection on human faces and voices. *Journal of Sex Research, 9*, 227–243.

Read, C., Buder, E. H., & Kent, R. D. (1992). Speech analysis systems; An evaluation. *Journal of Speech and Hearing Research, 35*, 314–332.

Reby, D. & Charlton, B. D. (2012). Attention grabbing in red deer sexual calls. *Animal Cognition, 15*, 265–270.

Reby, D., Charlton, B. D., Locatelli, Y., & McComb, K. (2010). Oestrous red deer hinds prefer male roars with higher fundamental frequencies. *Proceedings of the Royal Society of London B: Biological Sciences, 277*, 2747–2753.

Reby, D., & McComb, K. (2003). Anatomical constraints generate honesty: Acoustic cues to age and weight in the roars of red deer stags. *Animal Behaviour, 65*, 519–530.

Reby, D., McComb, K., Cargnelutti, B., Darwin, C., Fitch, W. T., & Clutton-Brock, T. H. (2005). Red deer stags use formants as assessment cues during intrasexual agonistic interactions. *Proceedings of the Royal Society of London B: Biological Sciences, 272*, 941–947.

Rendall, D., Owren, M. J., & Rodman, P. S. (1998). The role of vocal tract filtering in identity cueing in rhesus monkey (*Macaca mulatta*) vocalizations. *Journal of the Acoustical Society of America, 103*, 602–614.

Rendall, D., Owren, M. J., Weerts, E., & Hienz, R. D. (2004). Sex differences in the acoustic structure of vowel-like grunt vocalizations in baboons and their perceptual discrimination in baboons listeners. *Journal of the Acoustical Society of America, 115*, 411–421.

Rendall, D., Seyfarth, R., Cheney, D., & Owren, M. (1999). The meaning and function of grunt variants in baboons. *Animal Behaviour, 57*, 583–592.

Rendall, D., Vokey, J., & Nemeth, C. (2007). Lifting the curtain on the Wizard of Oz: biased voice-based impressions of speaker size. *Journal of Experimental Psychology. Human Perception and Performance, 33*, 1208–1219.

Riede, T., Arcadi, A. C., & Owren, M. J. (2007). Nonlinear acoustics in the pant hoots of common chimpanzees (*Pan troglodytes*): Vocalizing at the edge. *Journal of the Acoustical Society of America, 121*, 1758–1767.

Riede, T., Beckers, G. J. L., Blevins, W., & Suthers, R. A. (2004a). Inflation of the esophagus and vocal tract filtering in ring doves. *Journal of Experimental Biology, 207*, 4025–4036.

Riede, T., Bronson, E., Hatzikirou, H., & Zuberbühler, K. (2005a). Vocal production mechanisms in a non-human primate: Morphological data and a model. *Journal of Human Evolution, 48*, 85–96.

Riede, T., & Fitch, W. T. (1999). Vocal tract length and acoustics of vocalization in the domestic dog *Canis familiaris*. *Journal of Experimental Biology, 202*, 2859–2867.

Riede, T., Mitchell, B. R., Tokuda, I., & Owren, M. J. (2005b). Characterizing noise in non-human vocalizations: Acoustic analysis and human perception of barks by coyotes and dogs. *Journal of the Acoustical Society of America, 118*, 514–522.

Riede, T., Owren, M. J., & Arcadi, A. C. (2004b). Nonlinear acoustics in pant hoots of common chimpanzees (*Pan troglodytes*): Frequency jumps, subharmonics, biphonation, and deterministic chaos. *American Journal of Primatology, 64*, 277–291.

Riede, T., Suthers, R. A., Fletcher, N. H., & Blevins, W. E. (2006). Songbirds tune their vocal tract to the fundamental frequency of their song. *Proceedings of the National Academy of Sciences of the USA, 103*, 5543–5548.

Riede, T., & Zuberbühler, K. (2003a). Pulse register phonation in Diana monkey alarm calls. *Journal of the Acoustical Society of America, 113*, 2919–2926.

Riede, T., & Zuberbühler, K. (2003b). The relationship between acoustic structure and semantic information in Diana monkey alarm vocalization. *Journal of the Acoustical Society of America, 114*, 1132–1142.

Ryalls, J. H., & Lieberman, P. (1982). Fundamental-frequency and vowel perception. *Journal of the Acoustical Society of America, 72*, 1631–1634.

Sanvito, S., & Galimberti, F. (2003). Source level of male vocalizations in the genus *Mirounga*: Repeatability and correlates. *Bioacoustics, 14*, 47–59.

Sanvito, S., Galimberti, F., & Miller, E. H. (2007). Vocal signaling in male southern elephant seals is honest but imprecise. *Animal Behaviour, 73*, 287–299.

Scherer, K. R. (1986). Vocal affect expression: A review and a model for future research. *Psychological Bulletin, 99*, 143–165.

Schmidt-Nielsen, K. (1975). Scaling in biology: The consequences of size. *Journal of Experimental Zoology, 194*, 287–307.

Schneider, B., Cohen, E., Stani, J., Kolbus, A., Rudas, M., Horvat, R., et al. (2007). Towards the expression of sex hormone receptors in the human vocal fold. *Journal of Voice, 21*, 502–507.

Schön-Ybarra, M. (1995). A comparative approach to the non-human primate vocal tract: Implications for sound production. In E. Zimmermann & J. D. Newman (Eds.), *Current topics in primate vocal communication* (pp. 185–198). New York: Plenum Press.

Sébe, F., Duboscq, J., Ligout, S., Aubin, T., & Poindron, P. (2010). Early vocal recognition of mother by lambs: Contribution of low- and high-frequency vocalizations. *Animal Behaviour, 79*, 1055–1066.

Seyfarth, R. M., & Cheney, L. D. (1986). Vocal development in vervet monkeys. *Animal Behaviour, 34*, 1640–1658.

Simmons, L. W., Peters, M., & Rhodes, G. (2011). Low pitched voices are perceived as masculine and attractive but do they predict semen quality in men? *PLoS One, 6*(12), e29271.

Smith, D. R. R., Patterson, R. D., & Turner, R. (2005). The processing and perception of size information in speech sounds. *Journal of the Acoustical Society of America, 117*, 305–318.

Stoeger, A. S., Heilmann, G., Zeppelzauer, M., Ganswindt, A., Hensman, S., & Charlton, B. D. (2012). Visualizing sound emission of elephant vocalizations: Evidence for two rumble production types. *PLoS One, 7*(11), e48907.

Sundberg, J. (1975). Formant technique in a professional female singer. *Acoustica, 32*, 89–96.

Taylor, A. M., & Reby, D. (2010). The contribution of source-filter theory to mammal vocal communication research. *Journal of Zoology, 280*, 221–236.

Taylor, A. M., Reby, D., & McComb, K. (2008). Human listeners attend to size information in domestic dog growls. *Journal of the Acoustical Society of America, 123*, 2903–2909.

Taylor, A. M., Reby, D., & McComb, K. (2009). Context-related variation in the vocal growling behavior of domestic dogs, *Canis familiaris. Ethology, 115*, 905–915.

Taylor, A. M., Reby, D., & McComb, K. (2010). Size communication in domestic dog (*Canis familiaris*) growls. *Animal Behaviour, 79*, 205–210.

Taylor, A. M., Reby, D., & McComb, K. (2011). Cross modal perception of body size in domestic dogs (*Canis familiaris*). *PLoS One, 6*(2), e17069.

Titze, I. R. (1994). *Principles of voice production*. Englewood Cliffs, NJ: Prentice Hall.

Titze, I. R. (2008). Nonlinear source-filter coupling in phonation: Theory. *The Journal of the Acoustical Society of America, 123*, 2733–2749.

Titze, I., Fitch, W., Hunter, E., Alipour, F., Montequin, D., Armstrong, D., et al. (2010). Vocal power and pressure-flow relationships in excised tiger larynges. *Journal of Experimental Biology, 213*, 3866–3873.

Tokuda, I., Riede, T., Neubauer, J., Owren, M., & Herzel, H. (2002). Nonlinear analysis of irregular animal vocalizations. *Journal of the Acoustical Society of America, 111*, 2908–2919.

Townsend, S. W., Charlton, B. D., & Manser, M. B. (2014). Acoustic cues to identity and predator context in meerkat barks. *Animal Behaviour, 94*, 143–149.

Townsend, S. W., & Manser, M. B. (2011). The function of nonlinear phenomena in meerkat alarm calls. *Biology Letters, 7*, 47–49.

Vannoni, E., & McElligott, A. G. (2007). Individual acoustic variation in fallow deer (*Dama dama*) common and harsh groans: A source-filter theory perspective. *Ethology, 113*, 223–234.

Vannoni, E., & McElligott, A. G. (2008). Low frequency groans indicate larger and more dominant fallow deer (*Dama dama*) males. *PLoS One, 3*, e3113.

Vannoni, E., Torriani, M., & McElligott, A. G. (2005). Acoustic signalling in cervids: A methodological approach for measuring vocal communication in fallow deer. *Cognition, Brain and Behaviour, IX*, 551–565.

Volodin I. A., Volodina E. V., & Efremova K. O. (2009). Antelope, calling through the nose: structure of sounds and effect of sexual selection on the vocal behavior of the saiga (*Saiga tatarica*). *Zoologicheskii Zhurnal, 88*, 113–124.

Weissengruber, G. E., Forstenpointner, G., Peters, G., Kübber-Heiss, A., & Fitch, W. T. (2002). Hyoid apparatus and pharynx in the lion (*Panthera leo*), jaguar (*Panthera onca*), tiger (*Panthera tigris*), cheetah (*Acinonyx jubatus*) and domestic cat (*Felis silvestris f. catus*). *Journal of Anatomy, 201*, 195–209.

Wilden, I., Herzel, H., Peters, G., & Tembrock, G. (1998). Subharmonics, biphonation, and deterministic chaos in mammal vocalization. *Bioacoustics, 9*, 171–196.

Wilson, E. O. (1972). Animal communication. *Scientific American, 227*, 53–60.

Wolff, S. E., & Puts, D. A. (2010). Vocal masculinity is a robust dominance signal in men. *Behavioral Ecology and Sociobiology, 64*, 1673–1683.

Woods, R. H. (1893). Law of transverse vibrations of strings applied to the human larynx. *Transactions of the Royal Academy of Medicine in Ireland, 11*, 482–488.

Wyman, M. T., Mooring, M. S., McCowan, B., Penedo, M. C. T., & Hart, L. A. (2008). Amplitude of bison bellows reflects male quality, physical condition and motivation. *Animal Behaviour, 76*, 1625–1639.

Yin, S. (2002). A new perspective on barking in dogs (*Canis familiaris*). *Journal of Comparative Psychology, 116*, 189–193.

Yin, S., & McCowan, B. (2004). Barking in domestic dogs: Context specificity and individual identification. *Animal Behaviour, 68*, 343–355.

Chapter 9
Vocal Learning and Auditory-Vocal Feedback

Peter L. Tyack

Abstract Vocal learning is usually studied in songbirds and humans, species that can form auditory templates by listening to acoustic models and then learn to vocalize to match the template. Most other species are thought to develop vocalizations without auditory feedback. However, auditory input influences the acoustic structure of vocalizations in a broad distribution of birds and mammals. Vocalizations are defined here as sounds generated by forcing air past vibrating membranes. A vocal motor program may generate vocalizations such as crying or laughter, but auditory feedback may be required for matching precise acoustic features of vocalizations. This chapter discriminates limited vocal learning, which uses auditory input to fine-tune acoustic features of an inherited auditory template, from complex vocal learning, in which novel sounds are learned by matching a learned auditory template. Two or three songbird taxa and four or five mammalian taxa are known for complex vocal learning. A broader range of mammals converge in the acoustic structure of vocalizations when in socially interacting groups, which qualifies as limited vocal learning. All birds and mammals tested use auditory-vocal feedback to adjust their vocalizations to compensate for the effects of noise, and many species modulate their signals as the costs and benefits of communicating vary. This chapter asks whether some auditory-vocal feedback may have provided neural substrates for the evolution of vocal learning. Progress will require more precise definitions of different forms of vocal learning, broad comparative review of their presence and absence, and behavioral and neurobiological investigations into the mechanisms underlying the skills.

Keywords Auditory-vocal feedback • Compensation for noise • Lombard effect • Vocal convergence • Vocal imitation • Vocal learning • Vocal mimicry • Vocal plasticity

---

P.L. Tyack (✉)
Sea Mammal Research Unit, Scottish Oceans Institute, School of Biology, University of St. Andrews, St Andrews KY16 8LB, Scotland, UK
e-mail: plt@st-andrews.ac.uk

R.A. Suthers et al. (eds.), *Vertebrate Sound Production and Acoustic Communication*, Springer Handbook of Auditory Research 53,
DOI 10.1007/978-3-319-27721-9_9

9.1 Introduction: Definitions of Vocal Learning

Vocal learning is usually defined as the ability of an animal to modify the acoustic structure of sounds it produces based on auditory input. Many vertebrate species develop normal vocalizations even if they do not hear those of conspecifics, and are thought to inherit a motor program that generates the acoustic structure of each call. By contrast, some species such as humans and songbirds learn some signals from individuals with whom they interact, and these signals form a shared learned communication system. Janik and Slater (1997, p 59) define vocal learning as a modification of specific acoustic features of one's vocalizations, "as a result of experience with those of other individuals." This definition emphasizes learning to incorporate the calls of others into one's vocal repertoire. Compared to a more genetically constrained system, vocal learning can create a much more complex, open, and flexible communication system.

The classic animal model of vocal learning stems from work on oscine songbirds (Nottebohm 1970). The template model of song learning separates the process of forming an auditory memory or template through listening to a song (auditory learning) from the process of learning to produce that song by matching one's own vocal motor output to the stored template (sensorimotor learning) (Konishi 1965, 2004). For example, if a young male white-crowned sparrow (*Zonotrichia capensis*) is played recordings of the songs of an adult male conspecific at the right time, he will form memories or auditory templates of the songs he heard. As he matures and starts producing song-like sounds, he can slowly learn to match his own vocal motor output to the auditory templates, a pattern of development that Owren et al. (2011) call "reception-first" because of the need to hear the vocalization before learning to produce it. The requirement for vocal learning can be demonstrated by eliminating auditory input; if such a bird is deafened before vocal development, he will never produce normal song (Konishi 1965). A sparrow that retains normal hearing but that never has the opportunity to hear a model sound may develop song that is more normal, but that will not match wild-type songs of his population. In this case, the animal must learn to match his own vocal motor output against an auditory template that is inherited. This is often called an innate auditory template. Use of the term "innate" in this case means that it develops reliably in a species by inheritance. By contrast, many species of bird (Schleidt 1961; Konishi 1963; Nottebohm and Nottebohm 1971; Kroodsma and Konishi 1991) and mammal (Winter et al. 1973) can develop normal vocalizations without any auditory feedback. Vocalization in these species is thought to be structured by central pattern generators in the brain that control vocal motor output without requiring auditory input, a pattern of development that Owren et al. (2011) call "production-first." These species are thus classified as not having vocal learning. Note that species with vocal learning may also be able to produce some signals without auditory input. Laughter and crying in humans, for example, develop normally in hearing-impaired infants (Scheiner et al. 2006). Thus species with vocal learning may develop some vocalizations without reference to auditory input. This means that when one limits auditory input to test for its effects on vocal production, one must test all types of vocalization to fully search for vocal learning in a species.

We humans appreciate the flexibility of our language, but the evolutionary origins of human vocal learning are obscure, as no nonhuman primates show strong evidence for vocal learning. At least three taxa of birds are skilled at vocal learning, and among mammals we have evidence that bats, dolphins, elephants, humans, and seals are as well (Fitch and Jarvis 2013). Research on nonhuman primates and other mammals and birds has shown some minor modification of calls based on auditory input; these changes are similar in some ways to classic vocal learning and different in other ways. There are some additional ways in which auditory input affects vocalizations, such as changes by most birds and mammals to compensate for noise or interference (Brumm and Zollinger 2011), and specialized mechanisms such as Doppler compensation in some bats (Metzner et al. 2002). The lack of clear definitions of different forms of vocal learning has led to confusion and disagreement about which species show vocal learning and which do not. This scientific uncertainty interferes with understanding the evolution of vocal learning.

This chapter discusses a broad range of phenomena that have led to connections between auditory input and vocal motor output. Nottebohm and Liu (2010, p. 3) define vocal learning as modification of one's vocalizations "by reference to auditory feedback," a definition that would include any of these phenomena. In this chapter, this broadest category is called "auditory-vocal feedback." However, as discussed in the preceding text, other authors limit the term vocal learning to the development of calls that match those of other individuals with whom the learner has interacted. Following a distinction highlighted by Arriaga and Jarvis (2013) and using a terminology suggested by Fitch and Jarvis (2013), this chapter distinguishes between complex vocal learning, such as the classic cases of humans and songbirds, in which individuals can learn to produce new vocalizations by matching an auditory template formed from new sounds that they hear, from a more limited form of vocal learning, in which individuals use auditory input to modify the acoustic features of production-first vocalizations.

Janik and Slater (1997) also distinguish vocal production learning, which involves changing acoustic features of vocalizations, from learning a new use for a preexisting vocalization, or learning to comprehend the context in which a vocalization is produced. These distinctions are useful to discriminate vocal production learning from vocal usage or comprehension learning. However, this book focuses on sound production and this chapter only discusses vocal production learning, so here it is called by the more common name, vocal learning. Readers with any questions about this distinction should read Janik and Slater (1997).

9.2 Taxonomic Scope for Review of Vocal Sound Production

The word "vocal" derives from the human voice, but this chapter is a comparative review, so my definition of "vocal sounds" includes those from animals that produce sound by forcing air past membranes whose tension can be controlled. Frogs produce sound by moving air past vocal cords in the trachea into a vocal sac (Gans 1973), a mode of production that fits this definition of "vocal." The sound production organs of birds and mammals use air under pressure from the respiratory system to vibrate soft membranes in the airways. The sound production organ in birds is called the syrinx; that in toothed whales, the phonic lips (Cranford et al. 1996; Madsen et al. 2011); and that in most other mammals, the larynx. The membrane tension in all of these sound production organs is controlled by muscles that are typically well innervated, and some avian and mammalian species have articulatory control of acoustic filters that lie above the sound production organ itself. Control of pneumatic sound production in these taxa requires sophisticated simultaneous coordination of pressure in the lungs, tension in the vocal folds, and configuration of the acoustic filters. Neural control of the sound source and filters is often highly developed in many of these taxa (Fitch and Suthers, Chap. 1; Taylor et al., Chap. 8).

Mammals or birds with a production-first vocal development have been reported to develop species-typical vocalizations with no auditory feedback. However, in testing species whose vocal repertoires are less well known, researchers must also be careful not to miss a specific category of vocalization that may be learned. For example, if investigators just studied development of calls in oscine songbirds deprived of auditory feedback and missed testing song, they might fail to uncover evidence for vocal learning. In addition, evidence that production-first species never use auditory-vocal feedback to fine-tune their vocal production against an innate auditory template is not iron clad. Early studies that minimized the effect of deafening on vocal development are starting to be updated by studies that use more sophisticated acoustic analyses to show subtle differences. For example, Romand and Ehret (2004) demonstrated that although deafened kittens do develop species-specific mews, the mews of deafened or isolated kittens differed acoustically from those of normal kittens, suggesting the role of social experience and auditory feedback in fine-tuning mews. Thus auditory-vocal feedback may function to stabilize the acoustic structure of species-specific vocalizations in settings where this is important.

Janik and Slater (1997) differentiate vocal learning involving changes in the duration or level of a call from those that involve changes in frequency. They argue that simple control of exhalation can control duration and level, but that matching frequency requires more complex control. This chapter also argues for different levels of complexity of vocal learning, and agrees that matching of acoustic features involving coordination among respiratory, laryngeal, and articulatory muscles, like those leading to differences in frequency matching, clearly involve a more complex form of vocal learning than those that simply involve control of respiratory muscles.

However, a simple differentiation between duration/level parameters and frequency parameters may not suffice to distinguish simple versus complex forms of vocal learning. The intensity of vocal sounds is driven by increased pneumatic pressure from the lungs, and in the absence of counter-adjustments, increased pressure also leads to increased frequency in some species. The larynx in mammals and syrinx in birds can transfer more acoustic power at higher frequencies (Titze 1994). These biophysical relationships lead to a correlation between the source level and frequency of sounds in a diverse set of species (Au et al. 1995; Nemeth et al. 2013). Therefore, any evaluation of complexity of matching needs to account for these linkages between acoustic features of the sound production system.

As mentioned in Sect. 9.1, the primary distinction made in this chapter regarding complexity of vocal learning contrasts limited versus complex vocal learning. In the former, auditory feedback is used to modify existing features of a production-first call. In the latter, an auditory template of a sound is first learned from hearing a model (reception-first), and the animal then learns to match its vocal motor output to match the template. Either form may involve matching frequency and timing and amplitude to varying degrees of precision and complexity. Voice onset time provides an example from human speech of timing cues that may require complex vocal learning for a precise match. Other species such as sperm whales (*Physeter macrocephalus*) appear to learn the timing of brief transient vocalizations to create a complex repertoire of calls (Rendell and Whitehead 2005; Rendell et al. 2012). This chapter does not prejudge which acoustic features provide the complexity of vocal learning in different species, but rather distinguishes learning to modify existing calls from the ability to learn to develop new calls, which enables a more open-ended communication system.

9.3 Vocal Mimicry: Copying Sounds of Other Species

Some of the best evidence for complex vocal learning comes from animals under human care that imitated the sounds of humans or non-conspecifics. When an animal is introduced to an environment in which it is exposed to new sounds that are not part of its normal vocal repertoire, and when it makes precise imitations of these novel sounds, there are few alternative explanations than the animal has learned to produce the sounds it has heard. This process of learning through audition to create a new vocal motor pattern is clear evidence for complex vocal learning. It has been known for centuries that some songbird (Klatt and Stefanski 1974) and parrot (Pepperberg 2010) species kept by humans can copy the sounds of humans very precisely. Among mammals, there are cases of a harbor seal (*Phoca vitulina*; Ralls et al. 1985) and an Indian elephant (*Elephas maximus indicus*; Stoeger et al. 2012) imitating speech. African elephants (*Loxodonta africana*) have also been reported to imitate the sounds of a truck (Poole et al. 2005). Bottlenose dolphins (*Tursiops truncatus*) have also proven skilled at imitating synthetic computer-generated frequency modulation patterns (Richards et al. 1984). By contrast, intensive attempts to train nonhuman primates to imitate speech have failed to provide strong evidence for vocal imitation (Kellogg and Kellogg 1933; Hayes 1951; Hayes and Hayes 1952).

The ability of some nonhuman species to imitate speech is remarkable. Not only do these animals have to form auditory templates very different from those of their own species, but they also must adapt their sound production organs to produce sounds very different from those the organ usually produces. It has been argued that the relatively minor differences between the vocal tracts of anthropoid apes vs. humans may prevent apes from producing speech (Lieberman 1984), but birds such as mynah birds are able to adapt syrinx and beak to reproduce complex features of speech such as fundamental frequency contours, formants, and consonants such as fricatives and plosive bursts (Klatt and Stefanski 1974). Similarly, the male Asian elephant that imitated speech was able to match the first two format frequencies of his trainers quite precisely, even though the large size of his vocal tract normally generates much lower formant frequencies. He accomplished this match by inserting the tip of his trunk into his mouth just before imitating speech. This method of changing the vocal tract to produce unusual formants has not been reported for other elephants, nor does this elephant perform this action when making normal elephant vocalizations. This kind of creative manipulation of the vocal tract to change the frequency spectrum of a call is rare. Orangutans have been reported to hold leaves to their mouth, extending their vocal tract, to lower the frequency of a call (Hardus et al. 2009). There is scant evidence for nonhuman primates producing new vocalizations to match an auditory model, but there is some evidence involving nonvocal sounds produced by airflow past the lips. Orangutans have been reported to imitate human whistling spontaneously (Wich et al. 2009). An Asian elephant showed similar vocal creativity, putting her trunk against her mouth to produce whistles, a technique that was reported to have been learned from the elephant that originated the technique (Wemmer and Mishra 1982).

9.4 Evolution of Vocal Mimicry: Learning to Copy Sounds of Other Species

As discussed in Sect. 9.3, perhaps the best evidence for complex vocal learning comes from situations in which animals mimic human speech or artificial sounds synthesized by humans to incorporate acoustic features that differ from the subject's pre-exposure repertoire. This behavior of copying a sound that is not produced by a conspecific is called vocal mimicry (Baylis 1982). In a review of vocal mimicry, Kelley and Healy (2011) point out that about 20 % of songbird species have been reported to mimic sounds from non-conspecifics.

There is a certain irony that we humans, who pride ourselves on our abilities for vocal learning, require other species to match our speech for us to recognize their abilities for complex vocal learning. Vocal mimicry seems to reflect an unusual lack of constraint on vocal learning, and it seems likely that many species that have evolved skills for learning conspecific sounds might fail to imitate sounds of other species. On the other hand, social interaction appears to foster more open vocal development, for example, leading a white-crowned sparrow housed with singing males of the same and different species to be more likely to copy heterospecific song than if it just heard recorded songs (Baptista and Morton 1981). Kelley and Healy (2011) note the prevalence of mimicry among animals held in captivity, especially those from long-lived species with strong social bonds among fluid groups, and they suggest that animals that rely on copying conspecifics to maintain social relationships may broaden this pattern, copying human caregivers to maintain heterospecific social relationships in captivity.

The next section examines the specific settings in which some of the best evidence for mimicry by mammals and birds of human signals was observed. These include cases involving immediate mimicry of signals and cases suggestive of a temporal separation between formation of an auditory template of a sound and learning how to produce a sound that matches the template.

9.5 Separation Between Acquisition of an Auditory Template and Learning Through Auditory-Vocal Feedback: How to Produce a Sound that Matches the Template

There is evidence in two cases of mammalian mimicry of human speech for a separation between forming the auditory template and the vocal practice of matching motor output to the auditory template. The harbor seal that imitated speech was found as a newborn pup in May 1971 and was raised in the home of a local Maine resident, who named him Hoover, until August 1971, when he was transferred to the New England Aquarium and placed in a pool with other harbor seals (Hiss 1983). In 1976, he was first reported showing sexual behavior, and was first noted to make sounds "as if talking" (Ralls et al. 1985). This seal was first reported to produce a word from human speech in 1978, when an observer "wrote in the files 'he says "Hoover" in plain English. I have witnesses'" (Ralls et al. 1985, p. 1051). Hoover subsequently increased his repertoire of speech sounds, for which listeners could recognize a New England accent. The male Asian elephant that imitates human speech produces words in Korean that native speakers can classify accurately. He was raised for a long period in a setting with no other elephants where humans provided his only social interaction. It appears that he started to produce speech sounds at about 14 years of age, near his onset of sexual maturity (Stoeger et al. 2012). As in the classic songbird model, Hoover's auditory template appears to have formed early in life, and for both the seal and the elephant, the vocal motor learning phase did not take place until sexual maturity. Steroid hormones influence vocal learning circuits in songbirds (Brenowitz and Kroodsma 1996); perhaps the timing of vocal learning in these other species suggests a role for similar hormonal control.

9.6 Adult Animals that Rapidly Imitate Novel Sounds

This section discusses evidence from non-oscine bird and mammal species of capabilities for imitating new sounds as adults with a short interval between hearing a new sound and imitating it. Todt (1975) described a new method to train African grey parrots (*Psittacus erithacus*) to imitate human speech sounds. Adult parrots taught with this method learned new speech sounds, but the shortest interval between auditory presentation of a new sound and the parrot's first imitation was 8 h (overnight) and was generally 3 days. In this case, the mapping of auditory input to vocal motor output and matching of vocal output to the auditory template appears to take some time for consolidation. The lack of imitation until overnight suggests a role for sleep in the learning process, as has been observed for songbirds (Derégnaucourt et al. 2005).

Richards et al. (1984) report much more rapid imitation for a subadult female bottlenose dolphin (*Tursiops truncatus*) trained in an imitation task. The first step in the study was to define the pre-exposure repertoire of frequency-modulated whistle sounds produced by the subject and by the one other dolphin in the pool. The training procedure used operant conditioning in a series of successive training steps. First the dolphin was trained to produce any whistle sound after a *vocalize* signal was played. Then several model sounds (different from any in the pre-exposure repertoire) were introduced after the *vocalize* signal, and the dolphin was reinforced for responding immediately to the *vocalize* signal followed by the model signal by producing a whistle that matched first the duration and then frequency parameters of the model. Richards et al. (1984, p. 16) state that the subject "rapidly formed a generalized mimicry concept so that the presentation of any new model elicited an immediate attempt at imitation." Reiss and McCowan (1993) also report spontaneous vocal mimicry by captive bottlenose dolphins of computer generated frequency contours produced within 0.5 s of the end of the contour.

This behavior of captive dolphins imitating computer-generated models has a timing quite similar to that described for bottlenose dolphins copying whistles in the wild. Each individual bottlenose dolphin learns to produce an individually distinctive signature whistle (Janik and Sayigh 2013). Janik (2000) found that bottlenose dolphins in the wild may match the signature whistle of a dolphin with whom it is interacting. King et al. (2013) showed that the average latency between one bottlenose dolphin producing its signature whistle, and another matching the whistle was <1 s. The demonstrations of vocal learning using synthetic whistle-like sounds in captive dolphins seem to tap into a mode of whistle communication that involves rapid copying of conspecific sounds in the wild. Whistle matching in dolphins qualifies as complex vocal learning, but here, as in adult humans learning a new melody or word, the processes of auditory learning and sensorimotor learning are more rapidly and tightly coupled than in the classic descriptions of vocal learning. If they can produce copies that match a new model on the first attempt, this suggests an ability to map acoustic features as heard in the model directly onto acoustic properties of the vocal motor output.

9.7 Evidence that Mammals Learn to Produce Conspecific Sounds

Some songbird species have open-ended learning, and can continue to learn new songs as adults and add them to their song repertoire as they age (Beecher and Brenowitz 2005). In a few avian and mammalian species, such as the oscine yellow-rumped

Cacique (*Cacicus cela vitellinus*) and the humpback whale (*Megaptera novaeangliae*), not only do males retain the ability to learn new songs throughout their lifespan, but also the song of each region changes from month to month (Payne et al. 1983; Trainer 1989) and from year to year over decades (Payne and Payne 1985), with different members of a population tracking these changes in the song (Guinee et al. 1983). In the South Pacific, thematic material for humpback songs originates in waters off Australia, and spreads over several years to a series of breeding populations spread thousands of kilometers to the east (Garland et al. 2011). There is no way that this pattern of song change could occur except by vocal production learning, with whales continuously adjusting their songs based upon songs they hear. Delarue et al. (2009) report similar changes in the songs of bowhead whales (*Balaena mysticetus*) recorded in the Chukchi Sea.

A less dramatic pattern in which the young copy acoustic features of the calls of their mothers provides evidence for vocal learning in bats. Jones and Ransome (1993) studied the greater horseshoe bat (*Rhinolophus ferrumequinum*), which emits an echolocation call with most of its energy concentrated in a narrow frequency band. As females age from years 1 to 3, the frequency of this call increases, and thereafter decreases. The frequency of calls of young bats matches that expected for the age of their mothers, with calls of pups of older mothers lower than those of younger mothers. This pattern suggests that although bats inherit a template for the basic structure of their call, they learn to fine-tune the frequency of their call by comparison with the age-specific features of the call of their mothers. Such a pattern would fit the definition of limited vocal learning as defined in this chapter.

The Janik and Slater (1997) definition of vocal production learning emphasizes individuals learning acoustic properties of calls from conspecifics. However, as in the case of the bats, demonstrating vocal production learning is more difficult when the subject produces a sound similar to that of a conspecific than when it mimics a more exotic sound. Here one must show that the new sound was not part of the pre-exposure repertoire of the subject, and that the change is not produced by maturation of the vocal tract, or by a new context that elicits for the first time a species-specific call type that the subject could have produced all along. The whale and bat examples discussed in the preceding paragraphs meet these criteria, but it is difficult for many species whose vocalizations do not change in such distinctive ways.

9.8 The Role of Auditory Input in Mammalian Vocal Development

One approach around this problem is to study vocal development in animals that are deprived of auditory input. If such an animal develops normal vocalizations, then it must have production-first vocal development that does not require vocal learning. Different kinds of deprivation can isolate different kinds of vocal learning. For example, if an animal is completely deafened, it can neither form auditory templates from the sounds of others, nor can it learn to fine-tune its vocal motor output against an inherited auditory template. If an animal's hearing remains intact, but it is isolated from conspecifics, then it can fine tune its own vocalizations against innate auditory templates, but it cannot form new auditory templates based on the sounds of other individuals. In practice, it can be difficult to eliminate the possibility of an animal experiencing the sounds of others. For example, the hearing of birds may function in the egg, or of mammals in utero (Horner et al. 1987). Deafening and/or isolation can also cause more generalized deficits, so it can be difficult to prove that it was only the lack of auditory input that interfered with normal vocalization. In spite of these problems, Konishi's (1965) study of song in deafened or isolated white-crowned sparrows formed the basis of the template model of song learning.

We humans are a species that depends heavily on vocal learning for speech and singing a tune, so it comes as a surprise that vocal learning is not well developed in nonhuman primates. Frequently cited evidence against a role of auditory input in the vocal development of mammals comes from studies of squirrel monkeys (*Saimiri sciureus*). Winter et al. (1973) raised squirrel monkey infants with mothers that were normal or muted. They report that infants who heard no typical species-specific squirrel monkey sounds developed calls that were "virtually identical" with those of normal infants, and with no significant differences compared to the normal calls of adults. However, the sample size of normal monkeys was relatively small and the acoustic parameters, duration and fundamental frequency, were relatively simple by modern standards of acoustic analysis.

There has been recent interest in the question of whether mice (*Mus musculus*) may learn complex ultrasonic vocalizations. When adult male mice sense the presence of females, they produce complex sequences of repeated phrases of syllables that have been called songs (Holy and Guo 2005). Arriaga et al. (2012) found differences in songs of deafened versus normal mice, leading them to argue that male mice require auditory feedback to maintain normal songs, and they report that males altered their songs to match those of a cage mate, which they interpret as vocal imitation. By contrast, Hammerschmidt et al. (2012) studied the development of songs in deaf versus normal mice and found no difference, leading them to conclude that mice do not require auditory input for vocal development. Kikusui et al. (2011) demonstrated that two inbred strains of mice each had songs with different acoustic features. When they crossbred each strain, males raised with parents from the other strain developed songs similar to those of their genetic father, not the one whose sounds they heard. Both studies state they found no evidence for vocal learning of any sort. These diametrically opposing results suggest that the jury is still out on vocal learning in mice and that evidence for vocal learning in this species must be treated with caution. More research is needed in mice and many other species to test for differences in calls of animals with or without auditory input of normal calls to test whether species classed as non-learners may use auditory-vocal feedback to stabilize subtle acoustic features of vocalizations.

9.9 Weakness of Geographical Dialects as Evidence for Vocal Learning

One information source that has been argued as evidence for vocal learning is the existence of geographical dialects in vocalizations. If a young animal learns its vocalizations from its neighbors, this certainly can lead to geographical variation and vocal dialects. There is a large literature on song dialects in birds, and vocal learning can lead to convergence of songs within an area or population and divergence across areas or populations (Catchpole and Slater 2008). Janik and Slater (1997) cover evidence for vocal dialects in mammals, and they discuss many mechanisms that do not involve vocal learning that could lead to geographic variation in vocalizations. Geographical differences in vocalizations can be generated by genetic differences between populations, by differences in social structure that affect call usage, by differences in sound transmission in different environments, and by settling of animals in sites where they hear sounds similar to those they produce (Catchpole and Slater 2008). Given how well this topic has been covered elsewhere, this chapter focuses on a more fine-grained analysis of how calls converge among interacting individuals, which in my view provides better direct evidence for vocal learning.

9.10 Vocal Convergence as Animals Form a Group

Convergence of acoustic features of a vocalization among animals recorded before and after they form social bonds is a better test for vocal learning than evidence of geographical dialects. If the versions of the same call type from several animals differ from one another before exposure and become more similar after exposure to one another's vocalizations, then this provides stronger evidence for vocal learning than the simple demonstration of geographical variation. Note that vocal convergence may involve a call that developed without auditory input, but the fine-tuning of the call then does require feedback between auditory input and vocal motor output. The process of vocal convergence must allow fine-tuning of the auditory template as well as the converged vocalization. Vocal convergence thus qualifies as a form of limited vocal learning, by the definition given in Sect. 9.1. The fine-tuning of a species-specific vocalization to match that of other individuals helps to resolve the ethological question of why selection might favor allowing auditory input to modify the "correct" species-specific signal. Here the animal still maintains the species-typical call, but adds detail that indicates a social bond and membership in a social group. Vocal convergence has been reported for many species, including species for which the evidence of vocal learning is otherwise weak.

Some of the first evidence for vocal convergence in animals comes from birds. Most work on vocal learning in birds focuses on *songs* of oscine songbirds, but

Mammen and Nowicki (1981) worked on the *calls* of oscine black-capped chickadees (*Parus atricapillus*). They captured five winter flocks of chickadees and recorded the chick-a-dee call from each individual of each group. When they rearranged the members of three of these groups into different aviaries, they showed that the calls of members of each new aviary flock converged over a period of about a month. Nowicki (1989) studied this process of convergence, and showed that within the first week of being housed together, members of the group converged on calls matching the central tendency of features within the group rather than copying any one individual. Farabaugh et al. (1994) provide similar data for vocal convergence in the contact calls of the Psittaciform budgerigar (*Melopsittacus undulatus*). They raised two groups of three unrelated male budgerigars, each group in a different cage within the same room. None of the six males shared contact call types at the start of the study. The first time one male imitated the contact call of another was recorded 1 week after they were housed together. By 8 weeks, all birds in the same cage shared the same dominant call type, and this dominant call differed across the two cages. The lack of convergence of birds that could hear one another but did not interact directly suggests that actual social interaction made it more likely for calls to be imitated in this species than just hearing the calls. This indicates that the information affecting choice of models is not just auditory, but also involves social interaction. Some of the best evidence for vocal learning in hummingbirds comes from three male Anna's hummingbirds (*Calypte anna*) that showed convergence of song syllables when housed in the same room (Baptista and Schuchmann 1990).

There is evidence for vocal convergence in all of the mammalian taxa with good evidence for vocal learning: bats, cetaceans, elephants, humans, and seals. When Boughman (1998) transferred greater spear-nosed bat females (*Phyllostomus hastatus*) from one captive group to another, their call structure changed to become more similar to that of their new group mates. This pattern of fine-tuning of calls suggests that the bat case be viewed as limited vocal learning. Similar vocal convergence has been observed when bottlenose dolphins (*Tursiops* sp.) form social bonds. As male bottlenose dolphins mature, most form alliances that last for many years. Smolker and Pepper (1999) studied whistles produced by three males as they formed an alliance in the wild. Over the 4-year study, the distinctiveness of the whistles decreased as all three dolphins converged on a shared whistle type that rarely was produced before alliance formation. Watwood et al. (2004) were not able to study the process of convergence, but they confirmed that males in nine alliances produced whistles that were more similar to those of their own partners than to those of any of the other males. Fripp et al. (2004) provide evidence that dolphin calves in the wild model their signature whistles on those of community members, and Miksis et al. (2002) show that captive dolphin calves incorporate features of manmade signals as they develop their signature whistles. This matching of novel models suggests that vocal convergence in bottlenose dolphins may represent complex vocal learning, in which dolphins learn a novel or individual-specific model as opposed to a species-specific model. The evidence for vocal convergence in elephants involves a 23-year-old male African elephant (*Loxodonta africana*) that had spent 18 years housed with two female Asian elephants (*Elephas maximus*). This male produced sounds that were very different from normal sounds of African elephants, but that matched the chirp sounds produced by Asian elephants (Poole et al. 2005). This copying of a non-species-typical call differs from the other cases of conspecific vocal convergence; copying such a novel call qualifies as a case of complex vocal learning. Vocal convergence is well known in humans and is one aspect of a phenomenon called vocal accommodation (Street and Giles 1982; Giles 1984). Sanvito et al. (2007) showed on the breeding ground of southern elephant seals (*Mirounga leonina*) that agonistic calls of subordinate males tended to converge on those of the local dominant male.

In spite of the weak evidence for complex vocal learning among nonhuman primates, vocal convergence has been reported for several primate species: pygmy marmosets (*Cebuella pygmaea*; Snowdon and Elowson 1999), cotton-top tamarins (*Saguinus oedipus*; Weiss et al. 2001; Egnor and Hauser 2004), and chimpanzees (*Pan troglodytes*; Mitani and Gros-Louis 1998; Marshall et al. 1999; Crockford et al. 2004; Watson et al. 2015). Sugiura (1998) also demonstrated rapid matching in a study in which Japanese macaques matched acoustic features of coo calls played back to them. Vocal convergence has also been reported for taxa with little other evidence for vocal learning. For example, Briefer and McElligott (2012) raised young goat kids in groups. After 5 weeks together, half sibs in the same group had contact calls that were more similar than half sibs from different groups, and the calls converged over time.

In most of the cases of vocal convergence described in the preceding paragraphs, unrelated animals form a grouping that interacts socially. These individuals start with slightly different versions of a vocalization, and they converge on a common group-distinctive version of the vocalization. Owren et al. (2011) raise concerns that emotional responses to being housed with strangers may produce vocal patterns that look like vocal convergence but do not involve vocal production learning. However, looking at the broad range of vocal convergence studies, it is hard to imagine mechanisms other than vocal learning by which unrelated animals could develop vocalizations that are so distinctive across groups. Vocal convergence as a limited form of vocal learning has a broader taxonomic scope than complex vocal learning.

From a neural perspective, it is an open question whether fine-tuning specific acoustic parameters of an existing call type involves overlap with circuits for complex vocal learning or whether it involves a separate circuit. Humans and the avian taxa with reception-first vocal development have neural circuits in the telencephalon specialized for complex vocal learning, which are not as obvious in non-learners. The wider spread of vocal convergence than complex vocal learning suggests either that it uses different circuits or that it can be achieved with less obvious neural circuits. Owren et al. (2011, p 10) argue that "convergence and divergence phenomena may have little to say about reception-first vocal development." Arriaga and Jarvis (2013) suggest that fine-tuning the acoustic parameters of a call type whose central pattern generator (CPG) is located in the midbrain and/or brainstem would use neural circuits that modify the CPG based on cortical input and integrated auditory pathways. They propose that this involves different circuits from reception-first vocal learning, which involves forebrain circuits. On the other hand, learning to fine-tune a production-first call could involve overlap with the mechanisms used to produce a novel sound, with the primary differences involving how open the process selecting templates is to novel versus species-specific sounds, and how variable the system is for generating new vocalizations to match a template. Certainly if the signature whistles of bottlenose dolphins develop as reception-first calls that can match a novel model, then convergence of these signals may also be mediated by complex vocal learning circuits. Resolving these issues will require careful study of these neural circuits in species capable of vocal convergence and of complex vocal learning.

9.11 Neural Mechanisms that Enable Bats to Fine-Tune the Frequency of Echolocation Pulses Based on Auditory Input from Returning Echoes

Vocal convergence that involves fine-tuning the template for a production-first call based on auditory input from other conspecifics represents limited vocal learning. There is not much evidence on the neural basis for this form of limited vocal learning, but better evidence is available for a rapid and sophisticated form of auditory-vocal feedback used by echolocating bats to fine-tune their echolocation calls based on hearing their echoes. When a bat echolocates, it often needs to use auditory information about an incoming echo to adjust its next outgoing pulses. This puts a premium on auditory-vocal feedback rapid enough to function on time scales of a few tens of milliseconds. One of the most complex forms of feedback occurs for bats whose hearing is specialized to measure small changes in frequency over a narrow frequency band. The frequency band for which they can best discriminate frequencies (the "acoustic fovea") is narrow enough that differences in velocity between bat and prey can lead to Doppler shifts large enough to shift the echo returning from the prey to outside of the acoustic fovea. These bats shift the frequency of their outgoing pulses so that the Doppler-shifted echo remains in the acoustic fovea. Metzner (1989, 1993) reported on neural mechanisms for fine-tuning of vocal motor output based on auditory input studied in horseshoe bats with this Doppler-shift compensation. He found neurons within the midbrain that receive auditory input and that respond to vocal production and he proposed a mechanism for measuring the echo frequency and using this information to control the frequency of the outgoing pulse to compensate for Doppler shifts. Metzner (1996, p 252) studied the connectivity of this area, and proposed that it "serves as a link between the processing of auditory information and the control of vocalization and related motor patterns."

Studies of auditory-vocal feedback in bats show that midbrain and brainstem structures can rapidly fine-tune precise acoustic features of production-first sounds. Reliance on low levels of the brain, a few synapses from the relevant motoneurons, enables rapid processing. Vocal convergence of production-first sounds occurs over longer time scales, reducing the constraints on timing. However, vocal convergence is more complex than Doppler compensation in that it requires the formation of an auditory template, coupled with a mechanism to fine-tune a production-first vocal motor program so that the animal can reliably produce the fine-tuned call. We know that birds and humans use specialized auditory-vocal feedback mechanisms in the telencephalon to develop new reception-first calls. The bat work shows how auditory input can modulate pattern generators in the midbrain, but further research will be required to determine whether fine-tuning of production-first calls leading to vocal convergence can be achieved by similar feedback in the midbrain and brainstem, or whether this requires telencephalic pathways to support the learning and memory requirements of vocal learning. If vocal convergence does require telencephalic pathways, it will also be important to test whether this form of limited vocal learning involves the same circuits as complex vocal learning or not.

9.12 The Lombard Effect: Modifying Vocal Output in Response to Noise

The study of vocal convergence suggests that there may be a broader taxonomic range for this limited form of vocal learning than is typically credited for complex vocal learning, especially among mammals. But there may be an even more widely distributed connection between auditory input and vocal motor output. There is a problem faced by all animals that use sound to communicate, a problem that would select for a connection between auditory input and vocal motor output. This problem is communicating in varying noise conditions. Both in the ocean and in air, ambient noise levels vary over tens of decibels, leading to significant variation in the effective range of communication for a fixed signal level.

The best known compensation mechanism for noise is called the Lombard effect, named after the French otolaryngologist Etienne Lombard who discovered that humans speak more loudly when in the presence of loud noise (Brumm and Zollinger 2011). Brumm and Zollinger (2011) discuss the evidence of how broadly distributed the Lombard effect is among birds and mammals. The Lombard effect has been demonstrated among oscine songbirds, hummingbirds, and parrots, all of which are skilled at complex vocal learning; in the domestic fowl (Brumm et al. 2009), thought not to be capable of vocal learning; and crested tinamous, which belong to the most basal group of living birds. Hotchkin and Parks (2013) report that evidence for the Lombard effect has been found in every species of mammal in which it has been studied. The presence of the Lombard effect has not been as systematically explored among basal mammals as among birds, but the broad distribution and lack of evidence for absence of the Lombard effect in mammals suggest that it was likely present in the common ancestor of mammals. Brumm and Zollinger (2011) suggest that the Lombard effect has a very old history in birds and mammals, and they argue that either it independently evolved in both taxa or originated in a common ancestor and therefore could be 300 MY old and shared among amniotes.

The neural basis for the Lombard effect has not been studied in as much detail as song learning in oscine songbirds. Neurobiological studies in cats and squirrel monkeys locate the neuronal circuits for the Lombard effect in the brainstem. Nonaka et al. (1997) demonstrated the Lombard effect in decerebrate cats, showing that sufficient neuronal circuits exist within the brainstem to support the Lombard effect. Working with squirrel monkeys, Hage et al. (2006) report that neurons in the brainstem respond both to auditory input and vocal motor output, and Hage et al. (2006) suggest that these neurons mediate the Lombard effect by integrating audio-vocal information. This led Owren et al. (2011, p 7) to argue "the Lombard effect is mediated at the brainstem level, meaning its occurrence is likely uninformative with respect to the operation of higher level vocal control systems." By contrast, Eliades and Wang (2012) showed that the Lombard effect in common marmosets (*Callithrix jacchus*) is mediated by neurons in the auditory cortex that respond to the monkey's own vocalizations. When a monkey hears its own vocalization masked by noise, the neural response weakens, producing an error signal that predicts vocal intensity compensation in later vocalizations. This suggests that at least in primates, a parallel cortical circuit also affects modulation of call intensity based on auditory monitoring of one's own calls.

9.13 A Broader Look at Compensation for Noise in Animal Communication

Communicating in noise is a ubiquitous problem that may require modifying the outgoing signal depending on the noise present at the time of calling. The Lombard effect was the first such compensation mechanism studied in humans and animals, but the last few decades have seen a great expansion in evidence for compensation for noise in animal communication. Communication engineers recognize a suite of mechanisms that can be used to compensate for noise, including increasing the level of the signal, length of the signal, or redundancy of the signal. If the noise is limited to within a frequency band, then the frequency of the call can also be shifted outside of the noise band, just like switching a walkie-talkie to a quieter channel. If transient sounds, including calls of conspecifics, are interfering with communication, then the caller can wait for a quiet period to produce a call, or can shift the frequency of its own call away from the interfering frequency band.

All of the mechanisms identified by engineers to compensate for noise or interference listed in the preceding paragraph have been documented in animals that communicate with sound. One strategy involves waiting to produce a signal until the noise level reduces, or timing vocalizations to minimize overlap with competing transient sounds. Mechanisms for timing signals have evolved particular sophistication in animals when the signalers are competing for attention and the "noise" comprises competing transient signals from echoes or conspecific sounds (Greenfield 1994; Hall et al. 2006). However, modifying when one produces a call does not qualify as vocal learning and using auditory input just to decide when to call seems a particularly simple form of auditory-vocal feedback.

If the noise level is not changing rapidly enough, or if the animal cannot wait to get a signal through, then it can modify the acoustic structure of calls to compensate for the noise. Several animal taxa have been shown to increase the length of their calls in the presence of prolonged noise. Brumm et al. (2004) showed that a New World monkey, the common marmoset, lengthens the syllables of a call when exposed to white noise, which Egnor and Hauser (2006) also showed in studies of the cotton-top tamarin (*Saguinus oedipus*).

One of the predictions of communication theory (Shannon and Weaver 1963) is that the redundancy of signaling should increase as the channel becomes noisier. Some bird species increase the number of syllables in their calls or the bout duration of their songs with increasing noise: Japanese quail (*Coturnix coturnix japonica*; Potash 1972) and king penguins (*Aptenodytes patagonicus*; Lengagne et al. 1999). Chaffinches (*Fringilla coelebs*) sing longer bouts of the same song in noisy areas, increasing the redundancy of their singing (Brumm and Slater 2006b). Among mammals, humpback whales increased repetitions of phrases in their songs when they were exposed to a low-frequency sonar (Miller et al. 2000; Fristrup et al. 2003). These responses have been interpreted as compensation to increase the ability of receivers to detect and classify signals in a noisy channel. Turnbull and Terhune (1993) have shown that a harbor seal (*Phoca vitulina*) can detect a regular series of calls at a lower signal to noise ratio than a single call alone, providing support on the receiver side for this interpretation.

One of the most widespread compensation mechanisms reported for animals is shifting frequency to avoid band-limited noise. This frequency shifting is important to avoid interference from conspecific vocalizations, which are particularly likely to overlap in frequency. For example, some bats shift their echolocation calls away from the frequencies of conspecifics nearby in what is called a jamming-avoidance response (Ulanovsky et al. 2004).

The propulsion noise of ships in marine environments and road traffic noise in terrestrial environments both tend to be most intense at low frequencies, and these low frequencies also propagate best. Shipping noise has been increasing globally over the past century, and these long-term changes in noise have led to long-term increases in the frequencies of the contact calls of right whales (*Eubalaena* sp.) in the Atlantic Ocean, apparently to compensate for increasing low-frequency shipping noise (Parks et al. 2007).

Road noise from traffic varies in different locations, providing contrasts in noise that enable tests for how animals compensate. Slabbekoorn and Peet (2003) showed that male great tits (*Parus major*) recorded singing in quiet and noisy areas of cities showed a systematic correlation between amplitude of the noise and frequency of their songs. Halfwerk and Slabbekoorn (2009) conducted playback experiments to test whether great tits can rapidly switch the frequency of their songs outside of a noise band. When Halfwerk and Slabbekoorn (2009) played low-frequency noise, the subjects increased the frequencies of their low notes, and when they played high-frequency noise, the subjects decreased the frequencies of their high notes. This suggests that exposure to noise in a frequency band causes the birds to alter their singing pattern to emphasize energy outside of the noise band. There is evidence that some anurans (Parris et al. 2009; Cunnington and Fahrig 2010) and even an insect (bow-winged grasshoppers, *Chorthippus biguttulus*; Lampe et al. 2014) shift the frequency of their calls upward when in the presence of low-frequency noise. These results emphasize the taxonomic spread of mechanisms to compensate for noise, and how this selects for auditory-vocal feedback

One complication for these studies is that some species show a correlation between source level and frequency of calls: the harder they sing, the higher the frequency emphasis (Nemeth et al. 2013). These complications suggest the importance of measuring both level and frequency of calls as a function of noise. Potvin and Mulder (2013) set out explicitly to test whether birds elevate the frequency of their calls to avoid low-frequency noise or whether the rise in frequency is just a byproduct of calling more loudly. They exposed silvereyes (*Zosterops lateralis*) to sound playbacks of high- and low-frequency noise. When exposed to high-frequency noise, silvereyes reduced the minimum call frequency, but at the same time, they increased the average power of their calls. Cardoso and Atwell (2011) measured the intensity and frequency of songs of dark-eyed juncos (*Junco hyemalis*). They argue that oscine songbirds can to large measure control intensity and frequency independently, and that songbirds can adjust to low frequency noise either by singing louder, higher in frequency, or both. Understanding whether some frequency shifts are a byproduct of changes in source level must hinge on testing for taxon-specific linkages between acoustic features that result from the sound production mechanism, and on modeling the impact of changes in each feature on the active space of the calls in varying noise.

Reviewing the variety of noise-induced vocal modifications observed in wildlife, Hotchkin and Parks (2013, p 817) point out that "closely related species of mammals can exhibit very different vocal responses to noise." Some of this variability may have to do with tactical responses to variation in the timing and frequency spectra of the noise, some with differences in the communication tasks the animals are conducting, and some with taxon-specific linkages between acoustic parameters such as level and frequency. When humans speak in noise, the Lombard effect not only influences the intensity of the voice, but it also involves simultaneous modification of timing and frequency content (Lane and Tranel 1971). These changes in Lombard speech are thought to be linked through the biomechanical properties of the speech production system. Evidence for mechanisms to compensate for noise is so varied and ubiquitous that it suggests a strong selection pressure to evolve mechanisms that cover the specific problems faced by each taxon.

The mechanisms used by animals to compensate for noise appear to match specific problems. When faced with intermittent interference, many species time their calls to fall in the intervals between transient noises. When faced with band-limited noise, many species shift the frequency of their calls away from the dominant noise band. When faced with continuous wideband noise overlapping the call in frequency, many species call more loudly. Evidence that animals select which acoustic features to modify to most efficiently maintain effective communication in varying noise supports the view that compensation mechanisms are more complex than simple reflexes. Nuclei in the brainstem that coordinate respiration and larynx/syrinx could modulate the intensity of pressure, driving the intensity of vocalizations, leading to Lombard effect. Given correlation between level and frequency in some species, noise-stimulated changes in level could also lead to correlated changes in frequency. However, most of the other compensation mechanisms change the actual timing, repetition, or frequency of the signal. Modifying these features would appear to require modifying the actual pattern generated by a central pattern generator.

9.14 Modulation of Vocal Output Based on Auditory Input of Noise and Based on Balancing Benefits Against Costs and Risks of Different Effective Ranges of Communication

All animals that communicate must balance the costs and benefits of signaling. All signals involve some energetic cost, but a more significant cost in some settings is the risk that an unintended audience, such as a competitor, predator, or parasite, will detect your call. When male tungara frogs (*Physalaemus pustulosis;* Ryan et al. 1982) or field crickets (*Gryllus integer*; Cade 1975) produce calls to attract females, they also have a much higher risk of being killed by bat predators in the case of the frog or parasitic flies in the case of the cricket.

One underexplored aspect of noise compensation is the extent to which it helps a caller adjust its signal just to meet the detection requirements at the expected range of the audience, while reducing the risk that eavesdroppers will exploit the call to the caller's detriment. Brumm and Slater (2006a) have shown that zebra finches increase their song amplitude with increasing range to their intended receiver. And when male chickens (*Gallus gallus domesticus*) see a predator, they are less likely to produce an alarm call when not in the presence of an adult female (Evans and Marler 1992). However, few if any studies test whether animals respond to temporarily increased risk of eavesdropping by using noise compensation mechanisms to reduce the range at which their calls are detectable by the threat. Such a balance is implied in the presumed need to adjust calls to produce the correct range of detection. If animals were not responding to cost and risk, then why not just produce the loudest call possible? In fact, Zahavi and Zahavi (1997) argue in the context of reproductive advertisement displays that females may select a calling male based on his willingness to suffer the costs and risks of signaling. In this kind of setting, males may not adjust their signaling based upon noise, but just call as loudly as possible.

When the Lombard effect is induced in humans, it appears to be involuntary, and early descriptions sometimes called it a reflex. However Lane and Tranel (1971) argue that the Lombard effect is more complex than a reflex, and they emphasize that it is designed rather generally to adjust calling behavior so that the caller can better communicate in the presence of varying noise. In humans, the extent of the

Lombard effect depends on the importance of intelligibility as well as on the noise level (Lane and Tranel 1971). This differs from most reflexes that use feedback from internal sensory input to stabilize one simple form of motor output. The Lombard effect is much more complex in humans, who integrate information about noise and about the importance of getting the message across to modulate the level, length, and frequency of their vocalizations.

If animals evolved a mechanism to facilitate effective communication in varying noise, then it would function best by integrating all of the different kinds of information an animal needs to make the correct decision about adjusting the acoustic features of its signal. This view of noise compensation suggests that it is part of a broader set of mechanisms to modulate signals to reach the desired audience while reducing energetic costs and risks of detection by dangerous eavesdroppers. The signaler would have to integrate information from many different sources before making a decision about fine-tuning the acoustic properties of its call. If this integration is facilitated by information processing in the cortex, then the demonstration by Eliades and Wang (2012) that the Lombard effect in nonhuman primates also involves cortical circuits supports the idea that some animals as well as humans may integrate complex multimodal information to make decisions about how to adjust their vocalizations to balance the benefits, costs, and risks of vocalizing.

9.15 Neural Pathways for Learned and Unlearned Vocalizations in Birds and Mammals

The neural pathways for complex vocal learning have best been studied in oscine songbirds and humans. Comparisons of neural control of vocalization in these taxa vs. non-vocal learners emphasize the role of forebrain structures (telencephalon in birds (Jarvis 2007) and cortex in mammals (Jürgens 2009)) in producing learned vocalizations, while innate vocalizations are thought to be controlled by central pattern generators in the midbrain and/or brainstem (Wild 1997).

Over the last 40 years, neurobiologists have uncovered specialized areas of the songbird brain that process the information required for vocal learning. Different parts of the songbird brain are specialized to use auditory input to form templates, to use feedback from auditory input to improve the match between vocal output and the template, and to learn to produce a stable learned vocalization. Simpson and Vicario (1990, p 1541) "suggest that the learned features of oscine songbird vocalizations are controlled by a telencephalic pathway that acts in concert with other pathways responsible for simpler, unlearned vocalizations."

Jürgens (2009) argues that there are two separate neural pathways for the control of innate vocalizations versus learned vocalizations in mammals. Jürgens (2009) summarizes data suggesting a pathway for innate vocalizations involving the anterior cingulate cortex, the periaqueductal gray (PAG) in the midbrain, and areas of the reticular formation in the brainstem that have direct connections with phonatory motor nuclei. He argues that voluntary control of these vocalizations involves the anterior cingulate cortex, regulation of the initiation and intensity of innate vocalizations is performed in the midbrain PAG, and that most of the motor pattern generation involves the reticular formation of the lower brainstem. Many PAG neurons that correlate with vocalization fire before, not during, vocalization, and many respond to auditory, visual, or somatosensory input, suggesting that the PAG can mediate feedback to the vocal control system from other senses.

The muscles involved in sound production are also represented in the motor cortex, and Jürgens (2009) argues that learned vocal patterns are initiated by a second pathway involving the motor cortex. Jürgens (2002) argues that humans have a direct nerve fiber pathway from the facial motor cortical areas to the nucleus ambiguous in the brainstem, where motor neurons project to the intrinsic laryngeal muscles. He argues that this pathway is absent in nonhuman primates, and that it may be essential for cortical processes that enable vocal learning to provide fine motor control over the larynx. Birds with vocal learning also have similar projections directly from the telencephalon to motor neuron pools that innervate the syrinx, strengthening the evidence that direct connections from telencephalon are required for vocal learning to control fine details of vocal sound production.

Deacon (1998), by contrast, argues that many mammals including species thought to not be vocal learners may have had connections from the cortex to the nucleus ambiguous early in development, but that these weak cortical projections are not as important as the projection from PAG in species without complex vocal learning. Deacon (1998) argues that for human learned sounds to compete with the PAG pathway, these direct cortical pathways had to expand, leading to a larger, more distinct pathway for learned vocalizations that are driven under control from the premotor cortex rather than from PAG. Arriaga et al. (2012, p 107) reports "that mice have a cortical vocal premotor circuit that projects directly to vocal motoneurons in the brainstem." This projection is much weaker than that reported for humans and songbirds, providing support for Deacon (1998)'s more nuanced hypothesis regarding projections from the cortex to vocal motor neurons.

Both Jürgens (2002) and Deacon (1998) hypothesize that a direct cortical pathway to phonatory motor nuclei is essential for complex vocal learning, and they both predict that complex vocal learners have a direct tract from motor cortex to the brainstem nucleus that innervates the sound production organ. As mentioned at the start of this section, this has been measured in humans and songbirds, but tests are limited among other species with or without complex vocal learning. The mammals demonstrated to be complex vocal learners (seals, cetaceans, and elephants) offer unique opportunities for testing these hypotheses about the role of the cortex and about the need for direct connections from cortex to phonatory brainstem nuclei. If this argument also holds for the other mammals capable of producing novel signals, then it would predict strong pathways linking cortical circuits involved in producing vocalizations and motoneurons in the nucleus ambiguous that innervate the laryngeal muscles (or in the case of toothed whales, other motoneurons that innervate their unique sound production organ).

9.16 Evolution of Complex Vocal Learning

The study of vocal learning typically focuses on the role of learning in the development of complex birdsongs or human speech. When this is contrasted with evidence that vocalizations of other taxa can develop without auditory input from conspecifics, it is often concluded that the ancestral state in mammals and birds was lacking the character of vocal learning, and that it independently originated in the taxa where it has so dramatically been demonstrated. For example, Nottebohm (1972) argued that vocal learning evolved independently in songbirds, parrots, and hummingbirds from ancestors that lacked this trait.

Jarvis (2006, 2007) argues that all three avian vocal learning taxa share seven similar brain structures in the telencephalon. If the neural system controlling vocal learning evolved independently in the three vocal learning taxa of birds, then so many similarities in brain structures for vocal learning in the three avian taxa seems a remarkable coincidence. Jarvis (2007) compares vocal and auditory brain centers in the brains of humans and avian vocal learners, extending the argument for analogous brain structures for complex vocal learning to humans. He suggests that either these structures were used for some functions that made them particularly likely to be used for vocal learning or that a skill related to vocal learning was processed by these brain structures in a common ancestor.

The phylogeny of complex vocal learning suggests two or three independent origins in birds, and four or five among the mammals: humans, seals, cetaceans, elephants, and possibly bats (depending on whether vocal learning in bats is taken to be limited or complex; Fitch and Jarvis 2013). There may be similarities between the neural pathways for complex vocal learning in birds and mammals, but our lack of knowledge about these pathways in nonhuman mammals with complex vocal learning interferes with testing this idea. The data we have on the vocal learning abilities of different taxa are very spotty, and involve different criteria for learning. One critical component of a research program studying the evolution of vocal learning will be to identify a set of species likely to have or not to have each form of vocal learning, following explicit definitions. They will then need to be studied using comparable methods.

9.17 Did Complex Vocal Learning Evolve from Simpler Forms of Vocal Learning or Auditory-Vocal Feedback?

Much of the material reviewed in this chapter suggests that we need to question some assumptions about the evolution of vocal learning. There is something special about complex vocal learning systems that enable animals to learn new calls to form an open system of communication. However, rather than being an all-or-nothing capacity, there are a variety of ways that auditory input may alter vocal output, with varying taxonomic distributions and involving varying amounts of learning and flexibility.

Petkov and Jarvis (2012) make a similar argument for distinguishing a range of vocal learning capabilities, which may have differing taxonomic distributions.

Research on vocal learning must recognize the variety of ways that auditory input can affect vocal output. Here, following Fitch and Jarvis (2013), this chapter uses the term "complex vocal learning" for cases where animals can learn to imitate sounds that are not part of their species-specific vocal repertoire. This is essential for forming an open communication system, and may be fundamentally different from learning to fine-tune species-specific calls. The skills of learning to fine-tune species-specific calls, which Fitch and Jarvis (2013) call "limited vocal learning," is more widely distributed among mammals. This chapter also defines a broader range of "auditory-vocal feedback," which may not involve learning to develop a call based on an auditory template. If any of these forms of auditory-vocal feedback was shared among the ancestors of vocal learners and could form a substrate for the evolution of vocal learning, this could help to resolve the question of why the brain structures underlying independently evolved complex vocal learning appear to be so similar.

This chapter selected a definition for "vocal" that involves a sound production mechanism that may require auditory feedback to maintain stable acoustic features, and raises questions about the strength of evidence that hearing plays no role in the vocal development of many species identified as non-learners. The evidence reviewed on noise compensation suggests that most if not all mammals and birds have evolved the capability to modify acoustic parameters of their vocal output based on hearing the level and frequency range of ambient noise. Some bats have evolved similar mechanisms to modify their echolocation signals based on hearing echoes from previous signals. These phenomena appear to differ qualitatively from learning and memory mechanisms in which the learner forms or fine-tunes an auditory template and then learns to produce vocal sounds that match the template. Both limited and complex forms of vocal learning require auditory templates to be modifiable through audition, and require the ability to learn to match vocal motor output to a template. Thus it seems more likely that limited vocal learning is a better candidate than the simpler forms of auditory-vocal feedback for overlap in circuitry with complex learning. The distinction between complex versus limited vocal learning and the different forms of auditory-vocal feedback discussed in Sects. 9.11–9.13 suggests that they may involve neural circuits that do not fully overlap. However, it remains to be seen whether these neural circuits are completely independent, or whether some simpler forms may have served as precursors for the evolution of limited and/or complex vocal learning.

9.18 Selection Pressures for Auditory-Vocal Feedback and Vocal Learning

This chapter raises the question of whether simple forms of auditory-vocal feedback, which are widespread taxonomically, may have provided substrates for the evolution of more complex forms of vocal learning. This chapter does not propose a sharp definition separating auditory-vocal feedback from vocal learning. One critical distinction is that many of the simple forms of auditory-vocal feedback discussed here do not require memory, but vocal learning requires the formation of memories that enable the learning of auditory templates and the development of motor patterns producing vocalizations that match the template. As more evidence suggests a spectrum of complexity in vocal learning skills, more precise definitions will need to attend discussions of presence or absence of different forms of vocal learning in different taxa. Here I discuss a range of selection pressures that may be important for the evolution of vocal learning.

9.18.1 Stabilizing the Acoustic Structure of Vocalizations

When sound is produced by stridulation or by sonic muscles exciting an air sac, a central motor program may be able to yield a predictable sound. It is less clear how a stereotyped vocalization can be produced by passing air through vocal folds without some form of auditory-vocal feedback. For pneumatic sound production, producing a stereotyped vocalization may involve a sophisticated mix of adjustments of air pressure, tension on vocal folds, and shape of the vocal tract. Slight variation in the tension or mass of the folds or shape of the vocal tract may lead to variation in the signal that would be hard to detect using any sense other than audition. Thus one function of auditory-vocal feedback may be to stabilize the acoustic structure of species-specific vocalizations.

9.18.2 Compensation for Noise

Another early selective pressure for modifying vocal output based on auditory input may stem from the ubiquitous problem of adjusting calls to maintain the receiver's ability to detect and classify the call. Although data on the exact evolutionary origin of the Lombard effect are incomplete, it appears that soon after vertebrates developed pneumatic mechanisms to produce sounds, they evolved mechanisms to modify sounds to improve communication in increased noise. This is a ubiquitous problem, especially important when sender and receiver are more than a few meters apart and when the sender must balance the benefits of delivering a signal to an intended receiver against costs if the signal is intercepted by another animal that may be a threat. The demonstration that invertebrates may modify calls produced by stridulation to compensate for noise shows that this selection pressure likely has a broader taxonomic scope than that for stabilizing pneumatically produced sounds.

9.18.3 Vocal Convergence

The broad taxonomic distribution of vocal convergence suggests that once the connections between auditory input and vocal output were established, a common function was to fine-tune production of specific call types to match those in one's group or population. Group-distinctive vocalizations have been argued to mediate cohesion of groups in a wide variety of taxa. The accommodation literature argues that signal matching can function as an affiliative signal. When an animal learns vocalizations from territorial neighbors it can better detect intruders and can use vocal matching to direct specific signals to specific recipients, often as a threat (Vehrencamp 2001). A variety of ecological and social settings could select for these diverse functions of vocal convergence. It is hard to imagine other mechanisms for developing group-distinctive calls among unrelated individuals, so this may be an important selective pressure for vocal learning among species whose ecology creates a benefit for group-distinctive calls.

9.18.4 Echolocation

The evolution of echolocation may also provide selective pressures for specialized mechanisms that modify vocal output of echolocation clicks based on information received from incoming echoes. The use of echolocation requires very rapid feedback between auditory processing of echoes to regulate the timing and acoustic features of the next sonar pulses (Moss and Sinha 2003). As bats or dolphins approach a target, they may reduce the intensity of their outgoing pulse to stabilize the level of the echo (Madsen and Surlykke 2013). Bats with frequency modulated echolocation signals can detect differences in the delay between click and echo of as little as 60 microseconds, and detection of these delays is used to time outgoing clicks (Moss and Sinha 2003). Bats that analyze the Doppler shift of constant frequency echolocation signals are able to change the frequency of the outgoing signal so that the Doppler-shifted echo matches the best frequency for hearing (Metzner et al. 2002). The neurobiology of auditory processing and vocal motor control has been studied in echolocating bats, which are one of the nonhuman mammalian taxa known for vocal learning (Metzner and Schuller 2010). Bats have a mammalian auditory system, and the vocal motor pathway controlling the larynx is shared with many mammals, but they also have many specializations for rapid sensorimotor integration to provide rapid feedback between auditory processing of one click and producing the next. Echolocation has clearly thus provided a selection pressure in bats for specialized and rapid forms of modifying vocal output based on auditory input. The presence of vocal learning abilities in the two taxa specialized for echolocation, bats and toothed whales, suggests that echolocation may have selected for more sophisticated and rapid vocal learning skills in these taxa.

9.18.5 Sexual Selection

Sexual selection also can create selection pressures for using vocal learning to elaborate the acoustic structure of signals (e.g., increasing the song repertoire: Catchpole 1980, 1986), for males to copy the signals of more successful males or to match calls of neighbors to direct a threat (Vehrencamp 2001), or for males to copy acoustic features preferred by females (West and King 1988). As with nonacoustic displays, sexual selection can lead to the evolution of elaborate and striking advertisement displays. This elaboration can make the role of vocal learning in development of sexually selected songs particularly obvious. Many of the species shown to be capable of vocal learning—songbirds, seals, and whales—produce songs that are sexually selected advertisement displays. The bias toward males producing these songs, and the evidence for onset of learned signals by males at the time of sexual maturity suggest that sexual selection has influenced the evolution of vocal learning in songbirds, seals, whales, and perhaps elephants. But this does not necessarily imply that sexual selection provided the original selection pressure for links between auditory input and vocal motor output in the first place.

A contrasting view is suggested by the much broader distribution of other forms of auditory-vocal feedback, which suggests that ubiquitous problems of stabilizing vocalizations or of communication in noise may have provided earlier selection pressures for modifying vocal output based on auditory input. Vocal convergence appears to have a more limited taxonomic scope, but may also provide an early selection pressure for fine-tuning acoustic features of production-first vocalizations. Sexual selection in this case may have used some of these building blocks for the evolution of vocal learning mechanisms to generate more complex or more precisely matched vocal repertoires. However, the evidence cited in Sect. 9.15 which locates neural pathways for some forms of auditory-vocal feedback in the midbrain and brainstem, and those critical for vocal learning in the telencephalon are at odds with this hypothesis. Resolution of these contrasting views requires a broader taxonomic scope studying capabilities of noise compensation, vocal convergence, and vocal learning and their neural mechanisms.

9.18.6 Functions of Vocal Learning in Highly Social Taxa with Prolonged Periods of Dependency and Reliance on Knowledge of Older Members of the Group

Much has been written about similarities between birdsong and human speech (Doupe and Kuhl 1999; Fitch and Jarvis 2013). But there are many differences between sexually selected advertisement displays and speech. Although both communication systems may have sensitive periods with predispositions for learning the correct sounds, in many songbird species it is just the male that sings, and there is much less evidence for songs than speech having flexible associations between specific learned vocalizations and external referents.

Some of the vocal learning species that do not sing provide better evidence of these latter features. An African gray parrot trained to produce human speech sounds not only learned to produce words, but could also use these words to report on features of objects such as color, shape, and number (Pepperberg 1999). Similarly a bottlenose dolphin (*Tursiops truncatus)*, trained to mimic synthetic computer-generated sounds, learned to associate each sound with an arbitrary object such as a pipe, ball, or a frisbee. The dolphin then could produce the correct sound when shown one of these objects, so was able to vocally label an arbitrary object with a learned sound (Richards et al. 1984). Birdsong has provided fascinating parallels with speech in terms of sensitive periods, predispositions to learn, and the neurobiology of vocal development, but the abilities of species such as parrots and dolphins to learn arbitrary associations between learned vocalization and external referents provide a less explored and equally fascinating set of parallels with other aspects of speech.

Charvet and Striedter (2011) point out that oscine songbirds and parrots develop an unusually enlarged telencephalon through delayed maturation of this part of the brain. This requires a prolonged period of dependency, which is enabled by parental care. Charvet and Striedter (2011) argue that delaying telencephalic neurogenesis fosters the evolution of learned vocalization in humans, songbirds, and parrots. Following this argument, it is worth noting that among the most sophisticated vocal learners, there are a set of long-lived highly social animals with slow maturation and a prolonged period of dependence. Elephants and some toothed whales rival humans in having young dependent until their teenage years and with many adult females having a postreproductive period during which their reproductive effort is thought to be devoted to parental care (Marsh and Kasuya 1986; Tyack 1986; Lahdenperä et al. 2014). These older females retain knowledge of great importance to their groups (McComb et al. 2001; Brent et al. 2015). This social setting in which the young are dependent for years during which they can learn valuable information from caregivers may provide selective pressures for the use of vocal learning to develop more open forms of communication. The selection of songbirds as a model organism for vocal learning has enabled great progress in our understanding of the neural basis of the development of learned vocalizations, but I would argue that focus on communication in these other taxa may provide similarly valuable insights into the evolution of complex vocal learning in our own mammalian ancestors and into features of human communication for which we have few animal models.

9.19 Summary

The classic works on vocal learning have studied songbirds and humans to show how some species listen to an acoustic model and form an auditory template for the model. Then, often at a later date, they are able to listen to their own vocalizations, and slowly learn to produce sounds that match the template. If deprived of acoustic model to copy, these species learn to match their vocalizations to an inherited auditory template. This remarkable ability is usually contrasted with evidence that many other species can develop normal vocalizations with no auditory feedback at all.

However, there are a variety of ways that auditory input influences the acoustic structure of vocalizations that are more subtle than classic vocal learning. Some neurobiologists have argued that simple reflex-like auditory-vocal feedback taps completely different neural systems from vocal learning. But the last decades have revealed many different ways by which auditory input influences vocalizations. Some do not appear to involve learning and memory, some involve very rapid matching, and others involve matching on slow time scales similar to classic vocal learning. This chapter reviews these different forms of auditory-vocal feedback and asks whether some may use some of the same neural circuits and/or whether they may have provided neural substrates for the evolution of vocal learning.

This chapter defines vocalizations as sounds generated by passing air through vibrating lips, which then may be filtered by an upper respiratory path. The acoustic features of these kinds of vocal sounds are influenced by a complex combination of air pressure, tension, mass of the lips, and configuration of the vocal tract. A vocal motor program may be able to generate basic vocalizations such as crying or laughter, but auditory feedback may be required for matching more precise acoustic features of vocalizations.

This chapter discriminates limited vocal learning, in which an animal uses auditory input to fine-tune acoustic features of an inherited auditory template, from complex vocal learning, in which an animal can learn to produce a novel sound by learning a novel auditory template. The best evidence for complex vocal learning comes from birds and mammals that learn to imitate novel human-made sounds. Songbirds, parrots, seals, elephants, and dolphins all stand out for remarkable abilities of this kind of vocalization. Some bats and whales vary enough in their natural vocalizations, that one can track how conspecifics learn to match these variations.

A much broader range of mammals have been shown to converge in the acoustic structure of vocalizations when they are in a socially interacting group of conspecifics. If this convergence involves fine-tuning the acoustic structure of an inherited call based on hearing others, then it qualifies as limited vocal learning. It is not known whether this limited form of vocal learning uses different neural mechanisms than complex vocal learning.

Echolocating bats use rapid auditory-vocal feedback to modify outgoing echolocation signals based upon incoming echoes. They use midbrain and brainstem for a system that can rapidly modulate vocalizations in tens of milliseconds. These systems are capable of very precise matching, but need not involve learning or memory. This suggests that complexity and precision of matching does not necessarily imply higher development of vocal learning. On the other hand, these systems may facilitate the evolution of vocal learning, as the two echolocating taxa (bats and toothed whales) contain vocal learners.

All birds and mammals tested use auditory-vocal feedback to adjust their vocalizations to compensate for the effects of noise, and many species compensate for varying costs and benefits of communicating. Neurobiologists have shown that some of these mechanisms can be produced in the brainstem, and they often treat these mechanisms as reflex-like. However, many different options for compensation are available to animals. To make the correct decision about which to use, they must integrate information about noise, about their audience, and about risks that others may detect their signals to their disadvantage. In some nonhuman primates, these mechanisms involve cortical pathways as well as the brainstem, and it remains to be seen whether compensation mechanisms are completely independent of those used in vocal learning, or whether this ubiquitous capability might provide some substrates for the evolution of vocal learning.

Most work on the evolution of vocal learning assumes that the ancestral state involved no vocal learning, and therefore that taxonomically remote vocal learners must represent independent evolution of this trait. This interpretation is supported by the demonstration of specialized telencephalic nuclei in the brains of avian vocal learners, and of their absence in non-learners. However, the similarities in neural circuits used by the three avian taxa with vocal learning present a puzzle. If vocal learning originated independently, is it likely that all three taxa would use the same nuclei connected in the same way? One answer could be that vocal learning is not an all-or-none phenomenon, and that some of the less obvious forms of auditory-vocal feedback may provide shared substrates for the evolution of vocal learning.

This chapter has explored recent evidence for auditory-vocal feedback and for limited vocal learning to broaden the scope of how we think about vocal learning and to suggest new approaches to studying the evolution of vocal learning. Progress will require more precise definitions of different forms of vocal learning, broad comparative review of their presence and absence, and behavioral and neurobiological investigations into the mechanisms underlying the skills.

Acknowledgments P. L. T. thanks Henrik Brumm, Bjorn Merker, Arthur Popper, Peter Slater, and Rod Suthers for very helpful comments on an earlier draft of this chapter. This work received funding from the MASTS pooling initiative (The Marine Alliance for Science and Technology for Scotland) and their support is gratefully acknowledged. MASTS is funded by the Scottish Funding Council (grant reference HR09011) and contributing institutions.

References

Arriaga, G., & Jarvis, E. D. (2013). Mouse vocal communication system: are ultrasounds learned or innate? *Brain and Language, 124*, 96–116.

Arriaga, G., Zhou, E. P., & Jarvis, E. D. (2012). Of mice, birds and men: the mouse ultrasonic sound system has some features similar to humans and song-learning birds. *PLoS ONE, 7*(10), 1–15.

Au, W. W. L., Pawloski, J. L., Nachtigall, P. E., Blonz, M., & Gisner, R. C. (1995). Echolocation signals and transmission beam pattern of a false killer whale (*Pseudorca crassidens*). *The Journal of the Acoustical Society of America, 98*(1), 51–59.

Baptista, L. F., & Morton, M. L. (1981). Interspecific song acquisition by a white-crowned sparrow. *The Auk, 98*(2), 383–385.

Baptista, L. F., & Schuchmann, K.-L. (1990). Song learning in the Anna hummingbird (*Calypte anna*). *Ethology, 84*, 15–26.

Baylis, J. (1982). Avian vocal mimicry: Its function and evolution. In D. E. Kroodsma & E. H. Miller (Eds.), *Acoustic communication in birds* (pp. 51–83). New York: Academic.

Beecher, M. D., & Brenowitz, E. A. (2005). Functional aspects of song learning in songbirds. *Trends in Ecology & Evolution, 20*(3), 143–149.

Boughman, J. W. (1998). Vocal learning by greater spear-nosed bats. *Proceedings of the Royal Society B: Biological Sciences, 265*, 227–233.

Brenowitz, E. A., & Kroodsma, D. E. (1996). The neuroethology of birdsong. In D. E. Kroodsma & E. H. Miller (Eds.), *Ecology and evolution of acoustic communication in birds* (pp. 269–281). Ithaca, NY: Cornell University Press.

Brent, L. J., Franks, D. W., Foster, E. A., Balcomb, K. C., Cant, M. A., & Croft, D. P. (2015). Ecological knowledge, leadership, and the evolution of menopause in killer whales. *Current Biology, 25*, 1–5.

Briefer, E. F., & McElligott, A. G. (2012). Social effects on vocal ontogeny in an ungulate, the goat, *Capra hircus*. *Animal Behaviour, 83*(4), 991–1000.

Brumm, H., Schmidt, R., & Schrader, L. (2009). Noise-dependent vocal plasticity in domestic fowl. *Animal Behaviour, 78*(3), 741–746.

Brumm, H., & Slater, P. J. B. (2006a). Animals can vary signal amplitude with receiver distance: evidence from zebra finch song. *Animal Behavior, 71*, 699–705.

Brumm, H., & Slater, P. J. B. (2006b). Ambient noise, motor fatigue, and serial redundancy in chaffinch song. *Behavioral Ecology and Sociobiology, 60*, 475–481.

Brumm, H., Voss, K., Köllmer, I., & Todt, D. (2004). Acoustic communication in noise: Regulation of call characteristics in a New World monkey. *Journal of Experimental Biology, 207*(3), 443–448.

Brumm, H., & Zollinger, S. A. (2011). The evolution of the Lombard effect: 100 years of psychoacoustic research. *Behaviour, 148*, 1173–1198.

Cade, W. (1975). Acoustically orienting parasitoids: Fly phonotaxis to cricket song. *Science, 190*, 1312–1313.

Cardoso, G. C., & Atwell, J. W. (2011). On the relation between loudness and the increased song frequency of urban birds. *Animal Behaviour, 82*(4), 831–836.

Catchpole, C. K. (1980). Sexual selection and the evolution of complex songs among warblers of the genus Acrocephalus. *Behaviour, 74*, 149–166.

Catchpole, C. K. (1986). Song repertoires and reproductive success in the great reed warbler *Acrocephalus arundinaceus. Behavioral Ecology and Sociobiology, 19*, 439–445.

Catchpole, C. K., & Slater, P. J. B. (2008). *Bird song: Biological themes and variations*. Cambridge, England: Cambridge University Press.

Charvet, C. J., & Striedter, G. F. (2011). Developmental modes and developmental mechanisms can channel brain evolution. *Frontiers in Neuroanatomy, 5*, 1–5.

Cranford, T. W., Amundin, M., & Norris, K. S. (1996). Functional morphology and homology in the odontocete nasal complex. *Journal of Morphology, 228*, 223–285.

Crockford, C., Herbinger, I., Vigilant, L., & Boesch, C. (2004). Wild chimpanzees produce group-specific calls: A case for vocal learning? *Ethology, 110*, 221–243.

Cunnington, G. M., & Fahrig, L. (2010). Plasticity in the vocalizations of anurans in response to traffic noise. *Acta Oecologica, 36*(5), 463–470.

Deacon, T. W. (1998). *The symbolic species: The co-evolution of language and the brain*. New York: W. W. Norton.

Delarue, J., Laurinolli, M., & Martin, B. (2009). Bowhead whale (*Balaena mysticetus*) songs in the Chukchi Sea between October 2007 and May 2008. *The Journal of the Acoustical Society of America, 126*(6), 3319–3328.

Derégnaucourt, S., Mitra, P. P., Fehér, O., Pytte, C., & Tchernichovski, O. (2005). How sleep affects the developmental learning of bird song. *Nature, 433*(7027), 710–716.

Doupe, A. J., & Kuhl, P. K. (1999). Birdsong and human speech: Common themes and mechanisms. *Annual Review of Neuroscience, 22*, 567–631.

Egnor, S. E. R., & Hauser, M. D. (2004). A paradox in the evolution of primate vocal learning. *Trends in Neurosciences, 27*, 649–654.

Egnor, S. E. R., & Hauser, M. D. (2006). Noise-induced vocal modulation in cotton-top tamarins (*Saguinus oedipus*). *American Journal of Primatology, 68*(12), 1183–1190.

Eliades, S. J., & Wang, X. (2012). Neural correlates of the Lombard effect in primate auditory cortex. *The Journal of Neuroscience, 32*(31), 10737–10748.
Evans, C. S., & Marler, P. (1992). Female appearance as a factor in the responsiveness of male chickens during anti-predator behaviour and courtship. *Animal Behaviour, 43*(1), 137–145.
Farabaugh, S. M., Linzenbold, A., & Dooling, R. J. (1994). Vocal plasticity in budgerigars (*Melopsittacus undulatus*): Evidence for social factors in the learning of contact calls. *Journal of Comparative Psychology, 108*, 81–92.
Fitch, W., & Jarvis, E. (2013). Birdsong and other animal models for human speech, song, and vocal learning. In M. A. Arbib (Ed.), *Language, music, and the brain* (Strüngmann forum reports, Vol. 10, pp. 499–539). Cambridge, MA: MIT Press.
Fripp, D., Owen, C., Quintana-Rizzo, E., Shapiro, A., Buckstaff, K., Jankowski, K., et al. (2004). Bottlenose dolphin calves model their signature whistles on the whistles of community members they rarely hear. *Animal Cognition, 8*, 17–26.
Fristrup, K. M., Hatch, L. T., & Clark, C. W. (2003). Variation in humpback whale (*Megaptera novaeangliae*) song length in relation to low-frequency sound broadcasts. *The Journal of the Acoustical Society of America, 113*(6), 3411–3424.
Gans, C. (1973). Sound production in the Salientia: Mechanism and evolution of the emitter. *American Zoologist, 13*, 1179–1194.
Garland, E. C., Goldizen, A. W., Rekdahl, M. L., Constantine, R., Garrigue, C., Hauser, N. D., et al. (2011). Dynamic horizontal cultural transmission of humpback whale song at the ocean basin scale. *Current Biology, 21*, 687–691.
Giles, H. (1984). The dynamics of speech accommodation. *International Journal of the Sociology of Language, 46*, 1–155.
Greenfield, M. D. (1994). Cooperation and conflict in the evolution of signal interactions. *Annual Review of Ecology and Systematics, 25*, 97–126.
Guinee, L. N., Chu, K., & Dorsey, E. M. (1983). Changes over time in the songs of known individual humpback whales (*Megaptera novaeangliae*). In R. Payne (Ed.), *Communication and behavior of whales* (pp. 59–80). Boulder, CO: Westview Press.
Hage, S. R., Jürgens, U., & Ehret, G. (2006). Audio-vocal interaction in the pontine brainstem during self-initiated vocalization in the squirrel monkey. *European Journal of Neuroscience, 23*, 3297–3308.
Halfwerk, W., & Slabbekoorn, H. (2009). A behavioural mechanism explaining noise-dependent frequency use in urban birdsong. *Animal Behaviour, 78*(6), 1301–1307.
Hall, M. L., Illes, A., & Vehrencamp, S. L. (2006). Overlapping signals in banded wrens: Long-term effects of prior experience on males and females. *Behavioral Ecology, 17*(2), 260–269.
Hammerschmidt, K., Reisinger, E., Westekemper, K., Ehrenreich, L., Strenzke, N., & Fischer, J. (2012). Mice do not require auditory input for the normal development of their ultrasonic vocalizations. *BMC Neuroscience, 13*(1), 40.
Hardus, M. E., Lameira, A. R., Van Schaik, C. P., & Wich, S. A. (2009). Tool use in wild orang-utans modifies sound production: A functionally deceptive innovation? *Proceedings of the Royal Society of London B: Biological Sciences, 276*(1673), 3689–3694. doi:10.1098/rspb.2009.1027.
Hayes, C. (1951). *The ape in our house*. New York: Harper and Brothers.
Hayes, K. J., & Hayes, C. (1952). Imitation in a home-raised chimpanzee. *Journal of Comparative and Physiological Psychology, 45*, 450–459.
Hiss, A. (1983). Hoover. *The New Yorker, 3*, 25–27.
Holy, T. E., & Guo, Z. (2005). Ultrasonic songs of male mice. *PLoS Biology, 3*(12), 2177–2186.
Horner, K., Serviere, J., & Granier-Deferre, C. (1987). Deoxyglucose demonstration of in-utero hearing in the guinea pig foetus. *Hearing Research, 26*(3), 327–333.
Hotchkin, C., & Parks, S. (2013). The Lombard effect and other noise-induced vocal modifications: Insight from mammalian communication systems. *Biological Reviews, 88*(4), 809–824.
Janik, V. M. (2000). Whistle matching in wild bottlenose dolphins (*Tursiops truncatus*). *Science, 289*(5483), 1355–1357.

Janik, V. M., & Sayigh, L. S. (2013). Communication in bottlenose dolphins: 50 years of signature whistle research. *Journal of Comparative Physiology. A, Neuroethology, Sensory, Neural, and Behavioral Physiology, 199*(6), 479–489.

Janik, V. M., & Slater, P. J. B. (1997). Vocal learning in mammals. *Advances in the Study of Behavior, 26*, 59–99.

Jarvis, E. D. (2006). Selection for and against vocal learning in birds and mammals. *Ornithological Science, 5*, 5–14.

Jarvis, E. D. (2007). Neural systems for vocal learning in birds and humans: A synopsis. *Journal of Ornithology, 148*, S35–S44.

Jones, G., & Ransome, R. D. (1993). Echolocation calls of bats are influenced by maternal effects and changes over a lifetime. *Proceedings of the Royal Society B: Biological Sciences, 252*, 125–128.

Jürgens, U. (2002). Neural pathways underlying vocal control. *Neuroscience & Biobehavioral Reviews, 26*(2), 235–258.

Jürgens, U. (2009). The neural control of vocalization in mammals: A review. *Journal of Voice, 23*(1), 1–10.

Kelley, L. A., & Healy, S. D. (2011). Vocal mimicry. *Current Biology, 21*, R9–R10.

Kellogg, W. N., & Kellogg, L. A. (1933). *The ape and the child.* New York: McGraw-Hill.

Kikusui, T., Nakanishi, K., Nakagawa, R., Nagasawa, M., Mogi, K., & Okanoya, K. (2011). Cross fostering experiments suggest that mice songs are innate. *PloS ONE, 6*(3), 1–10.

King, S. L., Sayigh, L. S., Wells, R. S., Fellner, W., & Janik, V. M. (2013). Vocal copying of individually distinctive signature whistles in bottlenose dolphins. *Proceedings of the Royal Society of London B: Biological Sciences, 280*(1757), 20130053.

Klatt, D. H., & Stefanski, R. A. (1974). How does a mynah bird imitate human speech? *The Journal of the Acoustical Society of America, 55*, 822–832.

Konishi, M. (1963). The role of auditory feedback in the vocal behavior of the domestic fowl. *Zeitschrift für Tierpsychologie, 20*, 349–367.

Konishi, M. (1965). The role of auditory feedback in the control of vocalization in the white-crowned sparrow. *Zeitschrift für Tierpsychologie, 22*, 770–783.

Konishi, M. (2004). The role of auditory feedback in birdsong. *Annals of the New York Academy of Sciences, 1016*, 463–475.

Kroodsma, D. E., & Konishi, M. (1991). A suboscine songbird (eastern phoebe, *Sayornis phoebe*) develops normal song without auditory feedback. *Animal Behavior, 44*, 477–487.

Lahdenperä, M., Mar, K. U., & Lummaa, V. (2014). Reproductive cessation and post-reproductive lifespan in Asian elephants and pre-industrial humans. *Frontiers in Zoology, 11*(1), 54.

Lampe, U., Reinhold, K., & Schmoll, T. (2014). How grasshoppers respond to road noise: Developmental plasticity and population differentiation in acoustic signalling. *Functional Ecology, 28*(3), 660–668.

Lane, H., & Tranel, B. (1971). The Lombard sign and the role of hearing in speech. *Journal of Speech and Hearing Research, 14*, 677–709.

Lengagne, T., Aubin, T., Lauga, J., & Jouventin, P. (1999). How do king penguins (*Aptenodytes patagonicus*) apply the mathematical theory of information to communicate in windy conditions? *Proceedings of the Royal Society of London B: Biological Sciences, 266*, 1623–1628.

Lieberman, P. (1984). *The biology and evolution of language*. Cambridge, MA: Harvard University Press.

Madsen, P. T., Jensen, F. H., Carder, D., & Ridgway, S. (2011). Dolphin whistles: A functional misnomer revealed by heliox breathing. *Biology Letters, 8*(2), 211–213.

Madsen, P. T., & Surlykke, A. (2013). Functional convergence in bat and toothed whale biosonars. *Physiology, 28*(5), 276–283.

Mammen, D. L., & Nowicki, S. (1981). Individual differences and within-flock convergence in chickadee calls. *Behavioral Ecology and Sociobiology, 9*, 179–186.

Marsh, H., & Kasuya, T. (1986). Evidence for reproductive senescence in female cetaceans. *Reports of the International Whaling Commission, 8*, 57–74.

Marshall, A. J., Wrangham, R. W., & Arcadi, A. C. (1999). Does learning affect the structure of vocalizations in chimpanzees? *Animal Behavior, 58*, 825–830.
McComb, K., Moss, C., Durant, S. M., Baker, L., & Sayialel, S. (2001). Matriarchs as repositories of social knowledge in African elephants. *Science, 292*(5516), 491–494.
Metzner, W. (1989). A possible neuronal basis for Doppler-shift compensation in echo-locating horseshoe bats. *Nature, 341*(6242), 529–532.
Metzner, W. (1993). An audio-vocal interface in echolocating horseshoe bats. *The Journal of Neuroscience, 13*(5), 1899–1915.
Metzner, W. (1996). Anatomical basis for audio-vocal integration in echolocating horseshoe bats. *Journal of Comparative Neurology, 368*(2), 252–269.
Metzner, W., & Schuller, G. (2010). Vocal control in echolocating bats. In M. B. Stefan (Ed.), *Handbook of behavioral neuroscience* (Vol. 19, pp. 403–415). London: Elsevier.
Metzner, W., Zhang, S., & Smotherman, M. (2002). Doppler-shift compensation behavior in horseshoe bats revisited: Auditory feedback controls both a decrease and an increase in call frequency. *Journal of Experimental Biology, 205*, 1607–1616.
Miksis, J., Tyack, P., & Buck, J. (2002). Captive dolphins, *Tursiops truncatus*, develop signature whistles that match acoustic features of human-made model sounds. *The Journal of the Acoustical Society of America, 112*(2), 728–739.
Miller, P., Biassoni, N., Samuels, A., & Tyack, P. (2000). Whale songs lengthen in response to sonar. *Nature, 405*(6789), 903.
Mitani, J. C., & Gros-Louis, J. (1998). Chorusing and call convergence in chimpanzees: Tests of three hypotheses. *Behaviour, 135*, 1041–1064.
Moss, C. F., & Sinha, S. R. (2003). Neurobiology of echolocation in bats. *Current Opinion in Neurobiology, 13*(6), 751–758.
Nemeth, E., Pieretti, N., Zollinger, S. A., Geberzahn, N., Partecke, J., Miranda, A. C., et al. (2013). Bird song and anthropogenic noise: Vocal constraints may explain why birds sing higher-frequency songs in cities. *Proceedings of the Royal Society of London B: Biological Sciences, 280*, 20122798.
Nonaka, S., Takahashi, R., Enomoto, K., Katada, A., & Unno, T. (1997). Lombard reflex during PAG-induced vocalization in decerebrate cats. *Neuroscience Research, 29*(4), 283–289.
Nottebohm, F. (1970). Ontogeny of bird song. *Science, 167*(3920), 950–956.
Nottebohm, F. (1972). The origins of vocal learning. *American Naturalist, 106*, 116–140.
Nottebohm, F., & Liu, W.-C. (2010). The origins of vocal learning: New sounds, new circuits, new cells. *Brain and Language, 115*, 3–17.
Nottebohm, F., & Nottebohm, M. E. (1971). Vocalizations and breeding behavior of surgically deafened ring doves (*Streptopelia risoria*). *Animal Behavior, 19*, 313–327.
Nowicki, S. (1989). Vocal plasticity in captive black-capped chickades: the acoustic basis and rate of call convergence. *Animal Behavior, 37*, 64–73.
Owren, M. J., Amoss, R. T., & Rendall, D. (2011). Two organizing principles of vocal production: Implications for nonhuman and human primates. *American Journal of Primatology, 73*(6), 530–544.
Parks, S. E., Clark, C. W., & Tyack, P. L. (2007). Short- and long-term changes in right whale calling behavior: The potential effects of noise on acoustic communication. *The Journal of the Acoustical Society of America, 122*(6), 3725–3731.
Parris, K. M., Velik-Lord, M., & North, J. M. (2009). Frogs call at a higher pitch in traffic noise. *Ecology and Society, 14*(1), 25.
Payne, K., & Payne, R. (1985). Large scale changes over 19 years in songs of humpback whales in Bermuda. *Zeitschrift für Tierpsychologie, 68*, 89–114.
Payne, K., Tyack, P., & Payne, R. (1983). Progressive changes in the songs of humpback whales (*Megaptera novaeangliae*): A detailed analysis of two seasons in Hawaii. In R. Payne (Ed.), *Communication and behavior of whales* (pp. 9–57). Boulder, CO: Westview Press.
Pepperberg, I. M. (1999). *The Alex studies*. Cambridge, MA: Harvard University Press.
Pepperberg, I. M. (2010). Vocal learning in grey parrots: A brief review of perception, production, and cross-species comparisons. *Brain and Language, 115*, 81–91.

Petkov, C. I., & Jarvis, E. D. (2012). Birds, primates, and spoken language origins: Behavioral phenotypes and neurobiological substrates. *Frontiers in Evolutionary Neuroscience, 4*(12), 1–22.
Poole, J., Tyack, P., Stoeger-Horwath, A., & Watwood, S. (2005). Elephants are capable of vocal learning. *Nature, 434*(7032), 455–456.
Potash, L. M. (1972). Noise-induced changes in calls of the Japanese quail. *Psychonomic Science, 26*, 252–254.
Potvin, D. A., & Mulder, R. A. (2013). Immediate, independent adjustment of call pitch and amplitude in response to varying background noise by silvereyes (*Zosterops lateralis*). *Behavioral Ecology, 24*(6), 1363–1368.
Ralls, K., Fiorelli, P., & Gish, S. (1985). Vocalizations and vocal mimicry in captive harbor seals, *Phoca vitulina. Canadian Journal of Zoology, 63*(5), 1050–1056.
Reiss, D., & McCowan, B. (1993). Spontaneous vocal mimicry and production by bottlenose dolphins (*Tursiops truncatus*): Evidence for vocal learning. *Journal of Comparative Psychology, 107*(3), 301–312.
Rendell, L., Mesnick, S. L., Dalebout, M. L., Burtenshaw, J., & Whitehead, H. (2012). Can genetic differences explain vocal dialect variation in sperm whales, *Physeter macrocephalus*? *Behavior Genetics, 42*(2), 332–343.
Rendell, L., & Whitehead, H. (2005). Spatial and temporal variation in sperm whale coda vocalizations: Stable usage and local dialects. *Animal Behaviour, 70*(1), 191–198.
Richards, D. G., Wolz, J. P., & Herman, L. M. (1984). Vocal mimicry of computer-generated sounds and vocal labelling of objects by a bottlenosed dolphin, *Tursiops truncatus*. *Journal of Comparative Psychology, 98*, 10–28.
Romand, R., & Ehret, G. (2004). Development of sound production in normal, isolated, and deafened kittens during the first postnatal months. *Developmental Psychobiology, 17*, 629–649.
Ryan, M. J., Tuttle, M. D., & Rand, A. S. (1982). Bat predation and sexual advertisement in a neotropical frog. *American Naturalist, 119*, 136–139.
Sanvito, S., Galimberti, F., & Miller, E. H. (2007). Observational evidences of vocal learning in southern elephant seals: a longitudinal study. *Ethology, 113*, 137–146.
Scheiner, E., Hammerschmidt, K., Jürgens, U., & Zwirner, P. (2006). Vocal expression of emotions in normally hearing and hearing-impaired infants. *Journal of Voice, 20*(4), 585–604.
Schleidt, W. (1961). Operative Entfernung des Gehörorgans ohne Schädigung angrenzender Labyrinthteile bei Putenküken. *Experientia, 17*, 464–465.
Shannon, C., & Weaver, W. (1963). *Mathematical theory of communication*. Urbana, IL: University of Illinois Press.
Simpson, H. B., & Vicario, D. S. (1990). Brain pathways for learned and unlearned vocalizations differ in zebra finches. *The Journal of Neuroscience, 10*(5), 1541–1556.
Slabbekoorn, H., & Peet, M. (2003). Birds sing at a higher pitch in urban noise. *Nature, 424*, 267.
Smolker, R., & Pepper, J. W. (1999). Whistle convergence among allied male bottlenose dolphins (Delphinidae, Tursiops sp). *Ethology, 105*, 595–617.
Snowdon, C. T., & Elowson, A. M. (1999). Pygmy marmosets modify call structure when paired. *Ethology, 105*, 893–908.
Stoeger, A. S., Mietchen, D., Oh, S., de Silva, S., Herbst, C. T., Kwon, S., et al. (2012). An Asian elephant imitates human speech. *Current Biology, 22*(22), 2144–2148.
Street, R. L., & Giles, H. (1982). Speech accommodation theory: A social cognitive approach to language and speech behavior. In M. Roloff & C. R. Berger (Eds.), *Social cognition and communication* (pp. 193–226). Beverly Hills, CA: Sage.
Sugiura, H. (1998). Matching of acoustic features during the vocal exchange of coo calls by Japanese macaques. *Animal Behavior, 55*, 673–687.
Titze, I. R. (1994). *Principles of voice production*. Englewood Cliffs, NJ: Prentice Hall.
Todt, D. (1975). Social learning of vocal patterns and modes of their application in grey parrots (*Psittacus erithacus*). *Zeitschrift für Tierpsychologie, 39*, 178–188.
Trainer, J. M. (1989). Cultural evolution in song dialects of yellow-rumped caciques in Panama. *Ethology, 80*, 190–204.

Turnbull, S., & Terhune, J. (1993). Repetition enhances hearing detection thresholds in a harbour seal (*Phoca vitulina*). *Canadian Journal of Zoology, 71*(5), 926–932.

Tyack, P. L. (1986). Population biology, social behavior, and communication in whales and dolphins. *Trends in Ecology and Evolution, 1*, 144–150.

Ulanovsky, N., Fenton, M. B., Tsoar, A., & Korine, C. (2004). Dynamics of jamming avoidance in echolocating bats. *Proceedings of the Royal Society of London B: Biological Sciences, 271*(1547), 1467–1475.

Vehrencamp, S. L. (2001). Is song-type matching a conventional signal of aggressive intentions? *Proceedings of the Royal Society of London B: Biological Sciences, 268*, 1637–1642.

Watson, S. K., Townsend, S. W., Schel, A. M., Wilke, C., Wallace, E. K., Cheng, L., et al. (2015). Vocal learning in the functionally referential food grunts of chimpanzees. *Current Biology, 25*(4), 495–499.

Watwood, S., Tyack, P., & Wells, R. (2004). Whistle sharing in paired male bottlenose dolphins, *Tursiops truncatus*. *Behavioral Ecology and Sociobiology, 55*(6), 531–543.

Weiss, D. J., Garibaldi, B. T., & Hauser, M. D. (2001). The production and perception of long calls by cotton-top tamarins (*Saguinus oedipus*): Acoustic analyses and playback experiments. *Journal of Comparative Psychology, 115*, 258–271.

Wemmer, C., & Mishra, H. R. (1982). Observational learning by an Asiatic elephant of an unusual sound production method. *Mammalia, 46*, 556–557.

West, M. J., & King, A. P. (1988). Female visual displays affect the development of male song in the cowbird. *Nature, 334*, 244–246.

Wich, S. A., Swartz, K. B., Hardus, M. E., Lameira, A. R., Stromberg, E., & Shumaker, R. W. (2009). A case of spontaneous acquisition of a human sound by an orangutan. *Primates, 50*(1), 56–64.

Wild, J. M. (1997). Neural pathways for the control of birdsong production. *Journal of Neurobiology, 33*(5), 653–670.

Winter, P., Handley, P., Ploog, D., & Schott, D. (1973). Ontogeny of squirrel monkey calls under normal conditions and under acoustic isolation. *Behaviour, 47*, 230–239.

Zahavi, A., & Zahavi, A. (1997). *The handicap principle*. Oxford, England: Oxford University Press.

Chapter 10
Vertebrate Bioacoustics: Prospects and Open Problems

W. Tecumseh Fitch

Abstract Vertebrate bioacoustics has made great gains in the last two decades in terms of increased understanding of the functional morphology of the vocal tract: how sounds are produced by the larynx (or syrinx in birds) and then filtered in the vocal tract. Despite this fundamental progress, many unusual features seen in the vocal anatomy of particular vertebrates remain poorly understood. This results mainly from the fact that these potential vocal adaptations were described by classical comparative anatomists more than a century ago, long before a good understanding of the physics and physiology of vocal production was in place. Adding to this difficulty, many of the descriptions of anatomical peculiarities were published in non-English languages and often in hard-to-access journals. This chapter starts with a short review of the rise and fall of comparative anatomy as a leading branch of biology, focusing especially on vocal anatomy. It then provides a brief overview of the many known anatomical peculiarities that, although poorly understood, are thought to play some role in vocal production. Both morphology and possible function are considered, and any available empirical research is reviewed. The chapter covers most known vocal peculiarities including air sacs, vocal fold modifications, the syringeal bulla present in most ducks, or the elongated trachea seen in many bird species. Such unusual modifications of vocal anatomy will provide a rich and rewarding topic of future research in bioacoustics.

Keywords Animal bioacoustics • Animal communication • Comparative anatomy • Functional morphology • Laryngeal air sacs • Larynx • Syrinx • Tracheal elongation • Vocal tract

---

W.T. Fitch (✉)
Department of Cognitive Biology, University of Vienna,
14 Althanstrasse, Vienna 1020, Austria
e-mail: tecumseh.fitch@univie.ac.at

R.A. Suthers et al. (eds.), *Vertebrate Sound Production and Acoustic Communication*, Springer Handbook of Auditory Research 53,
DOI 10.1007/978-3-319-27721-9_10

10.1 Introduction

As this volume amply illustrates, the field of bioacoustics has made great progress in the last two decades, particularly with regard to a deeper understanding of how vertebrate vocalizations are produced, and their significance in a wider behavioral and evolutionary context. There were some previous islands of important progress, and even genius, in certain isolated domains: Paulsen's high-speed excised larynx experiments in the 1960s (Paulsen 1967), Martin's brilliant work on toad vocal production in the early 1970s (Martin 1971; Martin and Gans 1972), or Gaunt's research on the avian syrinx in the late 1970s and early 1980s (Gaunt et al. 1976; Gaunt and Gaunt 1985). But it was only in the 1990s that a research community began to develop and cohere, a community that asked similar questions and applied similar methods to diverse species, and then compared and contrasted the results. As the chapters in this volume amply demonstrate, this comparative approach to bioacoustics has become an increasingly unified and prominent subfield of biology. A strong virtue of comparative bioacoustics is its unified vertical scope, stretching from physical principles, through essentially physiological issues, right up to questions of cognition, neural control, evolution, and adaptive significance.

Despite these gains, this final chapter is written not to praise bioacoustics, but to bury it—bury it in new exciting questions that remain unanswered and in many cases unasked. Vertebrate bioacoustics comprises two major elements: vocal acoustics (the physics and physiology of vocal production) and comparative anatomy. These of course link to numerous other disciplines, especially "evo-devo" (evolutionary developmental biology), which explains how these structures arose, and animal behavior/animal cognition, which explores how they are perceived and put to use in communication. Although great progress has been made in our understanding of acoustic aspects of vertebrate vocalization, from production through propagation through perception and evolution, the connections to the anatomical domain remain relatively weak. An odd historical fact underlies this: The golden age of comparative anatomy (including of vocal organs like the larynx and syrinx) came and went before anyone understood how these organs actually worked mechanistically. The key functional insights were initiated, for the human larynx at least, in the domain of speech science in the 1940s and only became clear in the late 1950s. These insights took much longer to percolate over from the essentially engineering problems of early speech scientists to biologists exploring animal communication. As a result, by the time anyone knew how vocal organs actually worked, the scientists who had a rich grasp of their comparative anatomy were already long dead. Thus there was, and remains, a considerable gap between the richly detailed descriptions of differences between the larynges and especially syringes of different vertebrates from days past, and today's understanding of what functions these differences might subserve, mechanistically, or how and why they evolved. One goal of this chapter is to help close this gap, by illustrating some of the rich diversity of vocal anatomy documented by the classical comparative anatomists (much of it written in German). Because fish have already been ably reviewed in this volume, this chapter focuses on tetrapods.

Another field of connections that remain underexplored at present concerns the evolution and development of vocal organs and the vocal tract/upper respiratory tract more generally. The vocal tract is a dauntingly complex area of the body, and many of the key structures for vocal production and perception show a fascinating history of adaptation and exaptation. Certain aspects of these connections have already been well explored (cf. Fitch 2012). For example, Darwin suggested that the tetrapod lungs evolved from fish swim bladders, which acquired a new use when our aquatic ancestors emerged onto land. However, it is now clear that lungs actually evolved in basal fish (many of which still retain lungs today), and that the specialized swim bladders of teleost fish evolved from lungs (e.g., Alexander 1987; Arnason et al. 2001). Similarly the evolutionary story of the mammalian middle ear bones, which were exapted from articulatory jaw bones, has been well explored, thanks in part to a remarkable fossil record that preserves this transition (Allin 1975). Unfortunately, because the core vocal organs (the syrinx and larynx) are mostly cartilaginous and leave fossil traces very rarely, their origins and evolution are much hazier. There are other central questions for an evo-devo approach to vocal communication that deserve greater attention as well: Why is the avian syrinx so variable, and the mammalian larynx so conservative (Fitch and Hauser 2002)? Has the complexity of the syrinx in some clades (especially songbirds) been partly responsible for the explosion of species in those clades (Fitzpatrick 1988; Baptista and Trail 1992)? What are the precise homologies between vocal components across fish, amphibians, mammals, and birds—gene expression data may hold key insights that remain underexplored (cf. Bass and Chagnaud 2012). These and other connections represent equally important domains of future bridge building that, but because solutions seem at present far-off, they are not the focus of the current chapter.

This chapter has two main parts. The first provides a historical overview of the discipline of comparative anatomy with a focus on vocal anatomy. The goals are first to give those bioacousticians unfamiliar with this field of study a road map and highlight some of the key figures in the history of the discipline (most of whom had something to say about vocal anatomy), and second to highlight many of the anatomical issues relevant to vocal production that were raised, but never answered, by this field before the end of the its “golden age” in the early twentieth century.

The second part of the chapter then systematically reviews the many modifications of vocal anatomy that have been documented in the far-flung literature of comparative anatomy, as well as a review of the various functions for these anatomical peculiarities (both acoustic and other) over the years. The goal is again to provide a road map, but this time into the terra incognita of vertebrate vocal adaptations, most of which remain poorly understood, or even completely unstudied, today. In a sense, this second part tries to roughly chart out the territory that bioacoustics can hopefully move into and explore in the coming decades.

10.2 The Historical Development of Comparative Vocal Anatomy

A brief history of comparative anatomy shows how modern bioacoustics has ended up in its current quandary, also highlighting the authors who first described vocal traits of interest today. Comparative anatomy, like so many aspects of Western thought, begins with the Greeks, and with Aristotle in particular, whose *Historia Animalium* already documented many insights into the biology of diverse animal species at around 350 BC (Aristotle 350 BC). For example, Aristotle already knew that some birds must learn their song, or that dolphins have lungs rather than gills, and can use their respiratory tract to produce sounds. Unfortunately, the classical period ended with Galen's often erroneous descriptions of human anatomy, first circulated around 200 AD, which took hold and were treated as gospel for the next millennium. There were a few flashes of light during this long period of stasis. For example, Frederick II of Hohenstaufen was a Holy Roman Emperor who conquered Jerusalem, cultivated a learned multicultural court, and himself spoke six languages (including Arabic). Frederick published the first dedicated work on ornithology (a manual of falconry) around 1248 (Frederick II 1943; ca. 1248) where he not only described tracheal elongation in cranes for the first time, but also described in detail how to "play" a dead crane by blowing air into its trachea and then squeezing, like a bagpipe! Other isolated insights came from Leonardo da Vinci, who dissected both humans and animals and recorded many accurate observations in his notebooks—but little of this became widely known until long after his time.

The real birth of comparative anatomy occurred rather suddenly around 1600 (Cole 1949). Human anatomy had received a great impetus from the widely distributed work of Andreas Vesalius published in 1543, and Belon had already described the intranarial epiglottis of cetaceans by 1551. Ruini published a detailed and accurate anatomy of the horse, including its larynx, in 1598. Truly comparative work began with the nearly simultaneous publications of Fabricius (1600), who presented figures comparing the larynges of pigs, cattle, and sheep, and his student Casserius (1601), who dissected and described 20 mammalian species including those of humans and an ape. Together, these authors presented important insights into laryngeal function that remain valid to this day: Vesalius reported that cutting the recurrent laryngeal nerve silenced the voice, and Fabricius reports, based on the first excised larynx experiments, that the larynx itself (rather than lung movements) is the source of vocal sounds in mammals. Despite these virtues, these authors often had long speculative discourses, divorced from observation or experiment, that are of purely historical interest (a spectacular example is William Harvey's idea that the recurrent nerve, stretching down around the aorta like a pulley, acts as a sort of plucked string to produce the voice). In the two centuries that followed, apes and dolphins were of particular interest. For instance, Tyson (1680) dissected a porpoise and observed that the insertion of the cetacean larynx into the nasal cavity was quite different from most mammals, and in 1699 published a detailed anatomy of a chimpanzee (Tyson 1699). But it was not until 1778 that Camper described the vocal anatomy of an orangutan, and inferred that the large laryngeal air sacs would prevent speech by diverting the air from the vocal folds (Camper 1779). By this logic, it is unclear how orangutans can produce any vocalizations at all—an issue that Camper apparently did not consider.

The modern age of comparative anatomy begins with the founding of the great European natural history museums, for example, the Hunterian Museum in London (founded 1785 and partly absorbed by the British Museum of Natural History in 1881), or the Museum of Comparative Anatomy in Paris. These museums provided both large and diverse collections of specimens from around the world, along with employment for the many technicians and scientists necessary to prepare, study, and classify these specimens. Unfortunately, because most of these individuals had never even seen the living animals whose remains they studied, a regrettable tendency developed to derive detailed inferences about function from anatomy alone. Baron Cuvier (1769–1832) elevated this approach to a principle: that function can be inferred *only* from careful study of anatomical form, and that experimentation was in vain. The great comparative anatomist Richard Owen, a famous opponent of Darwin's evolutionary ideas, made excellent use of the large collections of the Hunterian and British museums to produce beautiful and still accurate descriptions of cheetahs, giraffes, anteaters, and other exotic species, but often ventured into this dangerous territory of functional speculation.

Although perhaps forgivable in 1600, this speculative tradition unfortunately continued into the twentieth century, especially in the work of Victor Negus (1949), who accompanied accurate and informative figures of the larynges of a wide diversity of species with speculations about function that verge on ridiculous. He asserts, for example, that laryngeal air sacs have no vocal function based on the observation that "certain animals with somewhat similar sacs are not noisy, while others with no dilatation are extremely vocal … the adult male Gorilla, with big sacs, is usually silent; the young of this species, and also young Chimpanzees, both with small or absent sacs, can be very noisy" (pp. 52–53). He seems hardly to have considered the idea that the acoustic role of air sacs might be more specific or subtle than making an animal "noisy" or "silent." This is not an isolated example: the entire book is organized around inferred functional principles that subsume accurate anatomical observations into a speculative (and often false) functional framework. This particular example is emphasized not only because Negus (1949) is sometimes treated as gospel (e.g., Lieberman 1984; Harrison 1995) but also because, unfortunately, most of the anatomical descriptions available today, particularly in English, suffer from this type of speculative inference of function based on overinterpretation of anatomical data. For example, Harrison's 1995 book repeats the argument from Negus quoted previously almost verbatim, and concludes that laryngeal air sacs, including huge ones like those present in great apes, are "relatively functionless" (Harrison 1995, p. 104), and Hilloowala (1976) continues Negus' contention that the "primary function" of the mammalian epiglottis is in aiding olfaction, speculating that primate air sacs compensate for the inability in herbivorous species to include the mouth as a resonant chamber (Hilloowala 1976). Similar speculations are typical in the literature on the avian syrinx, which featured a long debate about whether birds are more similar to oboes, trombones or organs—it turns out their functioning resembles none of these instruments (cf. Nowicki and Marler 1988). Many more examples of such speculative extrapolation can be found in even the recent anatomical literature.

Modern workers therefore cannot accept all of what they read in the anatomical literature at face value, and need to carefully separate anatomical observations from functional speculations. Fortunately, there is a large body of comparative anatomical literature on the vocal tract that mostly focuses on accurate description, often accompanied by beautiful figures, and covering the vertebrates with reasonable completeness (Göppert 1901; Gaupp 1904; Wiedersheim 1904). The "golden age" of comparative anatomy stretched from around Owen's time in 1860 (e.g., Gegenbauer 1874) to the first few decades of the twentieth century, and at this time German was considered to be *the* language in which comparative anatomy was published (even for English writers). Thus, unfortunately for many bioacousticians, this literature is written mostly in German, and in a style that can be difficult for even modern German speakers to decode. Unlike much of the English literature, the German literature also tends to have excellent historical and scholarly depth, citing earlier authors accurately and building on previous knowledge. But the linguistic and historical facts lead to the scientific problem that we face today: A large body of descriptive knowledge exists, but it is relatively inaccessible in our modern Internet-based knowledge economy and thus essentially buried to many readers. The best source of information for the mammalian larynx is Schneider (1964), but even this is hard to get, and written in German. It would be a monumental (but extremely valuable) task to translate and update this body of literature (cf. Gegenbauer 1878), but no one appears to have embarked on such an endeavor.

10.2.1 The Rise of Speech Science and Today's Dilemma

A detailed history of the rise of speech science was already given by Fitch and Suthers (Chap. 1) and will not be repeated here. Despite a long history of interest in how speech works, often focused on creating speaking machines (Dudley and Tarnoczy 1950), the creation of instrumentation such as the oscilloscope and spectrograph played a key role in the birth of modern speech science in the 1940s (Linggard 1985). Chiba and Kajiyama's breakthrough volume on the nature of formants and vowels (Chiba and Kajiyama 1941) led quickly to the clear formulation of the source-filter theory of speech production (Fant 1960, 2001). At roughly the same time, the nature of the physics and physiology underlying human vocal fold vibration was being clarified, and by the late 1950s van den Berg could confidently state the essential principles of the myoelastic-aerodynamic (MEAD) theory (van den Berg 1958). Thus, by the mid-1960s the basic theoretical edifice was present that allowed a clear understanding of human vocal production.

The application of these basic principles to animal communication took much longer, and is again reviewed in the introduction to this volume. Unlike the situation for human speech, in which the principles of the source-filter and MEAD theories quickly became common knowledge for the subsequent generations of speech scientists, application of these principles to other species happened piecemeal. Often small communities (e.g., the bat echolocation community) grasped these principles before they were applied more broadly. Thus it took almost four decades, from the 1960s to around 2000, before it became clear to a large percentage of the animal communication community that the same basic principles apply to most, if not all, tetrapods. The practical application of tools from speech science, such as linear predictive coding (LPC) analysis or analysis/resynthesis of animal vocalization for playback experiments, has really taken off only in the last decade. One can still find very basic confusions (e.g., between formants and harmonics) in some relatively recent literature (e.g., Geissler and Ehret 2002), and it is fair to say that animal bioacoustics still has not reached the broad consensus of techniques, approaches, and terminology that was already present in speech science by 1970. Nonetheless, an abundance of data gathered in the last decade strongly suggests that the principles governing human speech production apply, with a few exceptions, to other terrestrial vertebrates.

The paradox of modern comparative bioacoustics, then, is that we finally now have the conceptual and experimental tools available to understand the anatomical knowledge accumulated over many centuries by comparative anatomy, but this wealth of knowledge is rather inaccessible. It is hard work to unearth and interpret the understanding accumulated by these anatomists. The next section provides a personal summary of the many fascinating observations made during this early period (although my own knowledge of the literature is imperfect and incomplete) but hopefully this example will spur others to go deeper and further.

10.3 Understanding Vocal Diversity: Unsolved Problems

The remaining goal of this chapter is to provide an overview of the many anatomical features associated with the larynx and vocal tract that, although well-documented in the older anatomical literature, remain poorly understood acoustically. These include air sacs, tracheal diverticulae or pouches, syringeal bullae, or hollow cheekbones along with source modifications such as vocal membranes or pads. These anatomical peculiarities remain poorly understood or even mysterious from the viewpoint of their acoustic (or other) functions. In most cases, there may be nonvocal functions of such features, but for the most part the lack of scientific attention to these structures means that we simply do not know what they do or why they are present in some species or clades and not others. The treatment that follows errs on the inclusive, and probably includes some traits that will turn out to have only nonacoustic functions, in the hope that future research will clarify the issue.

Fig. 10.1 The normal tetrapod voice source is the larynx, illustrated here in mammals (a squirrel), anurans (a frog), and reptiles (a gecko). In anurans and reptiles, the vibrating tissues within the larynx are called vocal cords, and in mammals they are termed vocal folds. Note as well the large inflated air sac of the frog

10.3.1 Vocal Diversity in the Vocal Source

The "standard" vocal source among tetrapods is within the larynx, created by vibrations of either the vocal cords in reptiles and amphibians, or the vocal folds in mammals (Fig. 10.1). These structures are typically given different names because only in mammals are the vibrating tissues invested with muscle. Extirpation of the vocal cords or folds typically leads to muteness in anurans (reviewed in Weber 1976) and mammals (Griffin 1958).

In addition, in certain vertebrate clades novel sources have evolved (Fig. 10.2). As highlighted in Chap. 1, several vertebrate groups have evolved novel sound-producing sources. These include all of the sound producing mechanisms of fish (see Parmentier and Fine, Chap. 2), with the possible exception of lungfish, which use their primitive larynx to produce squeaks, burps, and "vocal sounds" (M'Donnell 1860). Other fish sounds are produced nonlaryngeally by a wide variety of novel mechanisms of diverse mechanical function and anatomical origin (cf. Parmentier and Fine, Chap. 2 and Ladich and Fine 2006). Among tetrapods, the avian syrinx and odontocete nasal sac system are examples of novel vocal sources.

The avian syrinx is a novel synapomophy of the entire class Aves. All birds possess at least a simple syrinx, and no living nonbirds have one (presumably the syrinx evolved among the feathered dinosaurs that gave rise to living birds). Toothed whales have also evolved a novel organ, the nasal sac complex, which seems to

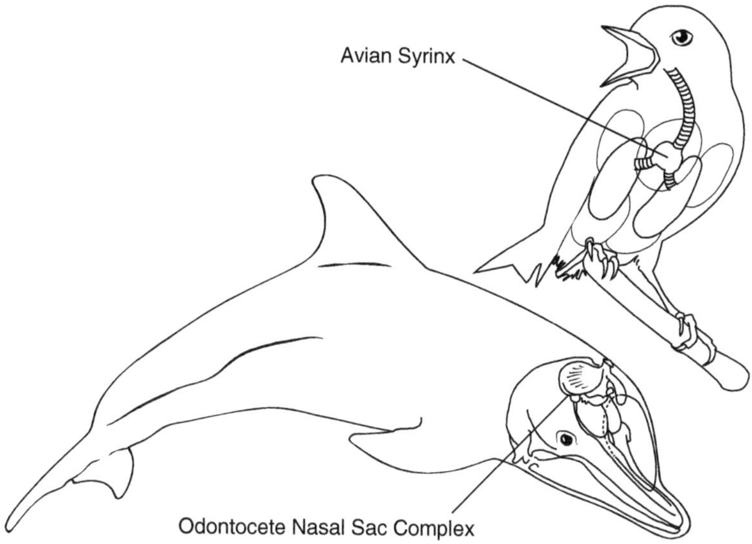

Fig. 10.2 Two examples of novel vocal sources: the syrinx of birds and the nasal sac complex of odontocete cetaceans

provide the main (or even sole) source of their sounds, again discussed earlier. Finally, those fish that make sounds use a bewildering diversity of mechanical sources to do so, often involving either vibrations of the swim bladder or stridulation of various bones (Bass 1990; Hawkins 1993).

Beyond these truly novel sources, several types of less extreme change can be observed in living tetrapods. Perhaps the most striking sort are enlarged larynges. Multiple clades, including humans, exhibit sexual dimorphism of the larynx (Kelley 1988; Titze 1989), but in some species this is taken to rather remarkable extremes. The best known is the huge hypertrophy of the larynx and hyoid bone in howler monkeys of the genus *Alouatta*. These monkeys have by far the largest relative larynx size among primates (Fig. 10.3). Howlers live, mostly arboreally, in dense forest environments, and produce very loud territorial vocalizations, often as a group (Altmann 1959; Whitehead 1995). Male howlers in particular have evolved a greatly expanded larynx, with very long vocal folds, and an air sac system including a midline subhyoid airsac that extends into a very large bullate hyoid bone (Kelemen and Sade 1960; Schön Ybarra 1986). The existence of this "super larynx" has been known since Darwin cited it as a potential example of sexual selection (Darwin 1871), but there is no experimental work to clarify how the vocal folds vibrate or what the function of the air sac system might be. Howler monkeys are particularly suited to a comparative approach (e.g., Dunn et al., 2015) because there are many species, and their phylogeny is now well understood based on modern molecular methods (Schneider et al. 1993). The various howler species exhibit considerable

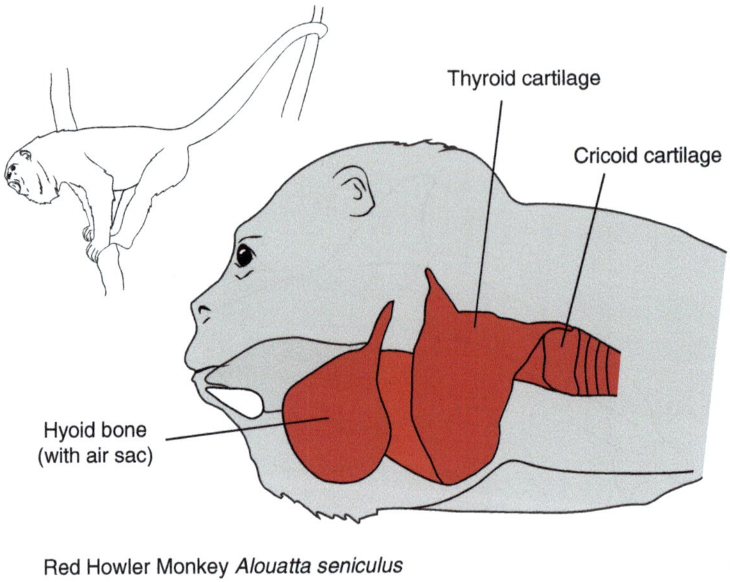

Fig. 10.3 Howler monkeys have an enlarged larynx and a bullate hyoid bone into which the subhyoid air sac inserts (the illustration is based on a red howler monkey, *Alouatta seniculus*)

variation in the size of the vocal structures both within a species (with males being much larger than females) and between species, and howler bioacoustics is a topic very well worth pursuing.

Several other mammals have hypertrophied larynges (Frey and Hofmann 2000; Frey et al. 2011). But the prize for largest larynx of all, relative to body size, must go to the hammerhead bat (*Hypsignathus monstrosus*), whose larynx is so enlarged that it fills the entire thoracic cavity, displacing the lungs themselves down into the abdomen (Sprague 1943; Schneider et al. 1967). This species, shown in Fig. 10.4, clearly deserves its informal moniker as a "flying larynx." As in howlers, this huge larynx is sexually dimorphic and only the male exhibits an enlarged larynx. Although little is known about the behavior of this African forest species, the males form leks along riverbeds from which they utter their very loud, monotonous, and metallic call (Bradbury 1977); females fly along the river, apparently evaluating these calls. Because most females mate with just a few of the many displaying males, this is a species in which intense sexual selection appears to have driven the evolution of a truly astonishing vocal apparatus.

It is worth noting that humans have an unusually pronounced vocal dimorphism in the size of the adult male larynx, with male vocal folds about twice the length of those in women (Titze 1989). This greatly exceeds the typical body size difference between men and women (where men are roughly 20 % heavier), and thus indicates true vocal dimorphism. Enlargement appears to be under the control of steroid hormones that begin circulating at puberty (see Chap. 8, Taylor et al.). Despite a common assumption that such dimorphism is typical in animals, existing data do not

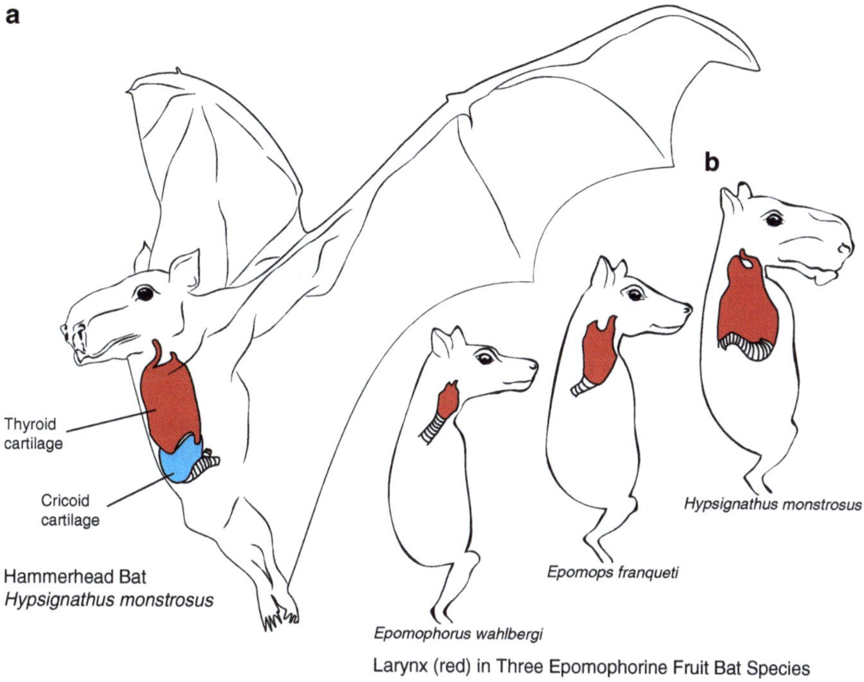

Fig. 10.4 (**a**) The hammerhead bat *Hypsignathus monstrosus* has the largest known larynx relative to body size: It fills the entire thoracic cavity, pushing the trachea and lungs down against the diaphragm. (**b**) The inset shows the less pronounced laryngeal enlargement of males of other bat species in the same subfamily, with the hammerhead bat shown again for comparison

support this intuitive idea (cf. Miller et al. 2008), for although a slightly larger source may be typical of males, it is only when this size difference exceeds that expected due to the (typically) slightly larger body size of males that we can accurately speak of a sexual dimorphism. Thus, despite the existence of numerous known examples of vocal sexual dimorphism, it remains unclear for the vast majority of species whether this is disproportionate to body size, and it thus remains unknown how typical, or uncommon, the vocal sexual dimorphism of our own species is.

10.3.1.1 Vocal Fold Variation

Numerous tetrapod species possess modifications of the vocal cords (anurans/reptiles) or vocal folds (mammals) within the larynx. The best known example is perhaps the tungara frog (*Physalaemus pustulosus*); these frogs have evolved a novel growth on their vocal cords that can be selectively engaged in vibration for a particular call type, the "chuck" (see Chap. 3, Colafrancesco and Gridi-Papp for more detail). Other anurans also have pads within their larynx (Weber 1977). There is extensive variation in the size and shape of anuran vocal cords, hyoid bones, and laryngeal cartilages (Trewavas 1932), but this interesting example of vocal diversity remains poorly documented and understood. It is probably adaptively relevant, given the importance of anuran vocalization in both mate choice and speciation (Blair 1958, 1964) and their apparently innate and species-specific calls (Gingras et al. 2013). Similar diversity in laryngeal morphology has been documented in geckos, the most vocal lizard clade, but its acoustic function has not been studied (Marcellini 1977; Moore et al. 1991; Rittenhouse et al. 1997; Russell et al. 2000).

The large roaring cats, members of the genus *Panthera*, all have massive upward extensions of the vocal folds that serve in the production of their powerful roar vocalizations (Hast 1986; Titze et al. 2010). Vocal fold length itself can also be increased, which requires concomitant changes in the laryngeal skeleton to contain these elongated folds. This accommodation explains why men have a thyroid notch that is lacking in women: This reflects the considerable changes in thyroid cartilage shape that occur in males at puberty (Sachs et al. 1973; Titze 1989). Similar moderate changes in larynx size, presumably related to increased relative vocal fold length, are found in several other mammals as well (Frey and Riede 2003; Frey et al. 2011). Finally, the angle of the vocal folds within the larynx, relative to the trachea, is highly variable (particularly within pinnipeds) and this may at least partially represent a way to fit elongated vocal folds, supporting low fundamental frequencies (f_0s), within the laryngeal skeleton (Schneider 1964, p. 70).

Another class of vocal fold adaptation works in the opposite direction: to allow very high-frequency, even ultrasonic, vocalizations. These are again upwardly directed extensions of the normal vocal folds, but here taking the form of very thin membranes. These are called vocal membranes (or in the older primate literature "vocal lips") and are common in primates (Starck and Schneider 1960; Mergell et al. 1999) and typical in echolocating bats (Griffin 1958). In bats these thin membranes are only about 6–8 μm in thickness (Griffin 1958), and have been shown by cutting to be responsible for high-frequency echolocation calls (Griffin 1958; Novick and Griffin 1961).

10.3.1.2 Variation in the Avian Syrinx

A fascinating type of vocal diversity that has received inadequate experimental attention is the remarkable variability in form, size, and presumed function of the avian syrinx. In songbirds (used here as a shorthand for the oscine passerines, which make up roughly half of all bird species), the syrinx is relatively consistent in form (cf. Chap. 5, Düring and Elemans) with the same sets of cartilages and vibrating membranes present in all songbirds. Nonetheless the songbird syrinx varies in its relative size, and also the degree of asymmetry and the role played by the two sides. For example, in some species virtually all song syllables seems to be produced by one side of the syrinx (Nottebohm 1971; Nottebohm and Nottebohm 1976); in others remarkable feats of alternation or coordination of sounds produced by the two sides of the syrinx are observed (Suthers 1999; Suthers et al. 2011). It may be that the songbird syrinx represents the pinnacle of complexity in a vocal organ on Earth today, and that once this songbird optimum was reached species vary mainly in the way that they "play" this instrument.

However, in nonoscine birds, syringeal diversity is the rule: so much so that the syrinx has played an important traditional role in avian classification and systematics (Garrod 1874–1878; Beddard 1898; Ames 1971). However, with few exceptions (e.g., some ducks, owls, and chickens) the functional relevance of this anatomical diversity remains completely unstudied, and even in those cases that have been studied remains poorly understood. Even a superficial perusal of the plates in Ames (1971) reveals a bewildering complexity and variability in all aspects of this organ. Given the importance of bird vocalizations as a topic for evolutionary and behavioral study among biologists, this anatomical richness would surely repay more detailed experimental study. This review only highlights a few areas of particular interest.

Perhaps the most striking set of variations are seen among the various duck species (members of the family Anatidae). Most male ducks possess a tracheal bulla—a swelling of the syringeal cartilages that may be hard or soft walled, and is remarkable for its diversity among relatively closely related species (Garrod 1875; Johnsgard 1961; Warner 1971). A few examples of anatid syringeal anatomy are illustrated in Fig. 10.5. Most authorities have presumed that because it creates a

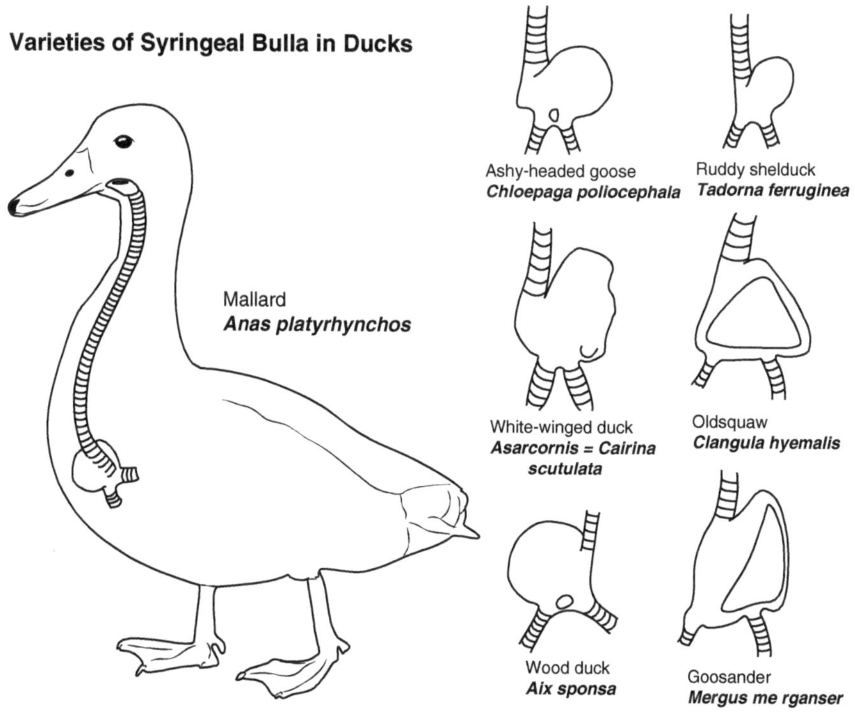

Fig. 10.5 Syringeal bulla in ducks. The syrinx with bulla is illustrated *in situ* in a mallard *Anas platyrhynchos*; to the *right* are anatomical variants of the bulla in six other duck species. The function of these bullae is unknown

large air space intimately associated with the syringeal membranes, the syringeal bulla plays some role in vocal variability among duck species. Waterfowl are renowned for their highly distinctive and variable courtship displays (Lorenz 1953; Johnsgard 1978), which are largely innate in males. Anatid syringeal diversity may provide a morphological means to diversify vocalizations reliably (cf. Miller et al. 2007). Given the ready availability of ducks (e.g., the common mallard *Anas platyrhynchos* has a well-developed syringeal bulla, and many other species are widely hunted or kept in parks and zoos), future research should focus on understanding how this well-documented form of syringeal variability translates into vocal acoustic diversity.

A second area of key difference concerns the general type of syrinx and its capacity for one- or two-voice phenomena. The songbird syrinx is clearly capable of independent control of two separate sets of vibrating vocal membranes (Suthers 1990; Suthers et al. 2011). In contrast, many other bird groups have a tracheal syrinx that has a pair of membranes directly facing each other across the tracheal lumen, in a configuration more reminiscent of the glottis in other tetrapods. These membranes bulge into the lumen and contact one another during vocalization (Goller and Larsen 1997; Larsen and Goller 2002). On its face, the anatomy in this case would not seem to support independently controlled vibration rates in the two membranes, and thus would appear incapable of two-voice phenomena. However, Zollinger and colleagues have recently documented complex vibratory regimes in one-half of a songbird syrinx (the mockingbird *Mimus polyglottos*) that are suggestive of two-voice phenomena, suggesting caution is warranted concerning this topic (Zollinger et al. 2008).

Modifications of the syringeal membranes themselves are another fascinating phenomenon. For example, male northern gannets (*Sula bassana*) have a peculiar protuberance within or beside the syringeal membrane, briefly described by Garrod: in "the lower portion of the windpipe of a male gannet … a pair of fatty bodies are developed just above the bifurcation of the bronchi, the like of which I have not elsewhere seen" (Garrod 1876, p. 335). Such a massive syringeal addition might provide a means of producing lower frequency fundamentals in gannet vocalizations, reminiscent of the vocal cords in tungara frogs (Gridi-Papp et al. 2006).

10.3.2 Morphological Diversity in the Vocal Tract

10.3.2.1 Tracheal Elongation

Because the avian sound source, the syrinx, lies at the base of the trachea, birds are unusual in having a suprasource vocal tract that includes the trachea. As a result, avian vocal tracts are disproportionately long compared to other mammals, and a moderate-sized bird such as a parrot or mynah has a vocal tract roughly as long as that of a human (Klatt and Stefanski 1974).

This fact about avian vocal production seems to underlie the repeated evolution of an elongated trachea in many bird lineages (Roberts 1880; Amadon 1969). As

Fig. 10.6 Tracheal elongation in birds. Illustrated are the different morphological variants shown in (from *left* to *right*) the crested guinea fowl (*Guttera pucherani edouardi*), the Eurasian spoonbill (*Platalea leucorodia*), the trumpeter swan (*Cygnus buccinator*) and (*top right*) the trumpet manucode (*Manucodia* = *Phonygammus keraudrenii*)

these clades are widely dispersed across the class Aves, and the precise form of tracheal elongation (TE) varies considerably between species, this is a trait that appears to have evolved convergently at least six times (Fitch 1999). In certain clades, including the cracids (guans and currasows), or in manucodes, the trachea forms loops that are relatively unconstrained by other aspects of the bird's anatomy (Amadon 1969; Clench 1978). In others, the elongated trachea invaginates into some portion of the bird's skeleton, such as the sternum in some swans and cranes (Roberts 1880), or the clavicle in the crested guinea fowl (Frith 1994). This diversity of forms of TE is illustrated in Fig. 10.6. TE is apparently found only in birds; the closest nonavian phenomenon seems to be the elongation of the bronchi seen in certain tortoises, which may allow for flexible extension and retraction of the neck (Siebenrock 1899). This phylogenetic exclusivity of TE to birds is consistent with the fact that only birds produce sounds with a vocal organ at the base of the trachea.

The many hypotheses that have been offered over the years to explain tracheal elongation have been previously reviewed (cf. Frith 1994; Fitch 1999). Most commentators have concluded that tracheal elongation has an acoustic function. In particular, by elongating the vocal tract, and thus lowering formant frequencies, tracheal elongation may serve to exaggerate the impression of size conveyed by vocalizations (Fitch 1999; Jones and Witt 2014), in a way directly analogous to the laryngeal lowering that has independently evolved in numerous mammalian clades (cf. Taylor et al., Chap. 8). Birds in general are probably sensitive to formant frequencies: many birds can imitate human speech (Nottebohm 1975; Pepperberg 2005) and/or learn specific patterns of harmonic amplitudes in conspecific song (Williams et al. 1989). Sensitivity to formants was tested in whooping cranes (*Grus americana*; Fitch and Kelley 2000), a species that has an elongated trachea that invaginates the sternum; TE is present but variable in most crane species. Whooping cranes showed an increased reaction to formants that had been synthetically altered while keeping all other cues fixed, showing that they perceive formants, and consistent with the "size exaggeration by formant lowering" hypothesis. A recent comparative analysis suggests that crane TE may have specifically developed in those crane species that practice long-distance migration, which favors small body size for energetic reasons (Jones and Witt 2014). To compensate for their decreased body size, migratory cranes appear to enhance their apparent size via tracheal elongation.

10.3.2.2 Air Sacs

Probably the most pervasive vocal tract peculiarity in vertebrates are air sacs attached to the larynx, trachea, or vocal tract. Such sacs occur in hundreds of vertebrate species, and are of such different form (even within limited clades, such as anurans or primates, where air sacs are typical but quite variable) that they appear to have evolved independently by convergent evolution many times. Unfortunately, it is difficult to find a single overview of all the different forms and types of vertebrate air sacs. Their presence or absence in certain clades such as primates is well documented enough that a relatively comprehensive overview is possible (Schneider 1964; Schön Ybarra 1995; Hewitt et al. 2002).

Anuran Air Sacs

Most frogs and toads possess elastic air sacs opening from the oral cavity (Liu 1935; Duellman and Trueb 1986); see Fig. 10.1 for an example. These may be either single (in the midline, opening from the floor of the mouth) or paired (lateral, opening from the mouth corners). Air sacs appear to be absent in species that do not produce sounds in air, such as African clawed frogs (genus *Xenopus*) and some *Rana* species (Wells 1977; Hayes and Krempels 1986). Anuran air sacs probably serve at least two distinct functions: respiratory efficiency via air rebreathing and sound propagation via impedance matching (Rand and Dudley 1993; Starnberger et al. 2014). The air rebreathing function can be readily observed in vocalizing frogs and also easily understood. Because frogs lack either a rib cage or diaphragm, they have a rather inefficient mechanism for filling the lungs with air, relying on buccal pumping (Gans et al. 1969; Gans 1970). This mechanism is also used by salamanders and does not differ qualitatively from the mechanism seen in lungfish (Bishop and Foxon 1968; Brainerd et al. 1993). Because filling the lungs with air requires many cycles of this buccal pumping, vocalizations involving full lung deflation would require a long silent pause between calls while refilling. But by closing the mouth and nostrils during vocalization, and inflating the vocal sac, the frog can re-inhale the same air and produce many vocalizations from the same lungful, as well as saving considerable metabolic expenditure via the energy stored by elastic recoil in the lungs and air sacs (cf. Dudley and Rand 1991). Because amphibians in general can achieve respiration through their skin, this rebreathing appears to lack any dire respiratory consequences (such as the CO_2 buildup that would occur in a human breathing into a bag).

Although rebreathing alone would provide a clear and adaptive function for anuran vocal sacs, they also appear to play a more specific impedance-matching role in many species (cf. Rand and Dudley 1993). Because the wavelengths of anuran advertisement calls are typically quite long, relative to body size, sound produced within the body is not efficiently radiated into the environment. But the inflatable air sac, by virtue of its large size and elastic walls, provides a more efficient radiating surface and thus increases the external amplitude of the vocalizations. Intriguingly, in some ranid frog species the external eardrum also plays a sound-radiating role, and thus is larger in males than in females (Purgue 1997).

One traditional hypothesis about air sacs has been convincingly refuted in several anuran species (Rand and Dudley 1993). This is the idea that the air sac acts as a coupled Helmholtz resonator, controlling vocal fold vibration frequency. By inducing frogs of four different species to vocalize in heliox, Rand and Dudley showed that airborne resonances have no appreciable influence on f_0, thus showing that source and filter are not coupled in these species (and by inference other anurans).

Although these two functions—rebreathing and impedance matching—appear to account for the ubiquity of air sacs in most anurans, air sacs may have further functions in some species (Starnberger et al. 2014). In numerous anurans the air sac is brightly colored, so that its inflation and deflation provides a clear visual signal (Preininger et al. 2013); in several species the importance of this visual signaling has been documented via experiments with robotic frogs with and without inflation (e.g., Narins et al. 2005). Furthermore, certain anuran species such as reed frogs have glandular patches on the air sac (Starnberger et al. 2014) that have been hypothesized to release chemical signals during vocalization (in addition to their acoustic and visual functions).

Laryngeal Air Sacs in Mammals

In mammalian larynges, air sacs extending outside of the cavity of the larynx are a quite common feature. These are of various types and can vary greatly in size, from pea-sized in some monkeys to large balloon-like structures in great apes (with a volume up to 6 L; Schön Ybarra 1995). A remarkable number of hypotheses, some

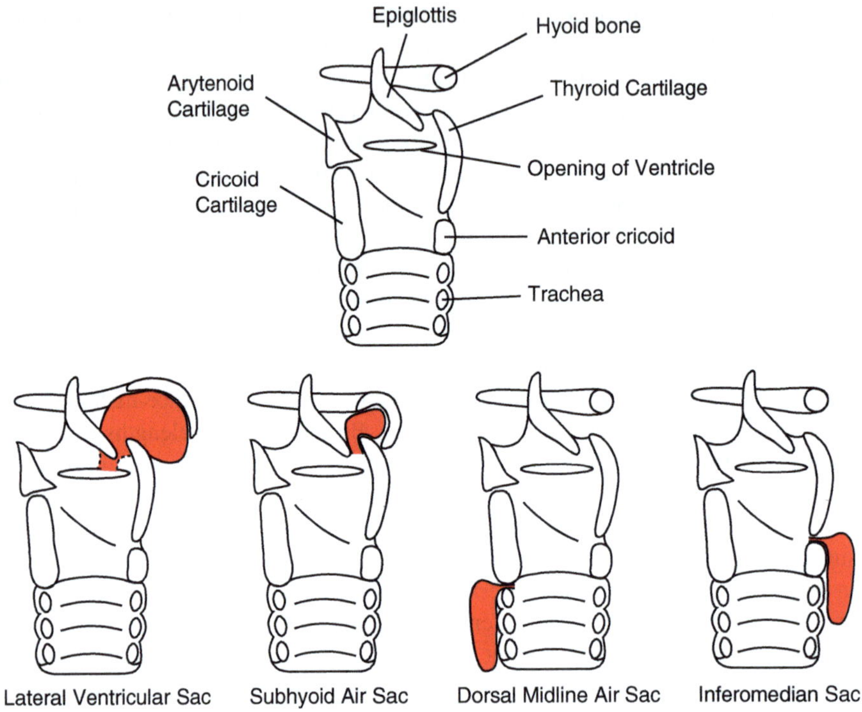

Fig. 10.7 Four types of primate air sac, schematic diagram. The top drawing illustrates the standard laryngeal conformation in primates lacking air sacs (e.g., in a human); the *red* form illustrates the various types of air sac seen in other primates (see text for details)

of them quite fanciful, have been suggested for the function of these air sacs, to be discussed after an anatomical overview. Different types of mammalian air sacs are schematically illustrated in Fig. 10.7.

There is great variety in the details of air sac anatomy, suggesting repeated independent evolution rather than conservation. Although good overviews of primate air sacs anatomy are available (Starck and Schneider 1960; Schön Ybarra 1995; Hewitt et al. 2002), the situation is less satisfactory for mammals more generally. Negus (1949) provides some nice figures for mammals more generally, and the best available overview, very briefly covering all mammals, is Schneider (1964). Because this literature is difficult to access, I document the great variety of air sacs here.

Probably the most common type of laryngeal air sacs are paired lateral ventricular sacs, which open into the larynx through the ventricle. The ventricle is a lateral cavity located between the vocal and ventricular folds, also termed the *cavum laryngis* or "Morgagni's sinus." Ventricles are quite variable across species in presence and size, and a considerable number of mammalian groups lack a ventricle entirely, including most ruminants, many carnivores, cetaceans, elephants, and marsupials (Schneider 1964). Even within a species, the size of the ventricle can be highly variable (Harrison

1995). In many groups the ventricle expands outside of the cartilaginous skeleton of the larynx to form ventricular air sacs, sometimes of great size. Examples include siamangs, howler monkeys, and all of the great apes (chimpanzees, bonobos, gorillas, and orangutans), where they were first reported by anatomists (Camper 1779; Starck and Schneider 1960). Smaller ventricular sacs are also seen in some rodents (marmots or squirrels). At least one pinniped, the South American sea lion (*Otaria byronia = flavescens*), has lateral air sacs but lacks a ventricle (Schneider 1964).

Another common form of air sacs is an unpaired ventral midline air sac, opening from a small midline cavity directly above the vocal folds termed the "fovea centralis." Like the ventricles, the presence or absence of a fovea centralis is variable among species, and several groups have a fovea centralis but lack air sacs. These include murid (mice, rats, etc.) and arvicolid (voles, lemmings, etc.) rodents, where the fovea centralis may play a role in ultrasound production via a whistle-like mechanism (Roberts 1975).

In a variety of species, this fovea is greatly enlarged to form an air sac that spills out of the laryngeal framework in various ways. For example, in many callitrichids (marmosets and tamarins) this ventromedial sac bulges out from between the thyroid and cricoid cartilages (Schneider 1964; Schön Ybarra 1995), termed an inferomedian sac. In contrast, in most cercopithecid monkeys (macaques, baboons, etc.) and in howler monkeys (genus *Alouatta*) this ventromedial sac instead extends up between the thyroid and epiglottis, termed a superomedian sac. Such an arrangement is also seen in tapirs (Negus 1949). In cercopithecids and howler monkeys, the air sac enters the globular hyoid bulla typical of this group, and is thus termed a subhyoid air sac. A related phenomenon are the subthyroid sacs seen some marsupials, including wallabies and phalangerid marsupials (phalangers, cuscuses, and possums), in which the thyroid cartilage itself is expanded and bullate, and the air sac fills this space (Sonntag 1921; Schneider 1964).

In reindeer ("caribou") *Rangifer tarandus*, alone among deer, a large air sac extends freely out between the thyroid and hyoid bone to balloon into the neck, and can have a volume of up to 4 L in large males (Frey et al. 2007b). Similar large extensible medial sacs are seen in the plains viscacha (*Lagostomus maximus*), a large fossorial rodent found in the plains of Patagonia (Zuckerkandl 1905). This air sac inflates before the production of low-frequency growls (Fitch, pers. obs., Eisenberg 1974).

A third type of air sac is apparently only found in primates: an unpaired dorsal midline air sac as found in some prosimians [including at least *Indri*, *Microcebus*, and lemur = *Varecia variegata* (Kollmann and Papin 1914)] and in spider monkeys (*Ateles* spp.). This dorsal sac emerges from between the cricoid cartilage and first tracheal ring, and runs down the dorsal trachea beneath the esophagus (Schneider 1964; Schön Ybarra 1995). These dorsal sacs are relatively small, only about 10–20 mm long in *Ateles*. This anatomical variety among distantly related mammals strongly suggests convergent evolution.

The largest air sac of all is certainly the gigantic ventromedial air sac found in the baleen whales (e.g., the blue whale *Balaenoptera musculus*): This extends out of the larynx and then stretches caudally along the trachea down to the bifurcation of the bronchi (Hosokawa 1950; Schneider 1964; Quayle 1991). This sac presumably plays a role in rebreathing air during the prolonged underwater vocalization bouts made by male baleen whales (Cummings and Thompson 1971; Payne and McVay 1971).

Very little is known about the development of air sacs. In humans, ventricular enlargement duplicating the anatomy of ape air sacs is not uncommon in the clinic, particularly in professions that involve generating sustained lung pressure such as wind instrument players or glass blowers (Harrison 1995). Negus suggested that producing loud, high-pressure vocalizations induces air sacs in nonhuman primates as well, but this hypothesis is falsified by the fact that small laryngeal air sacs are already present in utero in chimpanzees, before the animal has ever taken a breath of air (Schneider 1964, p. 81).

Other Forms of Air Sac

Other types of respiratory air sac are also occasionally observed in vertebrates. Outpouchings of the tracheal wall, called tracheal diverticula, are seen in multiple snake species, where they seem to play dual roles both in expansion of the body during defensive displays and to some extent in vocal production (Young 1992; Young et al. 2000). Hissing is a common vocal signal in reptiles (Mertens 1946; Gans and Maderson 1973; Auffenberg 1977). Young's experimental work indicates that hissing in snakes is produced simply by forceful exhalation through an open glottis: No laryngeal occlusion is needed. Thus, during hissing the trachea is fully coupled to the acoustic system. Young suggests that the low-frequency emphasis seen in certain snake species' hisses, termed "growls," result from low-pass filtering by the inflated tracheal sacs, which are coupled to the trachea via small holes and thus form an array of Helmholtz resonators. Other snakes also have expandable tracheal membranes, with single openings, but these may play only a neck-inflation role (Noble 1921; Young et al. 2000). Among other reptiles, chameleons apparently have a ventromedial laryngeal air sac though no details are available (Fig. 38 in Negus 1949; Lasiewski 1972). Alligator bellow production is accompanied by a prominent inflation of the region below the neck and jaw; because alligators have no vocal sacs per se, this probably reflects expansion of the upper esophagus or "crop" during vocalization through the glottis.

Emus have a unique form of tracheal air sac: a large (several liter) sac balloons outward from a rectangular opening in the tracheal rings at the base of the trachea, above the syrinx (Murie 1867; McLelland 1989). This tracheal sac can be easily seen to expand during the emu "drumming" display with the naked eye, so appears to play some vocal role (perhaps impedence matching). Expandable crops are also typical of some other bird species, and may also play a role in impedance-matching for low-frequency sounds, such as the bittern's "booming" (Chapin 1922).

Tracheal air sacs of a different sort, in this case hard-walled dilations of the upper tracheal rings, are seen in several clades of echolocating bats, in particular in the hipposerid and rhinolophid bats (Elias 1908; Suthers et al. 1988). These bats are nasal emitters, and produce constant-frequency echolocation calls that have a very strong second harmonic with a suppressed fundamental and higher harmonics. These bats have hard-walled pouches in both the nasal cavities and the upper trachea, that appear to play a role in this filtering (Hartley and Suthers 1988; Suthers et al. 1988), suppressing the fundamental and third harmonic. Interestingly, filling the tracheal pouches with dental cement leads to a change in the acoustic signal within the trachea, but not in the emitted signal, suggesting that the major function of these tracheal sacs may be to suppress reflections of the fundamental returning from the bronchial bifurcation and lungs (perhaps to avoid multiple tissue-borne vibrations reaching the cochlea).

10.3.2.3 Functions of Air Sacs

An abundance of hypotheses have been proposed for the function(s) of laryngeal air sacs, but so little experimental work has been done on them that this topic must remain, at present, highly speculative (Fitch and Hauser 2002). Most authors who have considered the topic concluded that the air sacs have some acoustic function (e.g., Hill and Booth 1957; Tembrock 1966; Schön Ybarra 1995). The earliest experimental work, reported by Jean-Pierre Gautier in 1971, involved puncturing the air sac of a guenon (*Cercopithecus* sp.) and observing the changes in its calls. Gautier reported that call intensity was decreased but the frequency was not changed, and suggested that the sac functions as an impedance matching system (as for frogs). However, inspection of the spectrograms in the paper suggests that there may also have been subtle changes in the spectrum. Similarly Hilloowala and Lass (1978) reported that surgical removal of the small subhyoid air sac from three rhesus macaques had no effect on formant frequencies.

In contrast, Bart de Boer has created computer models of primate air sacs that suggest that downward formant shifts in the filter function, relative to a vocal tract lacking air sacs, would be a major acoustic result of adding air sacs (de Boer 2008, 2009). This might serve as a method of exaggerating body size, similar to the laryngeal descent observed in humans, deer, and some other species (Fitch and Reby 2001; Fitch and Hauser 2002; McElligott et al. 2006; Charlton et al. 2011) and is consistent with observations of very low formants in some living primates (Haimoff 1983; Harris et al. 2006). Another acoustic possibility is that air sacs could enhance glottal efficiency via some form of constructive source-tract interaction (Fitch and Hauser 2002), but there is no evidence to support this idea [heliox experiments with primates would be an ideal way to test it, as in Koda et al. (2012)]. Thus, despite various plausible acoustic hypotheses for the functions of air sacs, the existing experimental and modeling data are inconsistent and inconclusive.

A second class of hypothesis involves respiratory functions. In addition to the rebreathing function of oral sacs in anurans, or in baleen whales discussed in section "Laryngeal Air Sacs in Mammals," several more speculative respiratory hypotheses have been suggested for air sacs in various clades. Negus (1949) suggested that ape air sacs act as storage sites for oxygen during vigorous activity. Because air sacs must be inflated with exhaled air that has already been processed by the lungs, sac air would be low in oxygen and high in CO_2 and thus of dubious respiratory value (Fitch and Hauser 2002). Air sacs are also found in some pinnipeds, where a gas storage function would potentially be of value during diving (Sleptsov 1940), but Fay (1960) doubted this possibility because the amount of additional oxygen stored in even large sacs would be trivial relative to dissolved blood O_2 in a diving pinniped.

A more plausible respiratory hypothesis was proposed by Hewitt and colleagues, who suggested that primates producing long series of rapidly repeated calls run a risk of hyperventilation, and that by rebreathing the "used" air from the air sacs, this danger is alleviated (Hewitt et al. 2002). Although such a hypothesis may be applicable to relatively large air sacs, such as those in great apes or some gibbons [e.g., the large and highly elastic air sacs of the siamang *Hylobates syndactylus* (Haimoff 1983)], its relevance for small rigid air sacs (e.g., in macaques) seems more dubious.

An interesting example of multifunctional air sacs is found in the Atlantic walrus (*Odobenus rosmarus*). Adult males have voluminous pharyngeal air sacs that open just dorsal to the larynx (Sleptsov 1940; Fay 1960). These sacs appear to produce the bizarre "bell" sound made by male walruses during courtship (Schevill et al. 1966; Stirling et al. 1987; Sjare et al. 2003), but the precise mechanism of production of this peculiar and unique sound is unknown. In addition to this presumed acoustic function, the pouches can be inflated as "life preservers," keeping the walrus afloat while sleeping at sea; Fay (1960) reports observing walruses sleeping at sea at least eight times, and the pharyngeal air sacs were invariably inflated, holding the head and shoulders out of the water. Finally, Sleptsov (1940) reported finding the sacs of two recently shot walruses filled with food (crustaceans and mollusks), and suggested a third function for these sacs as food storage systems. However, Fay (1960) suggested that Sleptsov's specimens had probably regurgitated stomach contents into the sacs while dying, and thought a food storage function unlikely. In any case, at least the "bell production" and "life preserver" hypotheses are both well supported, indicating at least two functions for walrus pharyngeal sacs. The possibility that air sacs may serve multiple functions should thus be seriously considered in other species.

A final issue worthy of note is that humans appear to have lost laryngeal air sacs after diverging from chimpanzees roughly 6 million years ago. This inference is clearly warranted by the comparative data: all other members of the (great ape + human) clade possess air sacs, strongly suggesting that they were present in our last common ancestor (Fitch 2000, 2010) and later lost. This comparative inference was reinforced by the recent discovery of the hyoid bone of an australopithecine, which has a bullate form nearly identical to that of a chimpanzee (Alemseged et al. 2006). This strongly suggests that laryngeal air sacs were still present in early hominins, and the absence of a bullate hyoid in later hominins such as Neanderthals suggest that they were lost in the genus *Homo* (Arensburg et al. 1989; Martínez et al. 2008). Unfortunately, the inference from the fossil hyoid alone remains insecure, as air sacs can be present in the absence of a bullate hyoid (e.g., in orangutans) and absent even when the hyoid is bullate (e.g., red colobus monkeys; Hill and Booth 1957).

Thus although we cannot be certain about precisely when our ancestors lost our air sacs, it is clear that we did. Understanding why this loss occurred requires a better understanding of the function of laryngeal air sacs in extant primates, particularly in great apes.

In summary, air sacs can clearly serve different functions in different clades. The best understood functions of air sacs are the dual rebreathing and impedence-matching functions in anurans; both may also apply in other species with large elastic air sacs (e.g., baleen whales). However, these hypotheses seem less likely for many types of laryngeal air sacs, especially those of relatively limited size (where rebreathing would have minimal benefits) or those that are ensconced in rigid bony walls (e.g., the subhyoid air sacs of cattharrhine monkeys, or the subthyroid sacs of some marsupials). Here, a frequency shifting function seems more likely.

10.3.3 Other Potential Vocal Adaptations

There are a number of other hollow cavities attached to the vocal tract that may potentially play a role in vocal production. Although it is clear that the primary function of some of the traits may be nonacoustic (e.g., for the elephant's trunk), their connection to the vocal tract means they will have acoustic consequences. In addition to the pharyngeal pouches mentioned previously in the walrus, pharyngeal pouches are also found in horses (Fish 1910), elephants (Miall and Greenwood 1878; Shoshani et al. 1998), and bears (Weissengruber et al. 2001). In horses, this pouch is termed a "guttural pouch" and has been hypothesized to play a role in brain cooling (Baptiste 1997; Baptiste et al. 2000); any potential acoustic role for these pouches is unexplored.

A rather bizarre pair of pouches that has at least a minor acoustic role is found in camels (*Camelus dromedarius*) and hooded seals (*Cystophora cristata*). In camels, the pouch is essentially an elastic inflatable extension of the soft palate, referred to by its Arabic name *dulaa* (Arnautovic and Abdul Magid 1974). This sac is inflated and everted from the mouth during courtship displays, almost like a child blowing bubblegum bubbles. During deflation the dulaa, which is reputedly more developed in males than females, makes a gargling, burbling sound (Fitch, pers. obs.).

The male hooded seal has a similar sac within its nasal passages (Fig. 58 in Negus 1949). Hooded seals give birth on the Arctic ice, where the male courts the nursing mother and engages in an impressive display. He first inflates his enlarged nose and then everts a bright-red nasal sac out of one nostril. This display has an acoustic component in which the inflated nasal cavities appear to play some role, albeit of low intensity (Terhune and Ronald 1973).

When it comes to enlarged or elongated noses that play at least some acoustic role, the trunk of both elephant species provides the most obvious example (see Narins et al., Chap. 7). Another example of an elongated nose is found in the male proboscis monkey (*Nasalis larvatus*), which, like an elephant, has the nostrils at the tip of the nose (so that nasal sounds pass through an elongated vocal tract) and may engage the nose during nasal “honk” vocalizations (Michael Owren, pers. comm.). Saiga antelope (*Saiga tatarica*) also have an enlarged, dome-shaped nasal passage, which has recently been described anatomically (Frey and Hofmann 1997) and is hypothesized to play a key role in vocalizations (Frey et al. 2007a). A similar nasal enlargement is seen in Guenther’s dikdik (*Rhyncotragus guetheri*; Frey and Hofmann 1996, 1997). A bizarre example of nasal elongation in an extinct species are the large hollow crests of some lambeosaurine dinosaurs, which housed elongated extensions of the nasal vocal tract and have been suggested to have an acoustic function such as body size exaggeration (Weishampel 1981, 1997).

A final example of a bizarre and poorly understood air cavity are the enlarged, inflated zygomatic arches of the paca (*Agouti paca*); these enlarged bony cavities have been hypothesized to play some resonant role in vocalizations and/or a mechanically produced jaw-chattering display (Hershkovitz 1955).

10.4 Conclusions

The list of vocal anatomical peculiarities presented in this chapter is almost certainly incomplete; even my 20 years of study is certainly not adequate to peruse the entire comparative anatomical literature, and there are probably many vocal adaptations that have never even been described. A rather surprising example of such an omission from the older literature is provided by the recent discovery of a permanently descended larynx seen in males of several deer species, including two quite common European species [red deer (*Cervus elaphus*) and fallow deer (*Dama dama*)]. Until 2001, a permanently descended larynx was widely believed to be unique to humans. Despite their being hunted for a millennium in Europe, and farmed for meat for a century, the permanent descent of the larynx in these two deer species went undescribed until (fueled by a new acoustic understanding of formant frequencies) it was documented and its acoustic effect of lowering formants described (Fitch and Reby 2001). Since this initial discovery, numerous other mammalian species have been found to have a permanently descended larynx (Weissengruber et al. 2002; Frey and Riede 2003; McElligott et al. 2006; Charlton et al. 2011). Apparently far from being uniquely human, this trait is not very unusual, and has evolved convergently at least four times. This nicely illustrates how an improved understanding of vocal production acoustics can open our eyes to anatomical phenomena (in this case lowered larynx position) that went unnoticed or at least unremarked by previous investigators. We can confidently expect that, now that basic principles of animal acoustic production are understood, more such discoveries await careful observers.

In summary, despite the truly impressive gains in our understanding of vertebrate vocal production in the last 20 years, as illustrated in the present volume, there is much more work to be done and many more anatomical puzzles to be solved. Virtually any of the isolated instances of vocal diversity cited above would be fuel for a very solid PhD thesis of four or five publications, and some of the most abundant vocal adaptations (such as air sacs, or syringeal diversity) would keep a good-sized scientific team busy for many years. As stressed in the introduction, many of the adaptations described in this chapter are barely known today, and their descriptions are often buried in inaccessible journals. Where the early anatomists commented on the potential function of vocal peculiarities, they often did so in ignorance of both the detailed behavior and communication of the species in question and the basic principles of vocal acoustics clarified in the last few decades. Thus, for the most part, vocal diversity of the sort discussed here is *terra incognita* with regard to its acoustic, physiological and behavioral functioning, and its evolutionary and adaptive relevance.

Vocal bioacoustics spans a remarkable diversity of disciplines and explanatory levels, from physics through anatomy to evolution, and for historical reasons has remained relatively untouched by the thriving field of functional morphology. Thus, there are few aspects of modern biology that appear so ripe for the picking, and promise such considerable and accessible gains in our understanding of the biology and evolution of communication more generally. In many cases (such as the role of air sacs, the broad relevance of formants in animal communication, the frequency of vocal dimorphism, or the role of hormones in developing vocal traits) advances in our comparative understanding would also be directly relevant to understanding human vocal production. This entire volume, including the present chapter, was written in the hope of inspiring just such progress, and if even a small percentage of its readers are thus inspired we can expect great gains in the coming decades.

References

Alemseged, Z., Spoor, F., Kimbel, W. H., Bobe, R., Geraads, D., Reed, D., et al. (2006). A juvenile early hominin skeleton from Dikika, Ethiopia. *Nature, 443*, 296–301.

Alexander, R. M. (1987). Swim bladder. In *McGraw-Hill encyclopedia of science and technology* (Vol. S, pp. 41–43). New York: McGraw-Hill.

Allin, E. F. (1975). Evolution of the mammalian middle ear. *Journal of Morphology, 147*, 403–438.

Altmann, S. A. (1959). Field observations on a howling monkey society. *Journal of Mammalogy, 40*(3), 317–330.

Amadon, D. (1969). Variation in the trachea of the Cracidae (Galliformes) in relation to their classification. *Natural History Bulletin of the Siam Society, 23*, 239–248.

Ames, P. L. (1971). The morphology of the syrinx in passerine birds. *Bulletin of the Peabody Museum of Natural History (Yale University), 37*, 1–194.

Arensburg, B., Tillier, A. M., Vandermeersch, B., Duday, H., Schepartz, L. A., & Rak, Y. (1989). A middle paleolithic human hyoid bone. *Nature, 338*, 758–760.

Aristotle. (350 BC). *The history of animals* (A. L. Peck, Trans. 1970 ed.). London: Heinemann.

Arnason, U., Gullberg, A., & Janke, A. (2001). Molecular phylogenetics of gnathostomous (jawed) fishes: Old bones, new cartilage. *Zoologica Scripta, 30*, 249–255.

Arnautovic, I., & Abdul Magid, A. M. (1974). Anatomy and mechanism of distension of the dulaa of the one-humped camel. *Acta Anatomica, 88*, 115–124.

Auffenberg, W. (1977). Display behavior in tortoises. *American Zoologist, 17*, 241–250.

Baptista, L. F., & Trail, P. W. (1992). The role of song in the evolution of passerine diversity. *Systematic Biology, 41*, 242–247.

Baptiste, K. E. (1997). Functional anatomy observations of the pharyngeal orifice of the equine guttural pouch (auditory tube diverticulum). *The Veterinary Journal, 153*, 311–319.

Baptiste, K. E., Naylor, J. M., Bailey, J., Barber, E. M., Post, K., & Thornhill, J. (2000). A function for guttural pouches in the horse. *Nature, 403*, 382–383.

Bass, A. H. (1990). Sounds from the intertidal zone: Vocalizing fish. *Bioscience, 40*, 249–267.

Bass, A. H., & Chagnaud, B. P. (2012). Shared developmental and evolutionary origins for neural basis of vocal acoustic and pectoral gestural signaling. *Proceedings of the National Academy of Sciences of the USA, 109*(Supplement 1), 10677–10684.

Beddard, F. E. (1898). *The structure and classification of birds*. London: Longmans and Green.

Bishop, I. R., & Foxon, G. E. H. (1968). The mechanism of breathing in the South American lungfish, *Lepidosiren paradoxa*: A radiological study. *Journal of Zoology (London), 154*, 263–271.

Blair, W. F. (1958). Mating call in the speciation of anuran amphibians. *American Naturalist, 92*(862), 27–51.

Blair, W. F. (1964). Isolating mechanisms and interspecies interactions in anuran amphibians. *Quarterly Review of Biology, 39*(4), 334–344.

Bradbury, J. W. (1977). Lek behavior in the hammer-headed bat. *Zeitschrift für Tierpsychology, 45*, 225–255.

Brainerd, E. L., Ditelberg, J. S., & Bramble, D. M. (1993). Lung ventilation in salamanders and the evolution of vertebrate air-breathing mechanisms. *Biological Journal of the Linnean Society, 49*, 163–183.

Camper, P. (1779). Account of the organs of speech of the orang outang. *Philosophical Transactions of the Royal Society of London B: Biological Sciences, 69*, 139–159.

Chapin, J. P. (1922). The function of the oesophagus in the bittern's booming. *Auk, 39*, 196–202.

Charlton, B. D., Ellis, W. A. H., McKinnon, A. J., Cowin, G. J., Brumm, J., Nilsson, K., et al. (2011). Cues to body size in the formant spacing of male koala (*Phascolarctos cinereus*) bellows: Honesty in an exaggerated trait. *Journal of Experimental Biology, 214*, 3414–3422.

Chiba, T., & Kajiyama, M. (1941). *The vowel: Its nature and structure*. Tokyo: Tokyo-Kaiseikan.

Clench, M. H. (1978). Tracheal elongation in birds-of-paradise. *Condor, 80*, 423–430.

Cole, F. J. (1949). *A history of comparative anatomy*. London: Macmillan Press.

Cummings, W. C., & Thompson, P. O. (1971). Underwater sounds from the blue whale, *Balaenoptera musculus. Journal of the Acoustical Society of America, 50*, 1193–1198.

Darwin, C. (1871). *The descent of man and selection in relation to sex* (1st ed.). London: John Murray.

de Boer, B. (2008). The acoustic role of supralaryngeal air sacs. *Journal of the Acoustical Society of America, 123*(5 Pt 2), 3732–3733.

de Boer, B. (2009). Acoustic analysis of primate air sacs and their effect on vocalization. *Journal of the Acoustical Society of America, 126*, 3329–3343.

Dudley, R., & Rand, A. S. (1991). Sound production and vocal sac inflation in the Túngara frog, *Physalaemus pustulosus* (Leptodactylidae). *Copeia, 1991*(2), 460–470.

Dudley, H., & Tarnoczy, T. H. (1950). The speaking machine of Wolfgang von Kempelen. *Journal of the Acoustical Society of America, 22*(2), 151–166.

Duellman, W. E., & Trueb, L. (1986). *Biology of the amphibians*. New York: McGraw-Hill.

Eisenberg, J. F. (1974). The functional and motivational basis of hystricomorph vocalizations. In I. W. Rowlands & B. J. Weir (Eds.), *The biology of hystricomorph rodents* (pp. 211–244). London: Academic.

Elias, H. (1908). Zur Anatomie des Kehlkopfes der Mikrochiropteren. *Morphologisches Jahrbuch, 37*, 70–118.

Fant, G. (1960). *Acoustic theory of speech production*. The Hague, Netherlands: Mouton & Co.

Fant, G. (2001). T Chiba and M Kajiyama: Pioneers in speech acoustics. *Quarterly Progress and Status Report, 42*(1), 59–60.

Fay, F. H. (1960). Structure and function of the pharyngeal pouches of the walrus (*Odobenus rosmarus* L.). *Mammalia, 24*, 361–371.
Fish, P. A. (1910). The exchange of air in the eustachian or guttural pouches of the horse. *American Journal of Physiology, 27*, 229–232.
Fitch, W. T. (1999). Acoustic exaggeration of size in birds by tracheal elongation: Comparative and theoretical analyses. *Journal of Zoology (London), 248*, 31–49.
Fitch, W. T. (2000). The evolution of speech: A comparative review. *Trends in Cognitive Sciences, 4*(7), 258–267.
Fitch, W. T. (2010). *The evolution of language*. Cambridge, England: Cambridge University Press.
Fitch, W. T. (2012). Evolutionary developmental biology and human language evolution: Constraints on adaptation. *Evolutionary Biology, 39*(4), 613–637.
Fitch, W. T., & Hauser, M. D. (2002). Unpacking "honesty": Vertebrate vocal production and the evolution of acoustic signals. In A. M. Simmons, A. N. Popper, & R. R. Fay (Eds.), *Acoustic communication* (pp. 65–137). New York: Springer.
Fitch, W. T., & Kelley, J. P. (2000). Perception of vocal tract resonances by whooping cranes, *Grus americana*. *Ethology, 106*(6), 559–574.
Fitch, W. T., & Reby, D. (2001). The descended larynx is not uniquely human. *Proceedings of the Royal Society of London B: Biological Sciences, 268*(1477), 1669–1675.
Fitzpatrick, J. W. (1988). Why so many kinds of passerine birds? A response to Raikow. *Systematic Zoology, 37*, 72–77.
Frederick II, H. R. E., 1194–1250. (1943/c. 1248). *The art of falconry, being the De arte venandi cum avibus* (C. A. Wood & F. M. Fyfe, Trans.). Stanford, CA: Stanford University Press.
Frey, R., Gebler, A., Fritsch, G., Nygrén, K., & Weissengruber, G. E. (2007a). Nordic rattle: The hoarse vocalization and the inflatable laryngeal air sac of reindeer (*Rangifer tarandus*). *Journal of Anatomy, 210*, 131–159.
Frey, R., & Hofmann, R. R. (1996). Evolutionary morphology of the proboscideal nose of Guenther's Dikdik (*Rhyncotragus guetheri* Thomas 1894—Mammalia, Bovidae). *Zoologischer Anzeiger, 235*, 31–51.
Frey, R., & Hofmann, R. R. (1997). Skull, proboscis musculature and preorbital gland in the Saiga antelope and Guenther's Dikdik (Mammalia, Artiodactyla, Bovidae). *Zoologischer Anzeiger, 235*, 183–199.
Frey, R., & Hofmann, R. R. (2000). Larynx and vocalization of the Takin (*Budorcas taxicolor* Hodgson, 1850—Mammalia, Bovidae). *Zoologischer Anzeiger, 239*, 197–214.
Frey, R., & Riede, T. (2003). Sexual dimorphism of the larynx of the Mongolian Gazelle (*Procapra gutturosa* Pallas, 1777) (Mammalia, Artiodactyla, Bovidae). *Zoologischer Anzeiger, 242*, 33–62.
Frey, R., Volodin, I. A., Volodina, E. V., & Juldaschev, E. T. (2011). Descended and mobile larynx, vocal tract elongation and rutting roars in male goitred gazelles (*Gazella subgutturosa* Güldenstaedt, 1780). *Journal of Anatomy, 218*(5), 566–585.
Frey, R., Volodin, I. A., & Volodina, E. V. (2007b). A nose that roars: Anatomical specializations and behavioural features of rutting male saiga. *Journal of Anatomy, 211*(6), 717–736.
Frith, C. B. (1994). Adaptive significance of tracheal elongation in manucodes (Paradisaeidae). *Condor, 96*, 552–555.
Gans, C. (1970). Respiration in early tetrapods: The frog is a red herring. *Evolution, 24*, 723–734.
Gans, C., De Jongh, H. J., & Farber, J. (1969). Bullfrog (*Rana catesbiana*) ventilation: How does the frog breathe? *Science, 163*, 1223–1225.
Gans, C., & Maderson, P. F. A. (1973). Sound producing mechanisms in recent reptiles: Review and comment. *American Zoologist, 13*, 1195–1203.
Garrod, A. H. (1874–1878). On some anatomical peculiarities which bear upon the major divisions of the passerine birds. *Proceedings of the Zoological Society of London,* 1874, 506–519.
Garrod, A. H. (1875). On the form of the lower larynx in certain species of ducks. *Proceedings of the Zoological Society of London, 1875*, 151–156.

Garrod, A. H. (1876). Notes on the anatomy of *Plotus anhinga. Proceedings of the Zoological Society of London, 1876*, 335–345.
Gaunt, A. S., & Gaunt, S. L. L. (1985). Syringeal structure and avian phonation. In R. F. Johnson (Ed.), *Current ornithology* (pp. 213–245). New York: Plenum Press.
Gaunt, A. S., Gaunt, S. L. L., & Hector, D. H. (1976). Mechanics of the syrinx in *Gallus gallus*: I. A comparison of pressure events in chickens to those in oscines. *Condor, 78*, 208–223.
Gaupp, E. (1904). Das Hyobranchialskelet der Wirbeltiere. *Ergebnisse der Anatomie und Entwicklungsgeschichte, 14*, 808–1048.
Gegenbauer, C. (1874). *Grundriss der vergleichenden Anatomie*. Leipzig, Germany: Wilhelm Engelmann.
Gegenbauer, C. (1878). *Elements of comparative anatomy* (F. J. Bell & E. R. Lankester, Trans.). London: Macmillan.
Geissler, D. B., & Ehret, G. (2002). Time-critical integration of formants for perception of communication calls in mice. *Proceedings of the National Academy of Sciences of the U S A, 99*, 9021–9025.
Gingras, B., Mohandesan, E., Boko, D., & Fitch, W. T. (2013). Phylogenetic signal in the acoustic parameters of the advertisement calls of four clades of anurans. *BMC Evolutionary Biology, 13*, e134.
Goller, F., & Larsen, O. N. (1997). *In situ* biomechanics of the syrinx and sound generation in pigeons. *Journal of Experimental Biology, 200*(16), 2165–2176.
Göppert, E. (1901). Beiträge zur vergleichenden Anatomie des Kehlkopfes und seiner Umgebung. In R. Semon (Ed.), *Zoologische Forschungreise in Australien und dem Malayischen Archipel* (Vol. 3, pp. 535–633). Jena, Germany: G. Fischer.
Gridi-Papp, M., Rand, A. S., & Ryan, M. J. (2006). Complex call production in the túngara frog. *Nature, 441*, 38.
Griffin, D. R. (1958). *Listening in the Dark*. New Haven, CT: Yale University Press.
Haimoff, E. H. (1983). Occurrence of anti-resonance in the song of the siamang (*Hylobates syndactylus*). *American Journal of Primatology, 5*, 249–256.
Harris, T. R., Fitch, W. T., Goldstein, L. M., & Fashing, P. J. (2006). Black and white colobus monkey (*Colobus guereza*) roars as a source of both honest and exaggerated information about body mass. *Ethology, 112*, 911–920.
Harrison, D. F. N. (1995). *The anatomy and physiology of the mammalian larynx*. New York: Cambridge University Press.
Hartley, D. J., & Suthers, R. A. (1988). The acoustics of the vocal tract in the horseshoe bat, *Rhinolophus hildebrandti. Journal of the Acoustical Society of America, 84*, 1201–1213.
Hast, M. H. (1986). The larynx of roaring and non-roaring cats. *Journal of Anatomy, 149*, 221–222.
Hawkins, A. D. (1993). Underwater sound and fish behaviour. In T. J. Pitcher (Ed.), *Behaviour of teleost fishes* (2nd ed., pp. 129–169). New York: Chapman & Hall.
Hayes, M. P., & Krempels, D. M. (1986). Vocal sac variation among frogs of the genus *Rana* from western North America. *Copeia, 1986*(4), 927–936.
Hershkovitz, P. (1955). On the cheek pouches of the tropical American paca, *Agouti paca* (Linnaeus, 1766). *Säugetierkundliche Mitteilungen, 3*, 67–70.
Hewitt, G., MacLarnon, A., & Jones, K. E. (2002). The functions of laryngeal air sacs in primates: A new hypothesis. *Folia Primatologica, 73*, 70–94.
Hill, W. C. O., & Booth, A. H. (1957). Voice and larynx in African and Asiatic Colobidae. *Journal of the Bombay Natural History Society, 54*, 309–321.
Hilloowala, R. A. (1976). The primate hyolaryngeal apparatus and herbivorous modifications. *Acta Anatomica, 95*, 260–278.
Hilloowala, R. A., & Lass, N. J. (1978). Spectrographic analysis of laryngeal air sac resonance in rhesus monkey. *American Journal of Physical Anthropology, 49*(1), 129–131.
Hosokawa, H. (1950). On the Cetacean larynx, with special remarks on the laryngeal sack of the Sei whale and the aryteno-epiglottideal tube of the Sperm whale. *Scientific Reports of the Whale Research Institute (Tokyo), 3*, 23–62.

Johnsgard, P. A. (1961). Tracheal anatomy of the Anatidae and its taxonomic significance. *Wildfowl Trust 12th Annual Report* (pp. 58–69).

Johnsgard, P. A. (1978). *Ducks, geese, and swans of the world.* Lincoln, NE: University of Nebraska Press.

Jones, M. R., & Witt, C. C. (2014). Migrate small, sound big: Functional constraints on body size promote tracheal elongation in cranes. *Journal of Evolutionary Biology, 27*, 1256–1264.

Kelemen, G., & Sade, J. (1960). The vocal organ of the howling monkey (*Alouatta palliata*). *Journal of Morphology, 107*, 123–140.

Kelley, D. B. (1988). Sexually dimorphic behaviors. *Annual Review of Neuroscience, 11*, 225–251.

Klatt, D. H., & Stefanski, R. A. (1974). How does a mynah bird imitate human speech? *Journal of the Acoustical Society of America, 55*(4), 822–832.

Koda, H., Nishimura, T., Tokuda, I. T., Oyakawa, C., Nihonmatsu, T., & Masataka, N. (2012). Soprano singing in gibbons. *American Journal of Physical Anthropology, 149*(3), 347–355.

Kollmann, M., & Papin, L. (1914). Etudes sur les Lemuriens I. Le Larynx et le Pharynx. Anatomie comparee et Anatomie microscopique. *Annales des Sciences Naturelles, Series 9, 19*, 227–317.

Ladich, F., & Fine, M. L. (2006). Sound generating mechanisms in fishes: A unique diversity in vertebrates. In F. Ladich, S. P. Collin, P. Moller, & B. G. Kapoor (Eds.), *Communication in fishes* (pp. 3–43). Enfield, NH: Science Publishers.

Larsen, O. N., & Goller, F. (2002). Direct observations of syringeal muscle function in songbirds and a parrot. *Journal of Experimental Biology, 203*, 25–35.

Lasiewski, R. C. (1972). Respiratory function in birds. In D. S. Farner & J. R. King (Eds.), *Avian biology* (Vol. 2, pp. 287–342). New York: Academic.

Lieberman, P. (1984). *The biology and evolution of language*. Cambridge, MA: Harvard University Press.

Linggard, R. (1985). *Electronic synthesis of speech*. Cambridge, England: Cambridge University Press.

Liu, C. E. C. (1935). Types of vocal sac in the Salientia. *Proceedings of the Boston Society of Natural History, 41*, 19–40.

Lorenz, K. (1953). *Comparative studies on the behaviour of the Anatinae*. London: Goodale Press.

M'Donnell, R. (1860). Observations on the habits and anatomy of the *Lepidosiren annectans. Natural History Review, 7*, 93–112.

Marcellini, D. (1977). Acoustic and visual display behavior of Gekkonid lizards. *American Zoologist, 17*, 251–260.

Martin, W. F. (1971). Mechanics of sound production in toads of the genus *Bufo*: Passive elements. *Journal of Experimental Zoology, 176*, 273–294.

Martin, W. F., & Gans, C. (1972). Muscular control of the vocal tract during release signaling in the toad *Bufo valliceps. Journal of Morphology, 137*, 1–28.

Martínez, I., Arsuaga, J.-L., Quam, R., Carretero, J.-M., Gracia, A., & Rodriguez, L. (2008). Human hyoid bones from the middle Pleistocene site of the Sima de los Huesos (Sierra de Atapuerca, Spain). *Journal of Human Evolution, 54*(1), 118–124.

McElligott, A. G., Birrer, M., & Vannoni, E. (2006). Retraction of the mobile descended larynx during groaning enables fallow bucks (*Dama dama*) to lower their formant frequencies. *Journal of Zoology, 270*, 340–345.

McLelland, J. (1989). Larynx and trachea. In A. S. King & J. McLelland (Eds.), *Form and function in birds* (Vol. 4, pp. 69–103). New York: Academic.

Mergell, P., Fitch, W. T., & Herzel, H. (1999). Modeling the role of non-human vocal membranes in phonation. *Journal of the Acoustical Society of America, 105*, 2020–2028.

Mertens, R. (1946). Die Warn- und Droh-Reaktion der Reptilien. *Abhandlungen der Senckenbergischen Naturforschenden Gesellschaft, 471*, 1–108.

Miall, L. C., & Greenwood, F. (1878). The anatomy of the Indian elephant. Part III. *Journal of Anatomy and Physiology, 13*(1), 17–50.

Miller, E. H., Seneviratne, S. S., Jones, I. L., Robertson, G. J., & Wilhelm, S. I. (2008). Syringeal anatomy and allometry in murres (Alcidae: Uria). *Journal of Ornithology, 149*(4), 545–554.
Miller, E. H., Williams, J., Jamieson, S. E., Gilchrist, H. G., & Mallory, M. L. (2007). Allometry, bilateral asymmetry and sexual differences in the vocal tract of common eiders *Somateria mollissima* and king eiders *S. spectabilis*. *Journal of Avian Biology, 38*, 224–233.
Moore, B. A., Russell, A. P., & Bauer, A. M. (1991). Structure of the larynx of the tokay gecko (*Gekko gecko*), with particular reference to the vocal cords and glottal lips. *Journal of Morphology, 210*, 227–238.
Murie, J. (1867). On the tracheal pouch of the emu (*Dromaeus novae-hollandiae* Vieill.). *Proceedings of the Zoological Society of London, 1867*, 405–415.
Narins, P. M., Grabul, D. S., Soma, K. K., Gaucher, P., & Hödl, W. (2005). Cross-modal integration in a dart-poison frog. *Proceedings of the National Academy of Sciences of the U S A, 102*(7), 2425–2429.
Negus, V. E. (1949). *The comparative anatomy and physiology of the larynx*. New York: Hafner.
Noble, G. K. (1921). Snakes that inflate. *Natural History, 21*, 166–171.
Nottebohm, F. (1971). Neural lateralization of vocal control in a passerine bird. I Song. *Journal of Experimental Zoology, 177*, 229–262.
Nottebohm, F. (1975). A zoologists's view of some language phenomena with particular emphasis on vocal learning. In E. H. Lenneberg & E. Lenneberg (Eds.), *Foundations of language development: A multidisciplinary approach* (Vol. 1, pp. 61–103). New York: Academic.
Nottebohm, F., & Nottebohm, M. E. (1976). Left hypoglossal dominance in the control of canary and white-crowned sparrow song. *Journal of Comparative Physiology. A, Neuroethology, Sensory, Neural, and Behavioral Physiology, 108*, 171–192.
Novick, A., & Griffin, D. R. (1961). Laryngeal mechanisms in bats for the production of orientation sounds. *Journal of Experimental Zoology, 148*(2), 125–145.
Nowicki, S., & Marler, P. (1988). How do birds sing? *Music Perception, 5*, 391–426.
Paulsen, K. (1967). *Das Prinzip der Stimmbildung in der Wirbeltierreihe und beim Menschen*. Frankfurt, Germany: Akademische Verlagsgesellschaft.
Payne, R. S., & McVay, S. (1971). Songs of humpback whales. *Science, 173*, 583–597.
Pepperberg, I. M. (2005). Insights into vocal imitation in Grey parrots (*Psittacus erithacus*). In S. L. Hurley & N. Chader (Eds.), *Imitation, Human Development, and Culture* (pp. 243–262). Cambridge, MA: MIT Press.
Preininger, D., Boeckle, M., Freudmann, A., Starnberger, I., Sztatecsny, M., & Hödl, W. (2013). Multimodal signaling in the small torrent frog (*Micrixalus saxicola*) in a complex acoustic environment. *Behavioral Ecology and Sociobiology, 67*, 1449–1456.
Purgue, A. P. (1997). Tympanic sound radiation in the bullfrog *Rana catesbiana*. *Journal of Comparative Physiology. A, Neuroethology, Sensory, Neural, and Behavioral Physiology, 181*, 438–445.
Quayle, C. J. (1991). A dissection of the larynx of a humpback whale calf with a review of its functional morphology. *Memoirs of the Queensland Museum, 30*(2), 351–354.
Rand, A. S., & Dudley, R. (1993). Frogs in helium: The anuran vocal sac is not a cavity resonator. *Physiological Zoology, 66*(5), 793–806.
Rittenhouse, D. R., Russell, A. P., & Bauer, A. M. (1997). Laryngotracheal morphology of the *Uroplatus fimbriatus* complex (Reptilia: Gekkonidae): Yet another autapomorphic organ system in a group of highly aberrant geckos. *African Journal of Herpetology, 46*, 36–48.
Roberts, T. S. (1880). The convolution of the trachea in the sandhill and whooping cranes. *American Naturalist, 14*, 108–114.
Roberts, L. H. (1975). The rodent ultrasound production mechanism. *Ultrasonics, 13*(2), 83–88.
Russell, A. P., Rittenhouse, D. R., & Bauer, A. M. (2000). Laryngotracheal morphology of Afro-Madagascan geckos: A comparative survey. *Journal of Morphology, 245*, 241–268.
Sachs, J., Lieberman, P., & Erickson, D. (1973). Anatomical and cultural determinants of male and female speech. In R. W. Shuy & R. W. Fasold (Eds.), *Language attitudes: Current trends and prospects*. Washington, DC: Georgetown University Press.

Schevill, W. E., Watkins, W. A., & Ray, C. (1966). Analysis of underwater *Odobenus* calls with remarks on the development and function of the pharyngeal pouches. *Zoologica (New York), 51*, 103–106.

Schneider, R. (1964). Der Larynx der Säugetiere. *Handbuch der Zoologie, 5*(7), 1–128.

Schneider, R., Kuhn, H.-J., & Kelemen, G. (1967). Der Larynx des männlichen *Hypsignathus monstrosus* Allen, 1861 (Pteropodidae, Megachiroptera, Mammalia). *Zeitschrift für Wissenschaftliche Zoologie, 175*, 1–53.

Schneider, H., Schneider, M., Sampaio, I., Harada, M., Stanhope, M., Czelusniak, J., et al. (1993). Molecular phylogeny of New World monkeys (Platyrrhini, primates). *Molecular Phylogenetics and Evolution, 2*(3), 225–242.

Schön Ybarra, M. (1986). Loud calls of adult male red howling monkeys (*Alouatta seniculus*). *Folia Primatologica, 34*, 204–216.

Schön Ybarra, M. (1995). A comparative approach to the nonhuman primate vocal tract: Implications for sound production. In E. Zimmerman & J. D. Newman (Eds.), *Current topics in primate vocal communication* (pp. 185–198). New York: Plenum Press.

Shoshani, J., Agnew, D., Watson, G., Marchant, G. H., & Marsac, E. C. (1998). *The pharyngeal pouch: A unique receptacle in the throat of an elephant.* Paper presented at the Proceedings of the 23rd National Conference of the American Association of Zoo Keepers, Inc., Detroit, October 6–10.

Siebenrock, F. (1899). Über den Kehlkopf und die Luftröhre der Schildkröten. *Sitzungsberichten der kaiserlichen Akademie der Wissenschaften, Wien, 108*(1), 581–595.

Sjare, B., Stirling, I., & Spencer, C. (2003). Structural variation in the songs of Atlantic walruses breeding in the Canadian High Arctic. *Aquatic Mammals, 29*, 297–318.

Sleptsov, M. M. (1940). On adaptations of pinnipeds to swimming. *Zoological Journal (Zoologicheskii Zhurnal), 19*(3), 379–386.

Sonntag, C. F. (1921). The comparative anatomy of the koala (*Phascolarctos cinereus*) and vulpine phalanger (*Trichosurus vulpecula*). *Proceedings of the Zoological Society of London, 39*, 547–577.

Sprague, J. A. (1943). The hyoid region of placental mammals with especial reference to the bats. *American Journal of Anatomy, 72*, 385–472.

Starck, D., & Schneider, R. (1960). Respirationsorgane. In H. Hofer, A. H. Schultz, & D. Starck (Eds.), *Primatologia* (Vol. 3/2, pp. 423–587). Basel, Switzerland: S. Karger.

Starnberger, I., Preininger, D., & Hödl, W. (2014). Review: From uni- to multimodality. Towards an integrative view on anuran communication. *Journal of Comparative Physiology. A, Neuroethology, Sensory, Neural, and Behavioral Physiology, 200*(9), 777–787.

Stirling, I., Calvert, W., & Spencer, C. (1987). Evidence of stereotyped underwater vocalizations of male Atlantic walruses (*Odobenus rosmarus rosmarus*). *Canadian Journal of Zoology, 65*, 2311–2321.

Suthers, R. A. (1990). Contributions to birdsong from the left and right sides of the intact syrinx. *Nature, 347*, 473–477.

Suthers, R. A. (1999). The motor basis of vocal performance in songbirds. In M. D. Hauser & M. Konishi (Eds.), *The design of animal communication* (pp. 37–62). Cambridge, MA: MIT/Bradford.

Suthers, R. A., Hartley, D. J., & Wenstrup, J. J. (1988). The acoustic role of tracheal chambers and nasal cavities in the production of sonar pulses by the horseshoe bat, *Rhilophus hildebrandti. Journal of Comparative Physiology. A, Neuroethology, Sensory, Neural, and Behavioral Physiology, 162*, 799–813.

Suthers, R. A., Wild, J. M., & Kaplan, G. T. (2011). Mechanisms of song production in the Australian magpie. *Journal of Comparative Physiology. A, Neuroethology, Sensory, Neural, and Behavioral Physiology, 197*, 45–59.

Tembrock, G. (1966). *Akustische Kommunikation bei Säugetieren.* Darmstadt, Germany: Wissenschaftliche Buchgesellschaft.

Terhune, J. M., & Ronald, K. (1973). Some hooded seal (*Cystophora cristata*) sounds in March. *Canadian Journal of Zoology, 51*, 319–321.

Titze, I. R. (1989). Physiologic and acoustic differences between male and female voices. *Journal of the Acoustical Society of America, 85*(4), 1699–1707.

Titze, I. R., Fitch, W. T., Hunter, E. J., Alipour, F., Montequin, D., Armstrong, D. L., et al. (2010). Vocal power and pressure-flow relations in excised tiger larynges. *Journal of Experimental Biology, 213*(22), 3866–3873.

Trewavas, E. (1932). The hyoid and larynx of the Anura. *Philosophical Transactions of the Royal Society of London B: Biological Sciences, 222*, 401–527.

Tyson, E. (1680). *Phocaena, or the anatomy of porpess*. London: Benjamin Tooke.

Tyson, E. (1699). *Orang-Outang, sive Homo Sylvestris: or, the Anatomy of a pygmie compared with that of a monkey, an ape and a man*. London: Thomas Bennet and Daniel Brown.

van den Berg, J. (1958). Myoelastic-aerodynamic theory of voice production. *Journal of Speech and Hearing Research, 1*, 227–244.

Warner, R. W. (1971). The structural basis of the organ of voice in the genera *Anas* and *Aythya* (Aves). *Journal of Zoology, 164*, 197–207.

Weber, E. (1976). Alterations in the release calls of six European anura (Amphibia) after partial or total extirpation of the vocal cords. *Behavioural Processes, 1*(3), 197–216.

Weber, E. (1977). Effects of extirpation of laryngeal muscles and vocal cords on vocalizations of *Bufo viridis* (Amphibia, Anura). *Israel Journal of Zoology, 26*, 230–247.

Weishampel, D. B. (1981). Acoustic analysis of potential vocalization in lambeosaurine dinosaurs (Reptilia: Ornithischia). *Paleobiology, 7*(2), 252–261.

Weishampel, D. B. (1997). Dinosaurian cacaphony: Inferring function in extinct organisms. *Bioscience, 47*, 150–159.

Weissengruber, G. E., Forstenpointner, G., Kübber-Heiss, A., Reidleberger, K., Schwammer, H., & Ganzberger, K. (2001). Occurrence and structure of epipharyngeal pouches in bears (Ursidae). *Journal of Anatomy (London), 198*, 309–314.

Weissengruber, G. E., Forstenpointner, G., Peters, G., Kübber-Heiss, A., & Fitch, W. T. (2002). Hyoid apparatus and pharynx in the lion (*Panthera leo*), jaguar (*Panthera onca*), tiger (*Panthera tigris*), cheetah (*Acinonyx jubatus*), and domestic cat (*Felis silvestris* f. *catus*). *Journal of Anatomy (London), 201*, 195–209.

Wells, K. D. (1977). The social behaviour of anuran amphibians. *Animal Behaviour, 25*(3), 666–693.

Whitehead, J. M. (1995). Vox Alouattinae: A preliminary survey of the acoustic characteristics of long-distance calls of howling monkeys. *International Journal of Primatology, 16*, 121–144.

Wiedersheim, R. (1904). Über das Vorkommen eines Kehlkopfes bei Ganoiden und Dipnoern sowie über die Phylogenie der Lunge. *Zoologische Jahrbücher, Supplementheft, 7*, 1–66.

Williams, H., Cynx, J., & Nottebohm, F. (1989). Timbre control in zebra finch (*Taeniopygia guttata*) song syllables. *Journal of Comparative Psychology, 103*, 366–380.

Young, B. A. (1992). Tracheal diverticula in snakes: Possible functions and evolution. *Journal of Zoology (London), 227*, 567–583.

Young, B. A., Nejman, N., & Jaggers, J. (2000). Snakes that inflate: The functional morphology of a defensive behavior. *Zoology, 103*, 25–30.

Zollinger, S. A., Riede, T., & Suthers, R. A. (2008). Two-voice complexity from a single side of the syrinx in northern mockingbird *Mimus polyglottos* vocalizations. *Journal of Experimental Biology, 211*, 1978–1991.

Zuckerkandl, E. (1905). Über laterale rachentaschen bei *Lagostomus trichodactylus*. *Zeitschrift für Morphologie und Anthropologie, 9*, 138–144.

MIX
Papier aus verantwortungsvollen Quellen
Paper from responsible sources
FSC® C105338
FSC
www.fsc.org

If you have any concerns about our products,
you can contact us on
ProductSafety@springernature.com

In case Publisher is established outside the EU,
the EU authorized representative is:
Springer Nature Customer Service Center GmbH
Europaplatz 3, 69115 Heidelberg, Germany

Printed by Libri Plureos GmbH
in Hamburg, Germany